Springer Series in Biophysics

Volume 19

Series editor
Boris Martinac

More information about this series at http://www.springer.com/series/835

Richard M. Epand • Jean-Marie Ruysschaert
Editors

The Biophysics of Cell Membranes

Biological Consequences

Springer

Editors
Richard M. Epand
Biochemistry and Biomedical Sciences
McMaster University
Hamilton, ON, Canada

Jean-Marie Ruysschaert
Sciences Faculty
Université Libre de Bruxelles
Bruxelles, Belgium

ISSN 0932-2353 ISSN 1868-2561 (electronic)
Springer Series in Biophysics
ISBN 978-981-10-6243-8 ISBN 978-981-10-6244-5 (eBook)
DOI 10.1007/978-981-10-6244-5

Library of Congress Control Number: 2017952612

Printed on acid-free paper

This Springer imprint is published by Springer Nature
The registered company is Springer Nature Singapore Pte Ltd.
The registered company address is: 152 Beach Road, #21-01/04 Gateway East, Singapore 189721, Singapore

Preface

Chapters in this volume discuss several aspects of the physical properties of biological membranes and how these properties influence their functioning. The reviews emphasize the mechanisms that result in these changes in membrane properties and function.

One of the rapidly developing areas in membrane biophysics in recent years has been the role of transbilayer lipid asymmetry. It is known that the lipid composition of the bilayer of a biological membrane is very different for the two monolayers that compose this bilayer. Most studies of model membranes have employed membranes with identical composition for the two monolayers. There are technical difficulties in making model membranes with transbilayer asymmetry. Chapter 1 describes the methods that are being developed to facilitate the preparation of asymmetric model membranes and how the presence of this transbilayer lipid asymmetry affects the physical properties of the membrane. The maintenance of transbilayer lipid asymmetry is intimately connected with the rates of lipid flip-flop, *i.e.* the movement of lipid from one face of the bilayer to the opposite side. In model membranes devoid of protein, flip-flop rates of polar lipids are generally very slow. However, in biological membranes these rates can be accelerated by specific proteins, in some cases using an active transport mechanism, as well as through non-specific disordering of membrane packing compared with a pure lipid membrane. Chapter 2 discusses how the flipping rate is dependent on both the chemical structure of the lipid as well as on the physical state of the membrane. Results of studies of flip-flop rates obtained both from experiments as well as computer simulations are presented.

The two principle components of biological membranes are proteins and lipids. The function of membrane proteins is modulated by lipids, both by binding to specific lipid binding sites on the protein as well as by modulating the general biophysical properties of the membrane. Some of these properties, including the formation of supercritical fluids as well as long range interactions involving curvature stress, curvature elasticity and hydrophobicity play key roles in the coupling of lipids and proteins. The mechanisms of this modulation of membrane protein function through coupling with the physical properties of the membrane are reviewed in Chap. 3. Chapter 4 describes how mechano-sensitive channels can be gated by stretching of the bilayer(forces-from-lipids principle)and/or by the forces conveyed to the channel from the cytoskeleton/extracellular matrix(force-from filament). The final two Chapters deal with larger scale systems. Chapter 5 discusses mechanisms of changes in cell shape. The factors involved can include the cytoskeleton, membrane-bending proteins and membrane biophysical properties including a role for lipid domains in cell membranes. The final Chapter considers the liposome as a minimal cellular model that can be used to simulate diverse processes from the origin-of-life to a reconstituted biochemical pathway. The possibility of applying such systems for future biotechnological applications is also considered.

This volume thus summarizes, from diverse points of view, the nature of membrane biophysical properties and how these properties impinge on the various functions of a biological membrane.

Hamilton, ON, USA Richard M. Epand
Bruxelles, Belgium Jean-Marie Ruysschaert

Contents

Chapter 1
Preparation and Physical Properties of Asymmetric Model Membrane Vesicles

Johnna R. St. Clair, Qing Wang, Guangtao Li, and Erwin London

Abstract Model biomembrane vesicles composed of lipids have been widely used to investigate the principles of membrane assembly and organization. A limitation of these vesicles has been that they do not mimic the transbilayer lipid asymmetry seen in many natural membranes, most notably the asymmetry in the plasma membrane of eukaryotic cells. Recently, a number of approaches have been developed to prepare asymmetric membranes and study their properties. This review describes methods to prepare asymmetric model membranes, and the physical properties of asymmetric lipid vesicles. Emphasis is placed on the vesicles prepared by cyclodextrin-catalyzed exchange, which has proven to be a versatile and powerful tool, including for studies manipulating lipid asymmetry in living cells.

Keywords Membrane domains • Liquid ordered state • Sphingolipids • Phospholipids • Cyclodextrins

1.1 Lipid Asymmetry: Definition and Origin

When studying biological membrane organization and function, one important aspect to consider is lipid asymmetry. Lipid asymmetry refers to the difference in lipid composition in the outer (exoplasmic, exofacial) leaflet vs. the inner (cytoplasmic, cytofacial) leaflet of a membrane. Many cell membranes possess lipid asymmetry. In mammalian cells, the outer leaflet of the plasma membrane is enriched in sphingomyelin (SM), glycosphingolipids (GSL) and phosphatidylcholine (PC), while the inner leaflet is composed mainly of phosphatidylethanolamine (PE), and anionic lipids such as phosphatidylserine (PS) and phosphatidylinositol (PI) (Fig. 1.1) [1]. Cholesterol is present in both leaflets, but its distribution is still in dispute [2, 3].

J.R. St. Clair • Q. Wang • G. Li • E. London (✉)
Department of Biochemistry and Cell Biology, Stony Brook University, Stony Brook, NY, 11794-5215, USA
e-mail: erwin.london@stonybrook.edu

R.M. Epand, J.-M. Ruysschaert (eds.), *The Biophysics of Cell Membranes*, Springer Series in Biophysics 19, DOI 10.1007/978-981-10-6244-5_1

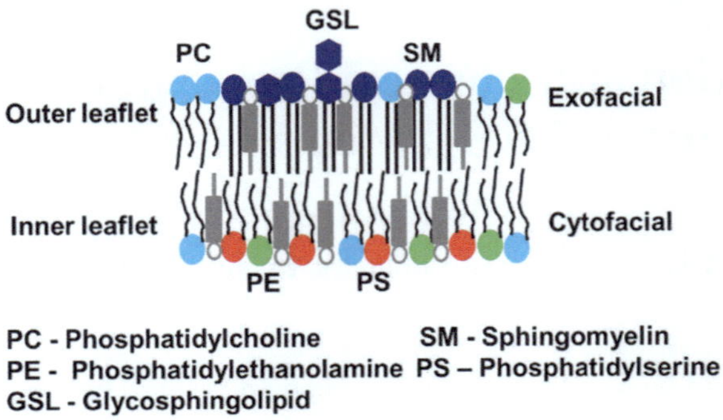

Fig. 1.1 Representation of lipid asymmetry in natural biomembranes. In mammalian cells the outer, or exofacial, leaflet is enriched in saturated acyl-chain sphingomyelin and in phosphatidylcholine, while the inner, or cytofacial, leaflet is composed primarily of phosphatidylethanolamine and phosphatidylserine. Cholesterol, shown in gray is present in both leaflets

In cells, lipid asymmetry is maintained by flippases and floppases, enzymes that control the movement of lipids across the bilayer, and by enzymes that synthesize and degrade lipids in one or the other leaflet [4]. Lipid flip-flop, or the transverse diffusion of lipids from one leaflet to another, counteracts asymmetry. The rate of spontaneous phospholipid flip-flop in the absence of proteins is generally slow, and can take days [5–9]. In contrast, cholesterol with its small and weakly polar headgroup, can cross the lipid bilayer in a minute or less [10].

1.2 Biological Function of Asymmetry

The full significance of lipid asymmetry remains elusive, but is known to be important in several biological processes. For example, the loss of PS asymmetry and the resulting display of PS in the outer leaflet of cell membranes is a signal which leads to the consumption of apoptotic cells by phagocytes [11], and is also a signal indicating that the membrane has been damaged, promoting blood clotting [12]. Some viruses even display PS in their outer leaflet to encourage engulfment by macrophages and achieve host infection [13, 14].

Asymmetry may also affect lipid-protein interaction. Transmembrane (TM) proteins typically have a higher positive charge at the cytofacial end of their TM helices relative to their exofacial end. This is known as the positive-inside rule [15]. Lipid charge asymmetry, with a higher negative surface charge at the inner leaflet/cytofacial surface, may help determine TM protein orientation, as well as influence the conformation of positively charged cytofacial juxtamembrane sequences [16]. Importantly, it has recently been observed that plasma membrane TM segments also have an orientational preference as judged from their abundance in natural sequences. The segments of TM helices having amino acids with smaller side chains exhibit a preference to be located in the outer leaflet relative to the inner leaflet [17].

1.3 The Physical State of Membranes: Membrane Domains and How They Might Be Influenced by Asymmetry

Another important aspect of membranes that is likely to be affected by asymmetry is lipid physical state. In a lipid bilayer composed of only phospholipids or sphingolipids there are two common physical states: the gel state and the liquid disordered state (Ld). As the name implies, lipids in the Ld state are disordered and loosely packed. Lipids in the gel state are tightly packed and solid-like, with much less lateral mobility/diffusion than Ld state lipids. In a pure lipid bilayer each individual lipid has its own characteristic melting, or transition temperature (Tm), at which the gel state will reversibly melt to form the Ld state. In lipid bilayers with mixtures of both high Tm lipids and low Tm lipids, gel and Ld states can co-exist in the same bilayer. The addition of cholesterol to gel state lipids generally changes the bilayer's physical state to what is known as the liquid ordered (Lo) state [18, 19]. Since sphingolipids and other lipids with saturated acyl chains have high Tm values, they readily form the Lo state. The Lo state has lipids that are tightly packed and ordered, as in the gel state, but with fast lateral diffusion, as in the Ld state. By light microscopy it is possible to observe co-existing Lo and Ld lipid phases in giant unilamellar vesicles (GUVs) composed of mixtures of high Tm lipids, low Tm lipids and cholesterol, for example, sphingolipids, unsaturated phospholipids and cholesterol [20, 21], and in giant plasma membrane vesicles (GPMV) derived from eukaryotic plasma membranes, which are rich in sphingolipids and cholesterol [22].

The lipid raft model posits that sphingolipid and cholesterol-rich Lo domains (lipid rafts) co-exist with Ld domains in living cells [18]. Lipid rafts may be an important feature of lipid organization in natural membranes. They have been proposed to play an important role in many cellular processes such as amyloid formation, protein and lipid sorting, cell signal transduction, and pathogen invasion [23–30].

If lipid rafts are an integral component of the machinery for transmitting information through the bilayer from the outside of the cell to the inside, then an interesting question arises: since the inner leaflet of the bilayer contains little to no raft-forming sphingolipids, then how can Lo domains form in the inner leaflet? The answer to this may be that outer leaflet lipids influence the physical properties of the inner leaflet: the physical properties of the two leaflets could be 'coupled' [31, 32]. This coupling could transfer information across membranes via the lipids themselves. For example, it is possible that domains are induced in the inner leaflet and these domains could then concentrate cytosolic-anchored proteins which have a high affinity for ordered domains, such as proteins anchored by saturated acyl chains [33, 34]. This could then lead to formation of specific protein-protein interactions (Fig. 1.2). Hints that interleaflet coupling may occur come from studies by the Mayor group which have shown that transmembrane interactions between outer leaflet long acyl-chain lipids and inner leaflet phosphatidylserine are crucial in generating actin dependent clustering of cytofacial lipid-anchored proteins [35]. It should be noted that this would represent a very different mechanism for signal

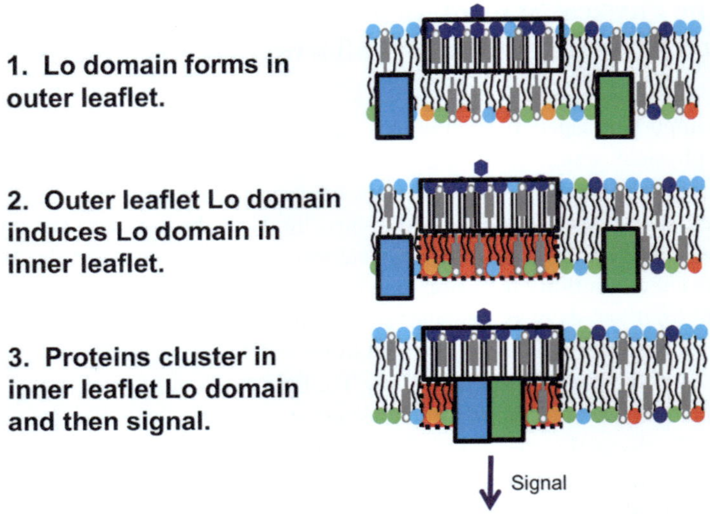

Fig. 1.2 Schematic illustration of interleaflet coupling giving rise to lipid-mediated signal transduction across membranes. Lo forming outer leaflet lipids are shown inducing Lo domains in the inner leaflet. This in turn may cause protein clustering in inner leaflet Lo domains which regulates signal transduction

transduction across membranes than that mediated by transmembrane proteins, and might provide a new target for biomedical applications.

1.4 Symmetric Model Membrane Vesicles

Techniques employing artificial membrane vesicles for the exploration of the principles governing natural membrane structure and function have been well-established. The most commonly used vesicles have lipid symmetry, i.e. the lipid composition in both leaflets is identical, or near-identical. Symmetric model membrane vesicles can be made from a variety of phospholipids and sphingolipids, with or without sterols, and in a variety of sizes. Multi-lamellar vesicles (MLV) can be prepared by adding buffer to a dried lipid film and then agitating. As the name implies MLV are many-layered, and they can be highly variable in size, with 8–15 concentric bilayers enclosed in a vesicle ranging from 0.5 to several microns in diameter [36]. Small unilamellar vesicles (SUV) are a single bilayer and are generally less than 50 nm in diameter. SUVs can be made by the sonication of dried lipids in an aqueous medium, or by dilution from ethanol [37, 38]. Generating SUVs with ethanol injection allows control over the size of SUVs by adjusting the concentration of lipid in ethanol before dilution [37]. Due to their small size, a feature of SUVs is a high level of curvature stress on the lipids and this has been shown to impact lipid packing [39]. Large unilamellar vesicles (LUV), commonly with a diameter of 100–200 nm, are relatively free from the curvature stress of SUVs. They can often be prepared by subjecting MLVs to multiple cycles of freeze-thaw followed by extrusion through membrane filters with pores of desired size, to obtain vesicles with a carefully controlled diameter [40]. Giant unilamellar vesicles (GUV) can approximate the size of living eukaryotic cells, ranging from 10,000 to 50,000 nm in diameter. GUV can be generated by electroformation [41], gentle hydration [42], and a number of oil-in-water techniques [43–51]. They are particularly useful for light microscopy studies.

Using NMR, interleaflet coupling was investigated in sonicated, symmetric SM SUV in one early pioneering study [52]. It was found that ion-induced changes in transition (melting) temperatures in outer leaflet could be partly transmitted to inner leaflet, indicating some level of interleaflet coupling, and it was proposed that interleaflet coupling could be a mechanism to transmit information across bilayers. Furthermore, it was found that the SM chain length influenced coupling, which was detected when SM acyl chain length was long. This could reflect some effect of acyl chain interdigitation.

1.5 Asymmetric Model Membranes: Planar Bilayers

Several different methods have been developed for the construction of asymmetric planar supported bilayers. The Langmuir/Blodgett (LB) [53] and Langmuir–Schaeffer (LS) [54] methods involve first depositing a monolayer of lipids onto a silicon substrate, and then adding a second monolayer by dipping the substrate again through an aqueous phase containing second leaflet lipids. Watanabe et al. recently modified supported planar bilayer techniques to develop a system with a capacity for simultaneously generating over 10,000 asymmetric supported lipid bilayers [55].

Fluorescence Interference Contrast Microscopy (FLIC) was used to compare the stability of asymmetry in planar supported bilayers constructed with several techniques [56]. Lipid asymmetry was measured in bilayers with and without cholesterol in Lo and Ld phases. Lipid asymmetry remained stable at 80% for up to 6 h, as judged by fluorescent probe partitioning in supported bilayers generated using the Langmuir-Blodgett/vesicle fusion (LB/VF) method.

Sum frequency vibrational spectroscopy also was used to examine the kinetics and thermodynamics of lipid asymmetry. In cholesterol-free supported planar LB/LS bilayers, asymmetry stability was limited to several hours and was variably dependent on headgroup structure [57], acyl chain length, lateral surface pressure [58], and acyl chain saturation [59]. Later work in LB/LS supported bilayers investigated the effects of sterol structural variation on flip-flop rates [60]. Vibrational spectroscopy revealed that in distearoyl PC (DSPC) bilayers the stability of lipid asymmetry varied considerably with sterol concentration and structure. The lack of stability seen in these studies may be due to the presence of bilayer defects specific to supported membranes [5].

Other efforts have concentrated on preparation of unsupported or cushioned planar bilayers in order to avoid the influence of a supporting substrate upon lipid behavior. Unsupported asymmetric planar bilayers can be made via the Montal-Mueller method [61]. This entails forming two separate lipid monolayers on each side of a moveable Teflon partition containing a small hole. Carefully moving the Teflon partition allows the formation of a bilayer at the hole in the partition where the monolayers make contact. Using asymmetric unsupported bilayers, the Keller group observed that in mixtures of diphytanoyl PC, dipalmitoylPC (DPPC) and cholesterol, Lo domains can form in one leaflet independently of the other and that by adjusting the lipid composition of one leaflet, domain formation in the opposing leaflet could be suppressed [62]. However, the use of hexadecane to prepare the monolayers raises the issue of residual solvent, which if present between the lipid leaflets could influence interleaflet coupling.

Avoiding issues of solvent and support effects by adapting the LB and LS methods, the Tamm and Naumann groups have used asymmetric bilayers cushioned by polymer tethered to lipids. The Tamm group generated a tethered double bilayer system and used single particle tracking to compare the mobility of the transmembrane protein syntaxin-1A in tethered vs. supported bilayers. The work demonstrated that there was no significant difference in syntaxin-1A mobility in tethered vs. supported bilayers [63]. Using the same approach the Naumann group examined the influence of lipid asymmetry on the sequestering and oligomerization behavior of integrins $\alpha v\beta 3$ and $\alpha 5\beta 1$ in bilayers [64]. The behavior of these TM proteins in asymmetric bilayers differed significantly from that in symmetric bilayers.

1.6 Asymmetric Lipid Vesicles

Lipid vesicles closely resemble natural biological membranes in that they are unsupported, fully hydrated and the bilayer is unbounded. Because of this, the results of work using lipid vesicles may be most informative for comparison to biological systems. Even more closely resembling natural biomembranes, would be asymmetric vesicles that reflected both the lipid diversity and asymmetry found in the membranes of living cells.

Early work in the development of asymmetric lipid vesicles examined spontaneous fluorescent lipid analog exchange between vesicle populations [65]. It was demonstrated that NBD-PC fluorescent lipid analogs could spontaneously transfer from the outer leaflet of 'donor' lipid vesicles to outer leaflet of 'acceptor' lipid vesicles. This would result in asymmetric vesicles in the sense that the fluorescent lipid would be restricted to a single leaflet after exchange. In some cases, as much as 50% of fluorescent donor lipids could be transferred to the acceptor vesicles. However, the method was limited in that it required use of fluorescent lipids in which one acyl chain was modified with a somewhat polar group.

1.7 Making Asymmetric Vesicles with Phospholipid Carrier Proteins

In other early work, the Zilversmit group used phospholipid exchange protein isolated from beef heart cytosol to exchange lipids between erythrocyte ghosts and PC/cholesterol liposomes [66]. In doing so they explored the natural lipid asymmetry in red blood cells and confirmed that lipid asymmetry is stable with sphingomyelin (SM) and PC largely limited to the outer leaflet. In studies using phospholipid exchange protein isolated from beef liver, the asymmetric transbilayer distribution of SM, PE, and PC was assessed [67]. The transfer of radiolabeled phospholipids from intact rat erythrocytes to unilamellar vesicles was measured as a function of time. It was found that all of the SM and more than half of the PC were located in the outer leaflet of the erythrocytes. Since nearly no PE was transferred, it appeared to be largely confined to the inner leaflet.

Also using lipid-exchange proteins, the Holloway group prepared asymmetric membranes to examine the depth of insertion of transmembrane protein cytochrome b5 in membranes. They generated model membranes in which only one leaflet contained phospholipids with brominated acyl chains, which quench tryptophan fluorescence. They found that cytochrome b5 orients in the bilayer with the tryptophan in its hydrophobic tail 7 nm from outer leaflet surface [68].

Lipids bound to bovine serum albumin (BSA) were used to generate asymmetry in rat liver endoplasmic reticulum vesicles for the study of lipid flip-flop rates [69]. A series of spin-labeled phospholipids were introduced into the outer leaflets of rat liver ER vesicles by binding them to BSA. To determine lipid flip-flop rates, ESR spectroscopy and kinetics assays were performed. Results showed that in all observed cases, phospholipid flip-flop was fast, with a half time of 20 min at 37 °C, and not dependent on lipid head group type. However, the spin-labelled phospholipids lacked long acyl chains, so the effects of acyl chain length on flip-flop rates could not be gauged.

Holzer et al. used pro-sterol carrier protein (pro-SCP2) to generate vesicles that were asymmetric with regard to negatively charged egg phosphatidylglycerol (PG) [70]. The degree of asymmetry in the vesicles was then measured using free-flow electrophoresis. Membrane curvature was found to be important for lipid transfer efficiency by pro-SCP2. By using both small donor and small acceptor lipid vesicles (50 nm diameter) exchange efficiency was increased by 55% over that achieved with larger vesicles.

1.8 Using pH Gradients to Make Vesicles with Anionic Lipid Asymmetry

The Cullis group induced asymmetry in vesicles containing negatively charged lipids by inducing a pH gradient across the bilayer. Following acidification of the exterior of PG-containing vesicles relative to the luminal cavity, 50% of PG

(5% of total outer leaflet lipids) moved to the inner leaflet in within 50 min [71]. Increased acyl chain saturation, chain length and inclusion of cholesterol significantly decreased this effect by decreasing membrane permeability and therefore the ability of PG to move across the bilayer. It was also shown that it was possible to reversibly manipulate the distribution of the negatively charged egg PG and egg phosphatidic acid (PA) in the bilayer [72]. EggPG and eggPA, but not the zwitterionic lipid dioleoylPE (DOPE) could be driven to the inner or outer leaflet, depending on the direction of the proton gradient. Cryoelectron microscopy was used to examine morphological changes in the bilayers of dioleoyl PG (DOPG)-containing LUVs with pH gradient-induced DOPG asymmetry. Flipping of DOPG from the outer to inner to leaflet or vice versa generated inversions and tubular protrusions [73]. This approach is limited to the movement of anionic lipids, and by the small percentages of anionic lipid that could be used (10% of total). Furthermore, the pH gradient necessary to induce movement of anionic lipids would not often be suitable for use in studies of protein conformation or function in asymmetric vesicles.

1.9 Making Asymmetric Vesicles with Water-in-Oil Techniques: Centrifugation Method

The Weitz group developed an oil-in-water technique to generate asymmetric giant unilamellar vesicles (GUV) [74]. In the method, an emulsion of dodecane, inner leaflet lipids and water was first prepared. This was layered over an intermediate phase consisting of dodecane and outer leaflet lipids, which was directly supported on an aqueous phase. As the droplets, which would have inner leaflet lipids in an inverted micelle type of structure, passed via centrifugation through the dodecane/outer leaflet layer into the aqueous phase, a monolayer of outer leaflet lipids was deposited on the outside of the inner leaflet monolayer vesicles, forming asymmetric bilayers. Using this technique, a controllable, high level (95%) of lipid asymmetry was reported.

Studies using vesicles prepared by oil-in-water methods to examine the effect of lipid asymmetry on the mechanical properties of bilayers showed that asymmetry significantly increases membrane rigidity. Two types of fluorescently-labelled asymmetric GUVs: dioleoyl PC $(\text{DOPC})_{\text{in}}$/1-palmitoyl-2-oleoyl PC $(\text{POPC})_{\text{out}}$ and $\text{POPC}_{\text{in}}/\text{DOPC}_{\text{out}}$ (in = inner leaflet lipid, out = outer leaflet lipid) were generated. Membrane rigidity was calculated by thermal fluctuation analysis of phase contrast micrographs. Results revealed that both types of asymmetric membranes exhibited a significantly higher bending rigidity compared to symmetric membranes consisting of either DOPC, POPC, or 1:1 DOPC:POPC [75].

The Takagi group has developed a method to control the size of water-in-oil asymmetric GUV. By adjusting the density of luminal contents with water-soluble molecules such as sucrose and glucose, the rate of transfer and therefore the size of vesicles could be controlled. Fluorescently labelled lipids were incorporated into the cholesterol-containing asymmetric GUVs and facilitated the observance of microdomain formation and confirmed lipid asymmetry [48].

1.10 Making Asymmetric Vesicles with Water-in-Oil Techniques: Droplet Interface Bilayers

Hwang and colleagues examined the effect of lipid asymmetry on the behavior of membrane proteins in droplet interface bilayers (DIB) [51]. In this method two types of aqueous droplets surrounded by inverted lipid monolayers with different lipids are deposited in an oil mixture. As the droplets meet, and monolayers come into contact, and an asymmetric bilayer forms at their junction. Employing monolayer droplets with differences in lipid charge, bilayers possessing a charge gradient were constructed and used to measure charge gradient-induced changes in the insertion and gating behavior of the outer membrane protein G (OmpG) from *Escherichia coli*. This method is limited to use with lipids that are soluble in oil, but allows fine control over mixtures of lipids for each monolayer.

1.11 Making Asymmetric Vesicles with Water-in-Oil Techniques: Microfluidic Approaches

The introduction of microfluidic techniques to the production of asymmetric lipid vesicles has enabled methods for the generation of water-in-oil GUV with controllable size, lipid asymmetry, and lumenal content. The Malmstadt group used a microfluidic flow-based layer-by-layer approach to produce asymmetric vesicles with biologically-relevant PS limited to a single leaflet [46]. First sucrose-loaded aqueous droplets were introduced to an oil stream containing dissolved inner leaflet lipids to develop a lipid monolayer. The droplets were then released into a second oil phase containing dissolved outer leaflet lipids floating over an aqueous phase. Following centrifugation, the droplets passed through the aqueous phase and were collected. Lipid asymmetry was confirmed via differential fluorescence quenching and selective labeling with biotinylated lipids. The vesicles were size-selectable because vesicles larger than 120 μm do not survive the centrifugation step and those smaller than 10 μm don't transfer through the second step. Lipid asymmetry was confirmed with about 85% asymmetry. It was necessary to identify and select oil-free vesicles visually.

The generation of asymmetric vesicles for drug delivery is a promising area of research. In a study of drug cytotoxicity, microfluidics were employed to generate GUVs with a cross-linked, chemically stable inner leaflet composed of trichloro(1H,1H,2H,2H–perfluorooctyl)silane (TPS) and an outer leaflet containing

DOPC. When loaded with the cancer therapeutic 5-fluorouacil, the asymmetric vesicles exhibited a pH-dependent threefold higher cytotoxicity relative to the free drug [76].

A high-throughput microfluidic process for generating water-in-oil GUV was recently developed [47, 77]. In the method, water, oil and inner leaflet emulsions were first formed. Outer leaflet lipids were then introduced via replacement of inner leaflet lipid solution with a second solution. Water-in-oil double emulsions were then formed and followed by the extraction of excess oil. Asymmetry and unilamellarity were confirmed with fluorescence quenching assays and transmembrane protein insertion assays. In a second report using this same fabrication strategy, the effects of lipid asymmetry on the mechanical properties of asymmetric bilayers vs. symmetric bilayers were examined. In cholesterol-free symmetric vesicles composed of dimyristoyl PC (DMPC), DOPC a 1:1 mixture of DMPC:DOPC, or asymmetric vesicles with a $DMPC_{in}/DOPC_{out}$ or $DOPC_{in}/DMPC_{out}$ composition results showed that the bending moduli and expansion moduli of asymmetric bilayers are different that those of symmetric bilayers.

1.12 Minimizing Residual Oil Contamination

A complication resulting from oil-in-water techniques used to generate asymmetric GUV is residual oil contamination. This residual oil likely impacts the physical properties and behavior of the bilayer. To circumvent this, methods to minimize oil contamination have been developed.

A microfluidics approach in which lipid monolayers were brought together in a fashion similar to that in droplet interface bilayers was employed to generate asymmetric GUV likely to have only minimal oil and then used to examine the effects of lipid composition and asymmetry on lipid vesicle fusion mediated by SNARE-family transmembrane proteins [45].

In another study, a high-throughput microfluidics method was developed to form asymmetric GUV with minimal oil contamination. In the study, cholesterol-rich asymmetric GUVs with ‘ultrathin shell’ (residual oil) were generated for the observation of lipid microdomain formation. Lipid asymmetry and the presence of lipid domains were observed with the use of fluorescently labelled lipids and fluorescent microscopy [44]. In another attempt to address the issues of solvent contamination, an asymmetric planar lipid bilayer was formed into lipid tubules which were then broken up by applying a microfluidic jet flow of aqueous buffer. Unilamellarity and residual solvent content was evaluated using confocal Raman scattering microscopy [78]. Although these methods may greatly reduce oil contamination, it is not clear whether the residual oil that remains would not perturb lipid behavior to a significant degree.

1.13 Early Studies Using Cyclodextrin Exchange to Alter Lipid Composition

A different strategy that is becoming more widely used is to prepare asymmetric lipid vesicles with cyclodextrins (CDs). CDs are water-soluble cyclic oligosaccharides composed of glucose monomers linked together by α-(1, 4) glycosidic bonds, and are typically composed of 6(α), 7(β), or 8(γ) units. They assume a conical barrel shape with a non-polar cavity and hydrophilic exterior. Hydrophobic guest molecules can complex with CDs. A wide range of functional groups can be conjugated to hydroxyl groups at the edges of CDs to modify their binding specificity [79, 80] (see Fig. 1.3). Changes the number of monomers comprising a CD alters the volume of the hydrophobic cavity, which can be varied to exclude or include larger or smaller guest molecules. For example, the hydrophobic cavity of αCDs is smaller than that of βCDs and therefore binds well to phospholipid acyl chains, but binds cholesterol either not at all or very poorly [80–83].

Cyclodextrins have long been used to remove sterols from cells, introduce new sterols into cells, or exchange one sterol in cells for another [84–88]. Similarly, CDs have been used to exchange sterols between model membrane vesicles [10, 89].

Early work exploring the properties of CDs as lipid carriers involved interactions of γ-CDs and fluorescent lipids. Carboxyethyl-γ-CD accelerated fluorescent lipid transfer from lipid vesicles to cells in culture [90]. Another study reported that the large hydrophobic cavity of γ-CD could bind more than one lipid acyl chain simultaneously [91]. In addition, stoichiometric measurements using fluorescent binding assays showed that the association constants of pyrene-labeled PC/γ-CD complexes decreased strongly with increasing acyl chain length [91].

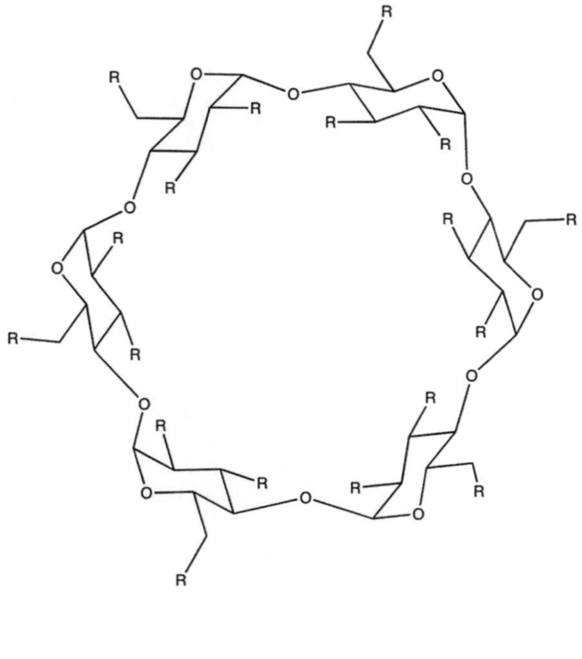

Fig. 1.3 α-cyclodextrin chemical structure. The 6-unit 1–4 linked α-D-glucopyranoside ring forms a barrel shape, with a hydrophobic interior cavity and a hydrophilic exterior surface. Functional groups can be conjugated to hydroxyl groups to modify their binding specificity. R = OH or modified OH

Other studies demonstrated that MβCD, among other cyclodextrins, could exchange phospholipids between artificial lipid vesicles, and at high concentrations even dissolve phospholipid vesicles [10, 92].

It was also shown that cyclodextrins can also cause the dissociation of small amounts of phospholipids from cells [86, 93].

1.14 Preparation and Properties of Asymmetric Vesicles Produced by MβCD-Catalyzed Exchange

The first study making use of CD to prepare highly asymmetric vesicles (AUV) employed MβCD to exchange lipids between the outer leaflets of two distinct vesicle populations [94]. The mixing of 'donor lipid'-loaded MβCD solutions (with excess donor lipid vesicles) and symmetric 'acceptor' SUV composed solely of the desired inner leaflet lipids was the key feature of the technique (Fig. 1.4). Two lipid exchanges were carried out: The first exchange catalyzed the installation of 'outer leaflet' lipids into acceptor SUV – effecting phospholipid asymmetry. Using this approach asymmetric SUV were prepared containing 75–100% SM in their outer leaflet, and DOPC, POPC, 1-palmitoyl-2-oleoyl PS (POPS) or a mixture of 1-palmitoyl-2-oleoyl PE (POPE) and POPS in their inner leaflet. A second,

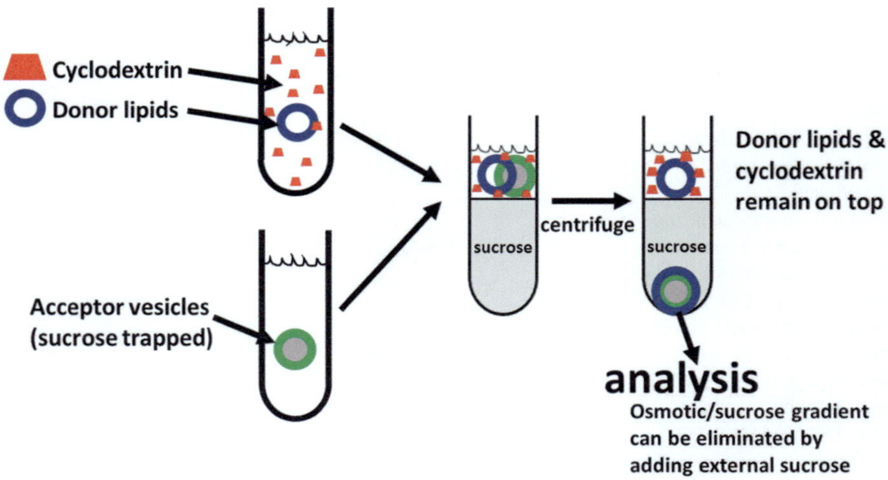

Fig. 1.4 A Scheme for Cyclodextrin-mediated Asymmetric Vesicle Preparation. Outer leaflet donor lipids in complex with cyclodextrin in solution (and some excess donor lipid vesicles) are incubated with sucrose-loaded inner-leaflet acceptor vesicles. The mixture is layered over sucrose at a lower concentration than that inside of the vesicles and ultracentrifuged. The asymmetric vesicle pellets and can be isolated from the other exchange reaction components. Notice that the resulting asymmetric vesicles will have an osmotic imbalance across the bilayer unless the osmotic strength of the external solution is adjusted after the vesicles have been isolated

optional exchange installed cholesterol to a final overall cholesterol concentration of approximately 25 mol%. The resulting asymmetric vesicles were purified from donor lipid mixtures by size chromatography. Fluorescence assays measured outer leaflet anisotropy and outer leaflet lipid charge and confirmed lipid asymmetry.

The thermal stability of ordered state SM domains in the outer leaflet of resulting AUV was found to be much higher than that in symmetric vesicles with the same overall composition, with a Tm approaching that of pure SM vesicles. This suggested that at the Tm of the outer leaflet, coupling to the inner leaflet composed of unsaturated phospholipids had broken down. Analogous properties were observed in asymmetric SUV with cholesterol (vs. corresponding symmetric SUV with the same lipid composition), except that the addition of cholesterol further thermally stabilized ordered domains in both symmetric and asymmetric vesicles. Experiments monitoring the orientation of hydrophobic helices suggested that membrane asymmetry could enhance formation of a transmembrane orientation within the bilayer.

A subsequent study developed a modified method to prepare plasma membrane-mimicking asymmetric LUV [95]. The lesser bilayer curvature of LUVs relieves the lipid packing stresses that exist in SUV. To do this, acceptor LUV were prepared with entrapped sucrose, and then isolated from donor lipids and MβCD after exchange by centrifugation. As with asymmetric SUV, nearly complete replacement of the outer leaflet lipid was achieved. Anisotropy assays revealed that the SM rich outer leaflet of asymmetric LUV melted at about the same temperature as symmetric LUV composed of pure SM, and at a much higher Tm than that of symmetric vesicles with the same overall lipid composition. This indicated that the weak interleaflet coupling at the temperature of outer leaflet melting seen in the prior study was not a result of the high membrane curvature in SUV.

MβCD was next used to prepare asymmetric GUVs for studies of interleaflet coupling with fluorescence correlation spectroscopy (FCS) [96]. In order to prevent the possibility that donor vesicles would stick to the GUV, residual donor vesicles were removed before mixing the donor lipid-MβCD mixture with acceptor GUV. This resulted in a lower, but still substantial, level of lipid exchange in the outer leaflet. Controls with fluorescent lipids demonstrated that stable lipid asymmetry was achieved in osmotically balanced vesicles. Osmotic balance across the bilayer is crucial because even a small imbalance has been shown to induce transient pore formation in GUV, which could potentially lead to a loss of asymmetry [97].

The introduction of SM into the outer leaflet of cholesterol-free DOPC GUV decreased lateral diffusion in the outer leaflet. Interleaflet coupling at room temperature was evaluated from the extent to which it also decreased lateral diffusion in the inner leaflet and was found to be stronger with inner leaflet brain PC (brain PC) than with DOPC.

In a follow-up study, the effect of varying acyl chain length on lateral lipid diffusion and membrane order in asymmetric GUVs and SUVs was further investigated [98]. As previously shown, MβCD-generated asymmetric GUV containing outer leaflet bSM and inner leaflet DOPC showed reduced lateral diffusion in the bSM outer leaflet, and little reduction in diffusion in the DOPC inner leaflet. However, when POPC or 1-oleoyl 2-myristoyl PC (OMPC) (both of which have one saturated and one unsaturated acyl chain) were used in place of DOPC, there was reduced lateral diffusion in both leaflets, with a similar level of reduction of diffusion in both leaflets, indicative of strong interleaflet coupling. Strong interleaflet coupling was further observed with the introduction of longer acyl chain milk SM or synthetic C24:0 SM into the outer leaflet opposing inner leaflet DOPC. This suggested that acyl chain length plays a role in interleaflet coupling. Interestingly, assays measuring inner leaflet lipid order in asymmetric SUV and GUV did not detect an increase in inner leaflet order when SM was introduced into the outer leaflet. For the SUV experiments, a recently developed method to selectively label inner lipids was used [99]. In this method inner leaflets are selectively labelled with fluorescent probes using low levels of MβCD and gel-filtration to remove outer leaflet fluorescent probes.

The difference in coupling with regard to diffusion and order is somewhat puzzling. One possible explanation is that order measurements are the average for an entire leaflet. It is possible that upon introduction of SM into the outer leaflet, an increase in order in inner leaflet ordered domains coupled to outer leaflet ordered domains rich in SM is "cancelled out" by a decrease in order in the remainder of the inner leaflet, which would be in contact with outer leaflet disordered domains rich in DOPC.

1.15 Studies of Lipid Flip-Flop in Asymmetric Vesicles

Next, a systematic study using MβCD exchange was undertaken to examine the effect of acyl chain structure upon the stability of asymmetry in SM_{out}/PC_{in} SUV in which the type of PC used was varied [9]. The level of outer leaflet exchange was quantified by comparing outer leaflet membrane order (assessed via fluorescence anisotropy of TMADPH added to the outer leaflet, TMADPH does not rapidly flip across the lipid bilayer) to that for a standard curve composed of symmetric bSM/PC vesicles with various % SM. Thin layer chromatography was used to measure overall lipid composition. Combining these two methods, the % of inner leaflet bSM, and thus the level of asymmetry, was calculated. In all cases exchange levels indicated, at a minimum, complete replacement of the outer leaflet PC with SM. For AUV containing PC with one saturated and one unsaturated acyl chain, asymmetry was virtually complete if the unsaturated chain had one or two double bonds. However, a small loss of loss of asymmetry was detected when the unsaturated acyl chain in the PC had four double bonds. AUV containing PCs with two monounsaturated acyl chains maintained full asymmetry for all except those with the shortest acyl chain studied (14 carbons), which again showed a small loss of asymmetry. In contrast, exchange-produced vesicles containing PCs with two polyunsaturated chains lost most (with two double bonds per chain) or all (with three or more double bonds per chain) of any asymmetry by the time asymmetry measurements were made, several hours after lipid exchange.

Measurement of transverse diffusion (flip-flop) rates for NBD-lipids in symmetric PC and SM/PC vesicles confirmed that the loss of asymmetry was associated with increased lipid flip-flop, and gave flip-flop values consistent with the literature for the effect of PC acyl chain structure upon lipid flip [100]. Acyl chain polyunsaturation may promote lipid flip-flop due to its effect of decreasing how tightly lipids are packed, and/or due to the increased polarity of the core of the bilayer, as C-C double bonds have increased polarity relative to single bonds [101]. Increased polarity would increase the solubility of lipid polar headgroups within the bilayer core.

A systematic study was also undertaken of the effects of phospholipid polar head group structure on the stability of bilayer asymmetry [8]. MβCD was used to generate asymmetric SUV with outer leaflet bSM and inner leaflet 1-palmitoyl-2-oleoyl-PG (POPG), DOPG, soy PI, tetraoleoylcardiolipin (tetraoleoyl CL), or POPS, in all cases alone or in combination with POPE. Inner leaflet dioleoyl PA (DOPA) was studied in a mixture with POPC. Proton NMR spectroscopy and observed changes in the thermal stability of SM-rich ordered domains helped detect an asymmetric distribution of SM and its decay over time. AUV containing a mixture of PE and anionic lipids maintained nearly full asymmetry that was stable for more than 1 day. However, most exchange vesicles containing anionic lipids without PE only showed partial asymmetry (which was assayed within hours after they were prepared) that decayed further after 1 day of storage. Asymmetry also decayed in 1 day for the vesicles containing PA. Interestingly, exchange vesicles with PS showed almost full asymmetry that was stable over 1 day even in when PE was not present in the inner leaflet. It is possible that lipids with a single charged anionic group undergo transverse diffusion across membranes more easily than lipids combined with multiple charge groups, such as PC, SM, PE and PS.

In a study using a modification of the lipid-exchange method to prepare asymmetric LUV without trapped sucrose (see below), 1H NMR was used to measure the flip-flop rates as a function of temperature using asymmetric vesicles in which chain-perdeuterated dipalmitoyl PC (DPPC-dC) was introduced into the outer leaflet of acceptor LUVs initially composed of headgroup-deuterated acceptor lipid (DPPC-dH) [5]. Flip-flop in gel phase bilayers was undetectable, while in the Ld state, from 50–65 °C, flip-flop half-time rates ranged between days to weeks. Intriguingly, flip-flop in asymmetric vesicles incubated at the main transition temperature (40 °C) for DPPC, was twice as rapid as in fully melted DPPC at 50 °C, indicative of defect-accelerated flip-flop in vesicles when gel and Ld phases co-exist.

1.16 Use of HPαCD to Prepare Asymmetric Vesicles Mimicking Plasma Membranes

Because MβCD binds well to cholesterol, it is difficult to control the levels of cholesterol in AUV during MβCD-mediated phospholipid exchange. To circumvent this, hydroxypropyl-α-cyclodextrin (HPαCD) was used instead. The hydrophobic cavity of the six glucose unit ring of HPαCD is too small to bind cholesterol, but can bind to the acyl chains of phospholipids. In a study comparing MβCD, HPαCD and HPβCD [81], it was found that while only MβCD could dissolve lipid vesicles, all three cyclodextrins could catalyze lipid exchange between vesicles.

HPαCD was used to construct plasma membrane-mimicking asymmetric LUVs with SM and/or POPC in the outer leaflet, POPE and POPS in the inner leaflet, and a wide range of cholesterol concentrations [82]. Efficient exchange (as high as 80–100% replacement of the outer leaflet as judged by thin layer chromatography) could be achieved, especially at high levels of cholesterol. Exchange of POPC donor lipid was somewhat more efficient than that of SM. Asymmetry was confirmed by chemical labelling of externally exposed POPE with TNBS, and by using a cationic peptide binding assay to measure the amount of externally-exposed anionic lipid. Asymmetry was stable for days.

1.17 Domain Formation in Asymmetric GUV Containing Cholesterol

HPαCD-catalyzed lipid exchange was used to prepare and investigate domain forming properties of asymmetric GUV in a following study [102]. Cholesterol-containing GUV with a mixture of SM and DOPC in their outer leaflets and DOPC in their inner leaflet were studied. Surprisingly, SM (and presumably cholesterol)-rich liquid ordered domains in the outer leaflet induced ordered domains in corresponding region of the inner leaflet, despite the strong tendency of DOPC, which has two unsaturated oleoyl chains, to remain in a relatively disordered state, even when mixed with cholesterol [103]. It was proposed that a lateral rearrangement of the inner leaflet was occurring, in which the regions of the inner leaflet in contact with the outer leaflet became enriched in cholesterol and depleted in DOPC. Supporting this, the NBD-DOPE used to probe inner leaflet domain formation, which like DOPC has two unsaturated oleoyl chains, was also depleted in the inner leaflet ordered domains in contact with outer leaflet ordered domains.

The contribution of SM acyl chain interdigitation to interleaflet coupling in the asymmetric GUV was investigated by comparing coupling when the outer leaflet contained egg SM, (predominately having C 16:0 acyl chains) vs. milk SM, which is rich in C 22–24 acyl chains that can interdigitate into the inner leaflet. With both types of SM, inner leaflet ordered domains formed. However, only when milk SM was used did inner leaflet ordered domains show probe partitioning properties similar to those in the outer leaflet, indicating stronger interleaflet coupling. This is consistent with the above-noted coupling seen in GUV lacking cholesterol.

1.18 Studies Using Cyclodextrin-Catalyzed Lipid Exchange to Study the Effect of Phospholipid Composition and Asymmetry Upon Membrane Protein Behavior

Transmembrane protein behavior can be influenced by lipid asymmetry in SUV. An asymmetric distribution of anionic lipids favored formation of a transmembrane configuration relative to a non-transmembrane (but membrane bound) state in which the peptide inserted only shallowly within the lipid bilayer [94].

The conformational behavior of the pore-forming toxin protein perfringolysin (PFO) was studied in AUV prepared using HPαCD to determine if lipid asymmetry could influence the conformation of this full length protein when it is membrane-inserted [21]. In a cholesterol concentration-dependent manner, PFO forms a large, homo-oligomeric β-barrel pore complex that spans the membrane bilayer of mammalian cells. PFO behavior was evaluated in asymmetric LUVs with a POPC outer leaflet, POPE-POPS inner leaflet and varying concentrations of cholesterol. Results were compared to those for PFO mixed with a variety of symmetric vesicles. This included symmetric vesicles with an overall lipid composition identical to that in the AUV, those containing only outer leaflet AUV lipids, and those composed of only the inner leaflet AUV lipids.

PFO binding and insertion were assessed by monitoring intrinsic Trp emission and the fluorescence intensity of an acrylodan-labeled residue. PFO oligomerization was measured with FRET and with SDS-agarose gel electrophoresis, as PFO oligomers do not unfold in SDS once the beta-barrel has formed. Pore-forming capacity was assayed by measuring the increase in externally-added BODIPY fluorescence emission intensity when BODIPY-labeled streptavidin was added to AUV with trapped biocytin that is released upon pore formation.

These assays demonstrated that the dependence of PFO conformation as a function of cholesterol concentration in AUV was different from that in symmetric vesicles of the same overall composition, and also different from that in symmetric vesicles that mimicked the outer or inner leaflets of the AUV. Interestingly, in the AUV a non-pore-forming, but transmembrane embedded, oligomeric conformation of PFO was observed at intermediate cholesterol concentrations. This may represent an intermediate stage in the PFO pore assembly process.

Cyclodextrin-mediated lipid exchange has also been used to investigate the effect of *in situ* changes in lipid composition upon the topography of lactose permease (LacY) reconstituted into model membrane vesicles. Following MβCD-induced enrichment of outer leaflet PE in proteoliposomes containing LacY in a partly inverted transmembrane conformation, LacY very quickly flipped to its native transmembrane conformation. Conversely, MβCD-mediated exchange introducing PG or CL into PE vesicles induced the flipping of LacY to form the partly inverted transmembrane orientation. The asymmetry of the lipids after the lipid exchange steps was not investigated [104, 105].

Membrane domain localization studies of nicotinic acetylcholine receptor (AChR) reconstituted in proteoliposomes found that localization of AChR within ordered lipid domains can depend upon lipid asymmetry [106]. Using both symmetric and asymmetric model systems containing cholesterol and different SM species, localization of AChR was assessed by FRET and by quantification of the distribution of AChR in detergent resistant and detergent soluble membrane fractions. AChR was reconstituted in symmetric vesicles composed of 1:1 POPC:Chol, or 1:1:1 POPC:Chol:SM. To prepare asymmetric vesicles, SM was exchanged into the outer leaflet of proteoliposomes containing AChR and POPC and cholesterol. In the case of 16:0 SM or 18:0 SM-containing symmetric vesicles, AChR was found to localize in Lo domains. No such Lo preference was found in symmetric vesicles containing bSM or 24:1 SM. However, in asymmetric models with bSM enrichment in the outer leaflet, AChR showed a strong preference for Lo domains.

1.19 Use of MαCD to Prepare Asymmetric Lipid Vesicles

Preparation of asymmetric LUV using MαCD was achieved very recently [as a portion of a study achieving lipid exchange in cells using MαCD, see below] [107]. Since MαCD does not interact well with cholesterol, this again allowed exchange of phospholipids without disturbing cholesterol levels in the acceptor vesicles from which the asymmetric LUV were prepared. MαCD was able to solubilize lipids at an even lower concentration than MβCD, indicating a strong interaction with lipids. In addition, like MβCD, MαCD had a relatively low specificity for lipid structure as judged by the concentration of MαCD needed to solubilize vesicles composed of various lipids. This made it possible to carry out exchange with various lipids at relatively low MαCD concentrations. Efficient exchange was obtained, but in contrast with HPαCD exchange was a bit more efficient for SM than POPC. Most importantly, the combination of strong phospholipid and sphingolipid interactions, coupled with negligible interactions with cholesterol, enabled applications of MαCD-induced lipid exchange to living cells, as described below.

1.20 Studies Using Cyclodextrin-Catalyzed Lipid Exchange to Prepare Asymmetric Planar Bilayers

Cyclodextrin-mediated exchange has also been extended to preparation of asymmetric planar lipid bilayers [108]. Fluorescence correlation spectroscopy was used to measure lipid mobility in each leaflet independently in asymmetric supported bilayers (aSLB). The method was compatible with protein reconstitution. DOPC/GPI-anchored placental alkaline phosphatase (PLAP) proteoliposomes were fluorescently-labelled and applied via a vesicle fusion method to a mica support. Leaflet-specific labeling of the supported leaflet was achieved by destroying NBD-DOPE fluorescence in the solution-exposed outer leaflet using sodium dithionite. Lipid asymmetry was induced using successive incubations with MβCD-loaded bSM. Cholesterol was introduced in additional exchange steps. Asymmetry was stable for several hours. Interestingly, no large scale lipid phase separation was observed in the aSLBs until asymmetry began to break down and bSM was present in both leaflets. Any domains present in the fully asymmetric aSLB were submicroscopic.

1.21 Modifying Cyclodextrin-Catalyzed Lipid Exchange for Study of Lipid Packing Using Small Angle Neutron Scattering

MβCD-catalyzed exchange has also been adapted for small-angle neutron scattering (SANS) studies [109]. This required modifying the protocol for MβCD-mediated exchange for LUV, in which trapping sucrose within the lumen of the vesicle aids AUV isolation, as the trapped sucrose interferes with SANS measurements. To avoid this, sucrose-loaded donor lipid vesicles preincubated with MβCD were mixed with sucrose-free acceptor vesicles. Following lipid exchange, the sucrose-laden donor vesicles were removed from the sucrose-free asymmetric LUV by centrifugation. Cyclodextrin with any bound lipid was then removed by filtration and washing. This allowed generation of asymmetric LUV in which one leaflet contained isotopically (deuterated)-labelled lipids (including fully labeled, or labeling the only the polar headgroup or acyl chains). Unlabeled POPC or DPPC were introduced into the outer leaflet of LUV containing partly or fully deuterated POPC. Similarly, deuterated POPC could be introduced into the outer leaflet of unlabeled POPC LUV. Lipid exchange levels and the degree of lipid asymmetry were assessed with 1H NMR and gas chromatography. Changes in bilayer thickness were assessed with SANS. It was found that DPPC-rich ordered (presumably gel) domains directly opposing the POPC leaflet exhibited reduced lipid packing density compared to typical gel phase lipids, indicating that an inner leaflet composed of Ld favoring lipids can partially fluidize outer leaflet ordered domains. This is indicative of a significant degree of interleaflet coupling.

1.22 Tuning Lipid Asymmetry to Control Phagocytosis of Model Membrane Vesicles

Liposomes can be easily generated that mimic apoptotic cells by containing PS in their outer leaflet. However, their utility in drug delivery is limited due to the speed at which they are detected by macrophages and phagocytosed. It would be useful to generate liposomes with encapsulated therapeutic agents that are protected from detection by the immune system until they have been able to bind their targets. In a recent study aiming to achieve this, engineered antibody-decorated asymmetric liposomes were developed to target pathogens to the immune system and to contain a built in time-delay switch dependent upon the flip-flop rates of PS [110].

Symmetric LUVs containing 1:1 DOPE:POPS and were incubated with MβCD pre-loaded with 1:2 bSM:DOPE (both donor and acceptor lipids contained a small amount of streptavidin-capped PE). Following outer leaflet lipid exchange, the resulting asymmetric vesicles were incubated with polyclonal anti-HIV-1 gp120 that was biotin-conjugated. These vesicles were then capable of binding HIV-1 virus-like particles. The slow flipping of PS to the outer leaflet of the vesicles over a period of 24–48 h then allowed the vesicles to then be targeted for destruction by macrophages. Varying the amount of bSM controlled the time delay before phagocytosis, with the highest outer-leaflet bSM content behaving similarly to LUVs without PS.

1.23 Extending Cyclodextrin-Induced Phospholipid Exchange to Mammalian Cells

Our understanding of the function of lipids in cellular membranes has fallen far behind that of other cellular molecules such as proteins or nucleic acids. This is largely due to the difficulty of manipulating lipid composition in living cells. It has been shown that membrane fatty acid composition, phospholipid composition, and cholesterol content can be modified in many different kinds of cultured mammalian cells by adding exogenous lipids in the form of liposomes or using lipids bound to carrier proteins. This topic has been comprehensively reviewed [111]. Although these methods alter membrane lipid composition, they only can make limited changes, and generally add lipids to cells, rather than substitute endogenous with exogenous lipids. Manipulating lipid composition using synthesis inhibitors can replace the original lipid composition of a cell with an altered one. However, inhibitor molecules act slowly. Furthermore, it is only possible to inhibit the synthesis of a subset of membrane lipids, most commonly sphingolipids or sterols [112, 113]. Mutations can be used to alter sterol structure [114, 115] and using a combination of mutations and introduction of new enzymes for wholesale metabolically re-engineering of lipid content can alter phospholipid composition drastically, but is difficult. To date phospholipid reengineering has been carried out successfully in *E. coli* [116, 117].

An alternative strategy is to use cyclodextrins to manipulate cell lipids *in vivo*. As noted above, MβCD has been widely used to manipulate plasma membrane sterol content in cultured cells. This can involve extracting cholesterol using empty MβCD, or loading cells with cholesterol by incubating them with MβCD/cholesterol complexes, or using MβCD/sterol complexes to substitute cholesterol in cells with another sterol [84, 85].

Use of cyclodextrins to alter other cellular lipids has been much less studied. Early studies concentrating upon fluorescently labeled lipids were noted above. In more recent work, a variety of unlabeled phospholipids was introduced into cells using MβCD [118]. One limitation of the approach was that the level of exogenous lipid introduced into cells was dependent upon acyl chain structure. Phospholipids with short 14 carbon acyl chains were mostly highly transferred into cells. It was unclear whether the introduction of lipids involved lipid exchange, because there was a net increase in cellular phospholipid in the transfer process. It was also found that additional cholesterol could to be introduced to compensate for MβCD-induced cholesterol depletion in cells. Interestingly, introduction of exogenous PE and PS with unnatural acyl chains to BHK21 hamster kidney cells led to rapid acyl chain remodeling, with substantial remodeling within 1 h, but remodeling was much more limited and slow when acyl chains matched or were close to those in endogenous lipids [119].

Recently, our group developed a method using MαCD to efficiently replace the plasma membrane outer leaflet phospholipids and sphingolipids in cells with exogenous phospholipids and sphingolipids, including unnatural lipids [107]. One advantage of MαCD relative to MβCD, is its small cavity, which is too small to interact with sterols. As a result, MαCD will not significantly alter cholesterol levels during phospholipid/sphingolipid exchange. To carry out exchange, donor lipids (SM or PC or a mixture of the two) in the form of multilamellar vesicles were incubated with a concentration of MαCD high enough to dissolve at least most of the vesicles at the lipid concentrations used (40 mM MαCD was sufficient for use with 1.5–3 mM lipid in most experiments). The mixture was then incubated with cultured human cells, generally at 37 °C for 1 h. This protocol extracted 70–80% of endogenous SM but only very low (10–15%) amounts of PC and PE from the cultured cells. There was minimal PS, PI, or cholesterol extracted from the cells. Similar results were obtained when an exchange was carried out with an unnatural C17:0 SM as the donor lipid, and the lipid composition in the cells before and after exchange was evaluated by mass spectrometry. Exchange was rapid; under the experimental conditions half-time at 37 °C was 15–20 min both for the removal of endogenous SM and introduction of exogenous fluorescently labeled lipid. The maximal SM exchange level was not affected when MαCD concentration, exogenous lipid concentrations, exogenous lipid compositions or cell types were varied. Given these observations and the fact that it has been reported that a significant fraction of SM is not localized in the plasma membrane [119] suggests almost complete replacement of plasma membrane outer leaflet sphingolipids was achieved. Efficient replacement of outer leaflet lipids was also consistent with the observation that up to 90% of cholera toxin B binding to cells was abolished after MβCD-induced lipid exchange, indicative of removal of the cholera toxin receptor glycosphingolipid GM1 from the plasma membrane outer leaflet upon exchange. In addition, carrying out exchange between model membrane vesicles, it was found that MαCD catalyzed exchange of different phospholipid or sphingolipids was relatively non-specific. Thus, it would be expected that all of the outer leaflet lipids (other than cholesterol) would be replaced upon exchange. This was confirmed by lipid analysis of the radiolabeled lipid removed from the cells upon exchange, which had a composition (SM ~ PC > PE, and little or no PS + PI) closely matching that previously reported for the outer leaflet of human red blood cells [1].

Mass spectrometry analysis of the difference in lipid composition before and after lipid exchange indicated that plasma membrane outer leaflet was enriched in a subset of PC species that are not highly unsaturated. It also showed that shorter acyl chain SM species were enriched in the plasma membrane outer leaflet, in agreement with prior studies [120].

All these results demonstrate that MαCD could be a useful tool to manipulate cell membrane lipid composition in an efficient manner. This may have many applications in addition to studies of lipid asymmetry. For example, the effect of altering plasma membrane outer leaflet lipid composition upon membrane protein function could be studied with this approach. In addition, the ability to introduce large amounts of unnatural phospholipids and/or sphingolipids into cells could aid analysis of lipid structure/function relationships.

Acknowledgement Support from NIH grant GM 112638 and GM 122493 is gratefully acknowledged.

References

1. Verkleij A, Zwaal R, Roelofsen B, Comfurius P, Kastelijn D, Van Deenen L (1973) The asymmetric distribution of phospholipids in the human red cell membrane. A combined study using phospholipases and freeze-etch electron microscopy. Biochim Biophys Acta Biomembr 323:178–193
2. Mondal M, Mesmin B, Mukherjee S, Maxfield FR (2009) Sterols are mainly in the cytoplasmic leaflet of the plasma membrane and the endocytic recycling compartment in CHO cells. Mol Biol Cell 20:581–588
3. Liu SL, Sheng R, Jung JH, Wang L, Stec E, O'Connor MJ, Song S, Bikkavilli RK, Winn RA, Lee D, Baek K, Ueda K, Levitan I, Kim KP, Cho W (2017) Orthogonal lipid sensors identify transbilayer asymmetry of plasma membrane cholesterol. Nat Chem Biol 13:268–274
4. Clark MR (2011) Flippin' lipids. Nat Immunol 12:373–375
5. Marquardt D, Heberle FA, Miti T, Eicher B, London E, Katsaras J, Pabst G (2017) 1H NMR shows slow phospholipid Flip-flop in gel and fluid bilayers. Langmuir 33:3731–3741
6. McConnell HM, Kornberg RD (1971) Inside-outside transitions of phospholipids in vesicle membranes. Biochemistry 10:1111–1120
7. Nakano M, Fukuda M, Kudo T, Matsuzaki N, Azuma T, Sekine K, Endo H, Handa T (2009) Flip-flop of phospholipids in vesicles: kinetic analysis with time-resolved small-angle neutron scattering. J Phys Chem B 113:6745–6748
8. Son M, London E (2013) The dependence of lipid asymmetry upon polar headgroup structure. J Lipid Res 54:3385–3393
9. Son M, London E (2013) The dependence of lipid asymmetry upon phosphatidylcholine acyl chain structure. J Lipid Res 54:223–231
10. Leventis R, Silvius JR (2001) Use of cyclodextrins to monitor transbilayer movement and differential lipid affinities of cholesterol. Biophys J 81:2257–2267

11. Li MO, Sarkisian MR, Mehal WZ, Rakic P, Flavell RA (2003) Phosphatidylserine receptor is required for clearance of apoptotic cells. Science 302:1560–1563
12. Lentz BR (2003) Exposure of platelet membrane phosphatidylserine regulates blood coagulation. Prog Lipid Res 42:423–438
13. Morizono K, Chen IS (2014) Role of phosphatidylserine receptors in enveloped virus infection. J Virol 88:4275–4290
14. Mercer J, Helenius A (2008) Vaccinia virus uses macropinocytosis and apoptotic mimicry to enter host cells. Science 320:531–535
15. von Heijne G (1992) Membrane protein structure prediction. Hydrophobicity analysis and the positive-inside rule. J Mol Biol 225:487–494
16. Slusky JS, Dunbrack RL Jr (2013) Charge asymmetry in the proteins of the outer membrane. Bioinformatics 29:2122–2128
17. Sharpe HJ, Stevens TJ, Munro S (2010) A comprehensive comparison of transmembrane domains reveals organelle-specific properties. Cell 142:158–169
18. Brown D, London E (1998) Functions of lipid rafts in biological membranes. Annu Rev Cell Dev Biol 14:111–136
19. Brown D, London E (1998) Structure and origin of ordered lipid domains in biological membranes. J Membr Biol 164:103–114
20. Korlach J, Schwille P, Webb WW, Feigenson GW (1999) Characterization of lipid bilayer phases by confocal microscopy and fluorescence correlation spectroscopy. Proc Natl Acad Sci 96:8461–8466
21. Lin Q, London E (2014) The influence of natural lipid asymmetry upon the conformation of a membrane-inserted protein (Perfringolysin O). J Biol Chem 289:5467–5478
22. Levental I, Grzybek M, Simons K (2011) Raft domains of variable properties and compositions in plasma membrane vesicles. Proc of Natl Acad Sci USA 108:11411–11416
23. Taylor DR, Hooper NM (2007) Role of lipid rafts in the processing of the pathogenic prion and Alzheimer's amyloid-beta proteins. Semin Cell Dev Biol 18:638–648
24. Williamson R, Usardi A, Hanger DP, Anderton BH (2008) Membrane-bound beta-amyloid oligomers are recruited into lipid rafts by a fyn-dependent mechanism. FASEB J 22:1552–1559
25. Cuadras MA, Greenberg HB (2003) Rotavirus infectious particles use lipid rafts during replication for transport to the cell surface in vitro and in vivo. Virology 313:308–321
26. Gulbins E, Kolesnick R (2003) Raft ceramide in molecular medicine. Oncogene 22:7070–7077
27. Lyman MG, Curanovic D, Enquist LW (2008) Targeting of pseudorabies virus structural proteins to axons requires association of the viral Us9 protein with lipid rafts. PLoS Pathog 4:e1000065
28. Murphy SC, Hiller NL, Harrison T, Lomasney JW, Mohandas N, Haldar K (2006) Lipid rafts and malaria parasite infection of erythrocytes. Mol Membr Biol 23:81–88
29. Riethmuller J, Riehle A, Grassme H, Gulbins E (2006) Membrane rafts in host-pathogen interactions. Biochim Biophys Acta 1758:2139–2147
30. Korade Z, Kenworthy AK (2008) Lipid rafts, cholesterol, and the brain. Neuropharmacology 55:1265–1273
31. Collins MD (2008) Interleaflet coupling mechanisms in bilayers of lipids and cholesterol. Biophys J 94:L32–L34
32. Kiessling V, Wan C, Tamm LK (2009) Domain coupling in asymmetric lipid bilayers. Biochim Biophys Acta 1788:64–71
33. Pinaud F, Michalet X, Iyer G, Margeat E, Moore HP, Weiss S (2009) Dynamic partitioning of a Glycosyl-phosphatidylinositol-anchored protein in Glycosphingolipid-rich microdomains imaged by single-quantum dot tracking. Traffic 10:691–712
34. Lenne PF, Wawrezinieck L, Conchonaud F, Wurtz O, Boned A, Guo XJ, Rigneault H, He HT, Marguet D (2006) Dynamic molecular confinement in the plasma membrane by microdomains and the cytoskeleton meshwork. EMBO J 25:3245–3256

35. Raghupathy R, Anilkumar AA, Polley A, Singh PP, Yadav M, Johnson C, Suryawanshi S, Saikam V, Sawant SD, Panda A, Guo Z, Vishwakarma RA, Rao M, Mayor S (2015) Transbilayer lipid interactions mediate nanoclustering of lipid-anchored proteins. Cell 161: 581–594
36. Olson F, Hunt C, Szoka F, Vail W, Papahadjopoulos D (1979) Preparation of liposomes of defined size distribution by extrusion through polycarbonate membranes. Biochim Biophys Acta Biomembr 557:9–23
37. Kremer J, Van der Esker M, Pathmamanoharan C, Wiersema P (1977) Vesicles of variable diameter prepared by a modified injection method. Biochemistry 16:3932–3935
38. Dua J, Rana A, Bhandari A (2012) Liposome: methods of preparation and applications. Int J Pharm Stud Res 3:14–20
39. Bezrukov SM (2000) Functional consequences of lipid packing stress. Curr Opin Colloid In 5:237–243
40. Hope MJ, Nayar R, Mayer LD, Cullis PR (1993) Reduction of liposome size and preparation of unilamellar vesicles by extrusion techniques. Liposome Technol 1:123–139
41. Angelova MI, Dimitrov DS (1986) Liposome electroformation. Faraday Discuss Chem Soc 81:303–311
42. Akashi K-i, Miyata H, Itoh H, Kinosita K (1996) Preparation of giant liposomes in physiological conditions and their characterization under an optical microscope. Biophys J 71:3242–3250
43. Sugiura S, Kuroiwa T, Kagota T, Nakajima M, Sato S, Mukataka S, Walde P, Ichikawa S (2008) Novel method for obtaining homogeneous giant vesicles from a monodisperse water-in-oil emulsion prepared with a microfluidic device. Langmuir 24:4581–4588
44. Arriaga LR, Datta SS, Kim SH, Amstad E, Kodger TE, Monroy F, Weitz DA (2014) Ultrathin shell double emulsion templated giant unilamellar lipid vesicles with controlled microdomain formation. Small 10:950–956
45. Richmond DL, Schmid EM, Martens S, Stachowiak JC, Liska N, Fletcher DA (2011) Forming giant vesicles with controlled membrane composition, asymmetry, and contents. Proc Natl Acad Sci 108:9431–9436
46. Hu PC, Li S, Malmstadt N (2011) Microfluidic fabrication of asymmetric giant lipid vesicles. ACS Appl Mater Interfaces 3:1434–1440
47. Lu L, Schertzer JW, Chiarot PR (2015) Continuous microfluidic fabrication of synthetic asymmetric vesicles. Lab Chip 15:3591–3599
48. Hamada T, Miura Y, Komatsu Y, Kishimoto Y, Vestergaard M d, Takagi M (2008) Construction of asymmetric cell-sized lipid vesicles from lipid-coated water-in-oil microdroplets. J Phys Chem B 112:14678–14681
49. Ito H, Yamanaka T, Kato S, Hamada T, Takagi M, Ichikawa M, Yoshikawa K (2013) Dynamical formation of lipid bilayer vesicles from lipid-coated droplets across a planar monolayer at an oil/water interface. Soft Matter 9:9539–9547
50. Elani Y, Law RV, Ces O (2015) Protein synthesis in artificial cells: using compartmentalisation for spatial organisation in vesicle bioreactors. Phys Chem Chem Phys 17:15534–15537
51. Hwang WL, Chen M, Cronin B, Holden MA, Bayley H (2008) Asymmetric droplet interface bilayers. J Am Chem Soc 130:5878–5879
52. Schmidt C, Barenholz Y, Huang C, Thompson T (1978) Monolayer coupling in sphingomyelin bilayer systems. Letters Nature 271:775–777
53. Blodgett KB (1935) Films built by depositing successive monomolecular layers on a solid surface. J Am Chem Soc 57:1007–1022
54. Roberts G (2013) Langmuir-blodgett films. Springer, New York
55. Watanabe R, Soga N, Yamanaka T, Noji H (2014) High-throughput formation of lipid bilayer membrane arrays with an asymmetric lipid composition. Sci Rep 4:7076
56. Crane JM, Kiessling V, Tamm LK (2005) Measuring lipid asymmetry in planar supported bilayers by fluorescence interference contrast microscopy. Langmuir 21:1377–1388
57. Anglin TC, Conboy JC (2009) Kinetics and thermodynamics of flip-flop in binary phospholipid membranes measured by sum-frequency vibrational spectroscopy. Biochemistry 48:10220–10234

58. Anglin TC, Cooper MP, Li H, Chandler K, Conboy JC (2010) Free energy and entropy of activation for phospholipid flip-flop in planar supported lipid bilayers. J Phys Chem B 114:1903–1914
59. Brown KL, Conboy JC (2013) Lipid flip-flop in binary membranes composed of phosphatidylserine and phosphatidylcholine. J Phys Chem B 117:15041–15050
60. Allhusen JS, Kimball DR, Conboy JC (2016) Structural origins of cholesterol accelerated lipid flip-flop studied by sum-frequency vibrational spectroscopy. J Phys Chem B 120:3157–3168
61. Montal M, Mueller P (1972) Formation of bimolecular membranes from lipid monolayers and a study of their electrical properties. Proc Natl Acad Sci 69:3561–3566
62. Collins MD, Keller SL (2008) Tuning lipid mixtures to induce or suppress domain formation across leaflets of unsupported asymmetric bilayers. Proc Natl Acad Sci 105:124–128
63. Murray DH, Tamm LK, Kiessling V (2009) Supported double membranes. J Struct Biol 168:183–189
64. Hussain NF, Siegel AP, Ge Y, Jordan R, Naumann CA (2013) Bilayer asymmetry influences integrin sequestering in raft-mimicking lipid mixtures. Biophys J 104:2212–2221
65. Pagano RE, Martin OC, Schroit AJ, Struck DK (1981) Formation of asymmetric phospholipid membranes via spontaneous transfer of fluorescent lipid analogs between vesicle populations. Biochemistry 20:4920–4927
66. Bloj B, Zilversmit D (1976) Asymmetry and transposition rate of phosphatidylcholine in rat erythrocyte ghosts. Biochemistry 15:1277–1283
67. Crain RC, Zilversmit DB (1980) Two nonspecific phospholipid exchange proteins from beef liver. 1. Purification and characterization. Biochemistry 19:1433–1439
68. Everett J, Zlotnick A, Tennyson J, Holloway P (1986) Fluorescence quenching of cytochrome b5 in vesicles with an asymmetric transbilayer distribution of brominated phosphatidylcholine. J Biol Chem 261:6725–6729
69. Herrmann A, Zachowski A, Devaux PF (1990) Protein-mediated phospholipid translocation in the endoplasmic reticulum with a low lipid specificity. Biochemistry 29:2023–2027
70. Holzer M, Momm J, Schubert R (2010) Lipid transfer mediated by a recombinant pro-sterol carrier protein 2 for the accurate preparation of asymmetrical membrane vesicles requires a narrow vesicle size distribution: a free-flow electrophoresis study. Langmuir 26:4142–4151
71. Redelmeier T, Hope M, Cullis P (1990) On the mechanism of transbilayer transport of phosphatidylglycerol in response to transmembrane pH gradients. Biochemistry 29:3046–3053
72. Hope MJ, Redelmeier TE, Wong KF, Rodrigueza W, Cullis PR (1989) Phospholipid asymmetry in large unilamellar vesicles induced by transmembrane pH gradients. Biochemistry 28:4181–4187
73. Mui B, Döbereiner H, Madden T, Cullis P (1995) Influence of transbilayer area asymmetry on the morphology of large unilamellar vesicles. Biophys J 69:930
74. Pautot S, Frisken BJ, Weitz DA (2003) Engineering asymmetric vesicles. Proc of Natl Acad Sci USA 100:10718–10721
75. Elani Y, Purushothaman S, Booth PJ, Seddon JM, Brooks NJ, Law RV, Ces O (2015) Measurements of the effect of membrane asymmetry on the mechanical properties of lipid bilayers. Chem Commun 51:6976–6979
76. Zhang X, Zong W, Hu Y, Luo N, Cheng W, Han X (2016) A pH-responsive asymmetric lipid vesicle as drug carrier. J Microencapsul 33:663–668
77. Lu L, Doak WJ, Schertzer JW, Chiarot PR (2016) Membrane mechanical properties of synthetic asymmetric phospholipid vesicles. Soft Matter 12:7521–7528
78. Kamiya K, Kawano R, Osaki T, Akiyoshi K, Takeuchi S (2016) Cell-sized asymmetric lipid vesicles facilitate the investigation of asymmetric membranes. Nat Chem 8:881–889
79. Dodziuk H (2006) Cyclodextrins and their complexes: chemistry, analytical methods, applications. Wiley, Weinheim
80. Somogyi G, Posta J, Buris L, Varga M (2006) Cyclodextrin (CD) complexes of cholesterol–their potential use in reducing dietary cholesterol intake. Die Pharmazie- Int J Pharm Sci 61:154–156

81. Huang Z, London E (2013) Effect of cyclodextrin and membrane lipid structure upon cyclodextrin-lipid interaction. Langmuir 29:14631–14638
82. Lin Q, London E (2014) Preparation of artificial plasma membrane mimicking vesicles with lipid asymmetry. PLoS One 9:e87903
83. Szente L, Fenyvesi É (2017) Cyclodextrin-lipid complexes: cavity size matters. Struct Chem 28:479–492
84. Kim J, London E (2015) Using sterol substitution to probe the role of membrane domains in membrane functions. Lipids 50:721–734
85. Zidovetzki R, Levitan I (2007) Use of cyclodextrins to manipulate plasma membrane cholesterol content: evidence, misconceptions and control strategies. Biochim Biophy Acta Biomembr 1768:1311–1324
86. Ohtani Y, Irie T, Uekama K, Fukunaga K, Pitha J (1989) Differential effects of α-, β-and γ-cyclodextrins on human erythrocytes. Eur J Biochem 186:17–22
87. Christian A, Haynes M, Phillips M, Rothblat G (1997) Use of cyclodextrins for manipulating cellular cholesterol content. J Lipid Res 38:2264–2272
88. Legendre J, Rault I, Petit A, Luijten W, Demuynck I, Horvath S, Ginot Y, Cuine A (1995) Effects of β-cyclodextrins on skin: implications for the transdermal delivery of piribedil and a novel cognition enhancing-drug, S-9977. Eur J Pharm Sci 3:311–322
89. Niu S-L, Mitchell DC, Litman BJ (2002) Manipulation of cholesterol levels in rod disk membranes by methyl-β-cyclodextrin effects on receptor activation. J Biol Chem 277:20139–20145
90. Tanhuanpää K, Somerharju P (1999) γ-Cyclodextrins greatly enhance translocation of hydrophobic fluorescent phospholipids from vesicles to cells in culture IMPORTANCE OF MOLECULAR HYDROPHOBICITY IN PHOSPHOLIPID TRAFFICKING STUDIES. J Biol Chem 274:35359–35366
91. Tanhuanpää K, Cheng KH, Anttonen K, Virtanen JA, Somerharju P (2001) Characteristics of pyrene phospholipid/γ-cyclodextrin complex. Biophys J 81:1501–1510
92. Anderson TG, Tan A, Ganz P, Seelig J (2004) Calorimetric measurement of phospholipid interaction with methyl-beta-cyclodextrin. Biochemistry 43:2251–2261
93. Kilsdonk EP, Yancey PG, Stoudt GW, Bangerter FW, Johnson WJ, Phillips MC, Rothblat GH (1995) Cellular cholesterol efflux mediated by cyclodextrins. J Biol Chem 270:17250–17256
94. Cheng H-T, London E (2009) Preparation and properties of asymmetric vesicles that mimic cell membranes effect upon lipid raft formation and transmembrane helix orientation. J Biol Chem 284:6079–6092
95. Cheng HT, London E (2011) Preparation and properties of asymmetric large unilamellar vesicles: interleaflet coupling in asymmetric vesicles is dependent on temperature but not curvature. Biophys J 100:2671–2678
96. Chiantia S, Schwille P, Klymchenko AS, London E (2011) Asymmetric GUVs prepared by MbetaCD-mediated lipid exchange: an FCS study. Biophys J 100:L1–L3
97. Oglęcka K, Rangamani P, Liedberg B, Kraut RS, Parikh AN (2014) Oscillatory phase separation in giant lipid vesicles induced by transmembrane osmotic differentials. elife 3:e03695
98. Chiantia S, London E (2012) Acyl chain length and saturation modulate interleaflet coupling in asymmetric bilayers: effects on dynamics and structural order. Biophys J 103:2311–2319
99. Chiantia S, Klymchenko AS, London E (2012) A novel leaflet-selective fluorescence labeling technique reveals differences between inner and outer leaflets at high bilayer curvature. Biochim Biophys Acta Biomembr 1818:1284–1290
100. Armstrong VT, Brzustowicz MR, Wassall SR, Jenski LJ, Stillwell W (2003) Rapid flip-flop in polyunsaturated (docosahexaenoate) phospholipid membranes. Arch Biochem Biophys 414:74–82
101. Smith M, Jungalwala F (1981) Reversed-phase high performance liquid chromatography of phosphatidylcholine: a simple method for determining relative hydrophobic interaction of various molecular species. J Lipid Res 22:697–704
102. Lin Q, London E (2015) Ordered raft domains induced by outer leaflet sphingomyelin in cholesterol-rich asymmetric vesicles. Biophys J 108:2212–2222

103. Bakht O, Pathak P, London E (2007) Effect of the structure of lipids favoring disordered domain formation on the stability of cholesterol-containing ordered domains (lipid rafts): identification of multiple raft-stabilization mechanisms. Biophys J 93:4307–4318
104. Vitrac H, MacLean DM, Jayaraman V, Bogdanov M, Dowhan W (2015) Dynamic membrane protein topological switching upon changes in phospholipid environment. Proc Natl Acad Sci 112:13874–13879
105. Vitrac H, Bogdanov M, Dowhan W (2013) In vitro reconstitution of lipid-dependent dual topology and postassembly topological switching of a membrane protein. Proc Natl Acad Sci 110:9338–9343
106. Perillo VL, Peñalva DA, Vitale AJ, Barrantes FJ, Antollini SS (2016) Transbilayer asymmetry and sphingomyelin composition modulate the preferential membrane partitioning of the nicotinic acetylcholine receptor in lo domains. Arch Biochem Biophys 591:76–86
107. Li G, Kim J, Huang Z, Clair JRS, Brown DA, London E (2016) Efficient replacement of plasma membrane outer leaflet phospholipids and sphingolipids in cells with exogenous lipids. Proc Natl Acad Sci 201610705
108. Visco I, Chiantia S, Schwille P (2014) Asymmetric supported lipid bilayer formation via methyl-β-cyclodextrin mediated lipid exchange: influence of asymmetry on lipid dynamics and phase behavior. Langmuir 30:7475–7484
109. Heberle FA, Marquardt D, Doktorova M, Geier B, Standaert RF, Heftberger P, Kollmitzer B, Nickels JD, Dick RA, Feigenson GW (2016) Sub-nanometer structure of an asymmetric model membrane: Interleaflet coupling influences domain properties. Langmuir 32:5195–5200
110. Petazzi RA, Gramatica A, Herrmann A, Chiantia S (2015) Time-controlled phagocytosis of asymmetric liposomes: application to phosphatidylserine immunoliposomes binding HIV-1 virus-like particles. Nanomedicine 11:1985–1992
111. Spector AA, Yorek MA (1985) Membrane lipid composition and cellular function. J Lipid Res 26:1015–1035
112. Adada M, Luberto C, Canals D (2016) Inhibitors of the sphingomyelin cycle: sphingomyelin synthases and sphingomyelinases. Chem Phys Lipids 197:45–59
113. Delgado A, Casas J, Llebaria A, Abad JL, Fabrias G (2006) Inhibitors of sphingolipid metabolism enzymes. Biochim Biophys Acta Biomembr 1758:1957–1977
114. Aguilar PS, Heiman MG, Walther TC, Engel A, Schwudke D, Gushwa N, Kurzchalia T, Walter P (2010) Structure of sterol aliphatic chains affects yeast cell shape and cell fusion during mating. Proc Natl Acad Sci 107:4170–4175
115. Ott RG, Athenstaedt K, Hrastnik C, Leitner E, Bergler H, Daum G (2005) Flux of sterol intermediates in a yeast strain deleted of the lanosterol C-14 demethylase Erg11p. Biochim Biophys Acta Mol Cell Biol Lipids 1735:111–118
116. Dowhan W (2009) Molecular genetic approaches to defining lipid function. J Lipid Res 50:S305–S310
117. Wikström M, Kelly AA, Georgiev A, Eriksson HM, Klement MR, Bogdanov M, Dowhan W, Wieslander Å (2009) Lipid-engineered Escherichia Coli membranes reveal critical lipid headgroup size for protein function. J Biol Chem 284:954–965
118. Kainu V, Hermansson M, Somerharju P (2010) Introduction of phospholipids to cultured cells with cyclodextrin. J Lipid Res 51:3533–3541
119. Kainu V, Hermansson M, Somerharju P (2008) Electrospray ionization mass spectrometry and exogenous heavy isotope-labeled lipid species provide detailed information on aminophospholipid acyl chain remodeling. J Biol Chem 283:3676–3687
120. Koivusalo M, Jansen M, Somerharju P, Ikonen E (2007) Endocytic trafficking of sphingomyelin depends on its acyl chain length. Mol Biol Cell 18:5113–5123

Chapter 2
Spontaneous Lipid Flip-Flop in Membranes: A Still Unsettled Picture from Experiments and Simulations

Maria Maddalena Sperotto and Alberta Ferrarini

Abstract Biomembrane asymmetry, whose regulation is important for function, is maintained by the movement of lipids from one bilayer leaflet to the other (flip-flop). During the last decades a number of studies have been done to characterize this process, and it was found that it can be spontaneous or assisted by protein transporters. It can be accelerated or inhibited by various factors, e.g., it can be induced by mechanical stresses. It was also found that flip-flop rate and mechanism strongly depend on the molecular structure of the flipping lipid and on the composition and physical state of the membrane. Yet, large discrepancies exist among the data available in the literature, and a quantitative and comprehensive understanding of this process is still missing. This chapter reviews our current knowledge of the molecular aspects of spontaneous (or passive) flip-flop. An overview of experimental studies is presented, together with a summary of the state of the art of computer simulation studies, which enable a direct insight at the molecular level. The achievements and limitations of experimental and computational approaches are pointed out, as well as the challenges that remain to be addressed.

Keywords Lipid flip-flop • Biomembrane asymmetry • Transbilayer diffusion • Transport proteins • Labeled lipids • Lipid membrane simulations • Phospholipid dynamics • Cholesterol dynamics • Roto-translational diffusion

M.M. Sperotto
DTU Bioinformatics, Department of Bio and Health Informatics, Technical University of Denmark, Kemitorvet, Building 208, DK-2800 Kgs. Lyngby, Denmark
e-mail: maria@cbs.dtu.dk

A. Ferrarini (✉)
Department of Chemical Sciences, University of Padova, Via Marzolo 1, I-35131, Padova, Italy
e-mail: alberta.ferrarini@unipd.it

R.M. Epand, J.-M. Ruysschaert (eds.), *The Biophysics of Cell Membranes*,
Springer Series in Biophysics 19, DOI 10.1007/978-981-10-6244-5_2

Abbreviations

1D	mono-dimensional
2D	two-dimensional
AA	all-atom
AFM	atomic force spectroscopy
BSA	bovine serum albumin
C8PC	1-myristoyl-2-[9-(1-pyrenyl)nonanoyl]-phosphatidylcholine
C10PC	1-palmitoyl-2-[9-(1-pyrenyl)nonanoyl]-phosphatidylcholine
C12PC	1-lauroyl-2-[9-(1-pyrenyl)nonanoyl]-phosphatidylcholine
C8PE	1-myristoyl-2-[9-(1-pyrenyl)nonanoyl]-phosphatidylethanolamine
C12PC	1-lauroyl-2-[9-(1-pyrenyl)nonanoyl]-phosphatidylethanolamine
C8PG	1-myristoyl-2-[9-(1-pyrenyl)nonanoyl]-phosphatidylglycerol
C12PC	1-lauroyl-2-[9-(1-pyrenyl)nonanoyl]-phosphatidylglycerol
C8PA	1-myristoyl-2-[9-(1-pyrenyl)nonanoyl]-phosphatidic acid
C12PA	1-lauroyl-2-[9-(1-pyrenyl)nonanoyl]-phosphatidic acid
CG	coarse-grain
CL	cardiolipin
CM	hydrodynamic center of mobility
DAG	diacylglycerol
$diC_{18-2}PC$	dilinoleyl-phosphatidylcholine
DMPC	dimyristoyl-phosphatidylcholine
DOPE	dioleoyl-phosphatidylethanolamine
DOPG	dioleoyl-phosphatidylglycerol
DOPC	dioleoyl-phosphatidylethanolamine
DOPS	dioleoyl-phosphatidylserine
DPPC	dipalmitoyl-phosphatidylcholine
DPPS	dipalmitoyl-phosphatidylserine
DSPC	distearoyl-phosphatidylcholine
DSPS	distearoyl-phosphatidylserine
EPC	egg phosphatidylcholine
EPR	electron paramagnetic resonance
ER	endoplasmatic reticulum
GPI	glycosyl-phosphatidylinositol
GUV	giant unilamellar vesicles
HRC	human red cell (erythrocytes)
LB	Langmuir- Blodgett
LS	Langmuir-Schäfer
LUV	large unilamellar vesicles
MBCD	methyl-β- cyclodextrin
MD	molecular dynamics
NBD	nitrobenzoxadiazole
NMR	nuclear magnetic resonance

NR	neutron reflectometry
PA	phosphatidic acid
PC	phosphatidylcholine
PE	phosphatidylethanolamine
PG	phosphatidylglycerol
PI	phosphatidylinositol
PMF	potential of mean force
POPC	palmitoyl-oleoyl-phosphatidylcholine
PS	phosphatidylserine
SLB	planar supported lipid bilayer
SAPC	stearoyl-arachidonyl-phosphatidylcholine
SANS	small-angle neutron scattering
SFVS	sum-frequency vibrational spectroscopy
SM	sphingomyelin
SOPC	stearoyl-oleoyl-phosphatidylcholine
SUV	small unilamellar vesicles
TEMPO-DPPC	1,2-dipalmitoyl-*sn*-glycero-3-phospho(tempo)choline

2.1 Introduction

A biomembrane is a thin (around 5 nm) soft sheet, which is composed of a lipid bilayer matrix in which proteins are embedded or attached. At the macroscopic level, biomembranes are flexible, deformable, with a curvature energy comparable to $k_B T$ at room temperature, and fluid, i.e., the lipid-protein bilayer does not exhibit shear restoring forces. At the microscopic level membranes are a dynamic environment: lipids undergo a variety of motions over a broad range of time scales [43, 139], which include rotations around their longitudinal axis and intramolecolar motions (rotations around chain bonds), as well as lateral (within the plane of the membrane) and transversal (from one leaflet to the other) displacements. The characteristic rates depend on the kind of lipid and on the physical state of the membrane. The fastest are bond rotations, which occur on the picosecond scale, whereas typical times for lipid reorientations are of the order of a nanosecond in the liquid crystal phase. Lateral displacements are characterized by diffusion coefficients in the range of 0.1–1 $\mu m^2\, s^{-1}$. The transbilayer movement, called *flip-flop* [72], can be much slower, ranging from seconds to hours and even days, strongly dependent upon the nature of the lipid, the composition and the physical state of the membrane. There are also reports of shorter characteristic times, on the millisecond scale, for lipids with small head groups, such as cholesterol and diacyglycerols.

From the physical and chemical point of view, biomembranes are complex mixtures. A variety of tools, from spectroscopy to bioinformatics, have been used to prove that they contain a myriad of lipid species (see Fig. 2.1) [44], whose importance for the structure and function of the cell is still a matter of debate [143, 144].

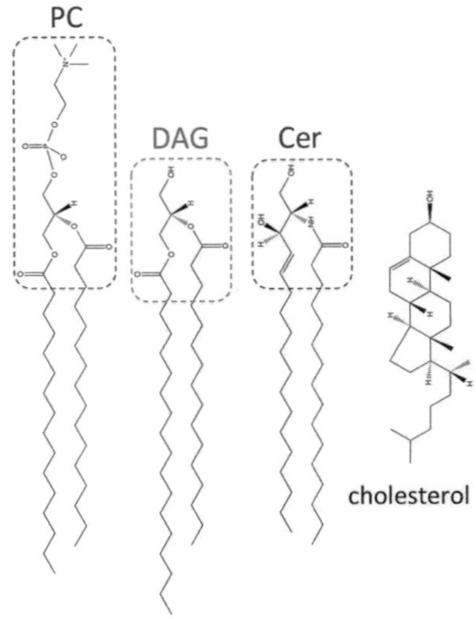

Fig. 2.1 Examples of lipid structures: dipalmitoyl-phosphatidylcholine (DPPC), 1-stearoyl-2-palmitoyl-*sn*-glycerol, C16-ceramide, cholesterol (from *left* to *right*)

The plasma membrane of eukarotic and prokaryotic cells and the membranes of inner organelles of eukaryotes are both laterally heterogeneous and asymmetric with respect to composition of the leaflets [37, 39, 153]. During the last 15 years, the occurrence of in-plane heterogeneous structures has been widely investigated [24, 100, 129]. Regarding the transversal asymmetry, since the early '70s [20] it has been known that the human erythrocyte membrane (HRC) is enriched in phosphatidylcholine (PC) and sphingomyelin (SM) in the outer (exoplasmatic or exofacial) leaflet, and in phosphatidylserine (PS), phosphatidylethanolamine (PE) and phosphateidylinositol (PI) in the inner (cytoplasmic) leaflet. The asymmetry is lower in biogenic membranes, i.e., the Golgi and the endoplasmic reticulum (ER), and this is likely related to their role at the beginning of the secretory pathway and their retention of a very low concentration of sterols, which would enable the formation of transient defects, so allowing small molecules to diffuse across them [21, 144]. The function, differentiation, and growth of cells depend on the regulation of lipid asymmetry [34, 156], whose alterations are reported to be associated with diseases, such as cancer, Alzheimer [25], and increased risk of atherosclerosis [154], with chronic ethanol consumption [152], and with apoptosis [135]. An example where the transversal asymmetry plays a role in the synaptic membrane. In the human brain, cholesterol constitutes approximately 25% of the total amount present in the body, and the cytofacial leaflet of synaptic plasma membranes contains approximately five- to six-fold more cholesterol than the exofacial one. Changes in cholesterol asymmetry may affect the formation and functioning of synaptic plasma membranes, with serious health implications [153].

Multiple mechanisms contribute to the onset and maintenance of asymmetry in biomembranes and flip-flop has a crucial role. Passive transbilayer diffusion is a spontaneous process, which tends to re-establish a uniform molecular distribution between the two leaflets. Thus, the existence of lipid asymmetry implies that flip-flop is somewhat hindered, or that there are mechanisms to maintain it. Experiments on reconstituted membranes by Kornberg and McConnell [72], who called flip-flop the spontaneous, i.e., thermally induced transbilayer motion of phospholipids, indicated that this process takes several hours. Later on Cullis and de Kruijff [33] deduced that in biomembranes the time scale was much shorter (less than microseconds) and attributed this to the formation of transient non-bilayer structures. It is now accepted that the maintenance of lipid asymmetry in biomembranes depends on the interplay of spontaneous flip-flop and the action of protein transporters, originally called *flippases*. Numerous experiments have been done to understand what triggers or facilitate flip-flop, characterize its time scales and determine how this process is affected by composition and physical state of the bilayer [22, 30, 36, 38, 65, 122, 126]. There is now general agreement on the importance of the lipid head group for the rate and the mechanism of spontaneous flip-flop. In particular, the process is orders of magnitude faster for species with small and/or neutral head groups, such as cholesterol and fatty acids. It is also recognized that flip-flop is strongly affected by the physical state of the bilayer and can be dramatically influenced by modifications in the chemical structure of the flipping lipid. However, the data available in the literature show striking discrepancies: for the same system, flip-flop rates ranging from the sub-second scale to days are reported. This is due to the intrinsic difficulty to detect the molecular process and to standardize the experimental procedures [38, 55, 85, 124, 132, 157]. Another reason is the sensitivity of flip-flop to chemical and physical changes; on the other hand, this sensitivity is likely crucial for the role of flip-flop in biomembranes.

This chapter presents a summary of our present knowledge of the molecular aspects of passive or spontaneous flip-flop, which occurs without the direct action of proteins. Protein-mediated flip-flop has been addressed by previous reviews [18, 93, 122, 126], and a recent one [111] offers an overview of the different types of transporters and mechanisms. A comprehensive a comprehensive account of the transmembrane dynamics of lipids and their biological implications is presented in Ref. [38]. In the next Section, our present knowledge of protein-assisted flip-flop is summarized. This is followed, in the third Section, by an overview of the experimental methods used to study flip-flop, where the main advantages and drawbacks of the different techniques are highlighted. This is useful for understanding the reasons behind the discrepancies among the results available in the literature. In the fourth Section, we summarize the state of the art of computational studies, which constitute a valuable complement to experimental investigations. Computer simulations can provide direct insight into the molecular aspects of flip-flop, with a resolution hardly accessible to measurements, and under conditions that may be difficult to achieve in experiments. A recent review addressed more technical aspects of computational studies of passive flip-flop [109] and a previous one focused on simulation studies of defect-mediated flip-flop [54]. In the present chapter, the achievements, potentiality and limitations of computational methods are pointed out, paying attention to their capability to provide useful kinetic and mechanistic insights on flip-flop, rather than on methodological aspects.

2.2 Spontaneous Versus Protein-Assisted Flip-Flop

In most experiments the rate of flip-flop is determined from the disappearance of a signal related to the unbalance of population between the two leaflets of a bilayer. So, it is customary to characterize the time scale of flip-flop by the half-time $t_{1/2}$, defined as the time required for the 50% decrease of the signal. If flip-flop is described as the unimolecular process:

$$\mathrm{T} \underset{k_{T\leftarrow B}}{\overset{k_{B\leftarrow T}}{\rightleftharpoons}} \mathrm{B} \tag{2.1}$$

where T (top) and B (bottom) indicate the position of the flipping lipid and $k_{B\leftarrow T}$, $k_{T\leftarrow B}$ are the rate constants for the translocation processes in the two opposite directions, the population unbalance between the two leaflets decays in time with the rate constant $k_{ff} = k_{B\leftarrow T} + k_{T\leftarrow B}$. This is related to the flip-flop half-time as $t_{1/2} = \ln 2/(k_{ff})$. In most cases $k_{B\leftarrow T} = k_{T\leftarrow B} = k_f$, thus $t_{1/2} = \ln 2/(2k_f)$.

Lipids with small head groups, such as cholesterol, fatty acids, diacylglycerols (DAGs) and ceramides, can spontaneously flip-flop at a relatively high rate. In contrast, the spontaneous flip-flop of phospholipids with charged/polar heads is very slow (the time scale may range from hours to days), because moving the heads from the water-membrane interface to the hydrophobic interior is energetically highly unfavorable. Tables 2.1 and 2.2 report half-times of spontaneous flip-flop ($t_{1/2}$) for various lipids, as obtained by different experiments.

In 1972, just a year after Kornberg and McConnell [72] had found that spontaneous phospholipid flip-flop is a very slow process, Bretscher [20] postulated the existence of transmembrane lipid-transporter proteins, which would accelerate the flip-flop rate to a biologically relevant scale. He proposed that the asymmetrical organization of phospholipids in HRC membranes was due to specific lipid enzymes, which he called phospholipid flippases. However, to start to acknowledge the ingenious idea by Bretscher it took a while longer. It was in 1984 that Seigneuret and Devaux [125] demonstrated for the first time the existence of phospholipid flippases in HRC membranes, using spin-labeled lipids. Their discovery was then confirmed in 1985 by Daleke and Huestis [35] who used non-labelled lipids, and in 1987 the existence of an erythrocite aminophospholipid transporter was demonstrated by Connor and Schroit [29] using fluorescent lipid analogues. It is now recognized that various transporters, called *flippases*, *floppases* and *scramblases*, contribute to maintain transmembrane asymmetry [111]. Flippases mediate the transbilayer diffusion against a concentration gradient from the exoplasmic to the cytosolic face, at the expense of ATP hydrolysis; the ATP-driven movement in the opposite direction is instead facilitated by floppases, whereas scramblases mediate the ATP-independent bidirectional transport. Due to these proteins, the flip-flop rate ranges from $100\,\mathrm{s}^{-1}$, in the case of ATP-driven transporter [27], to more than $10{,}000\,\mathrm{s}^{-1}$ for scramblases [50]. Regarding ATP-driven lipid transporters, these belong to the family of P4-ATPases or ABC transporters. Among the latter ones, there is a P-glycoprotein, called P-glycoprotein, called MDR1 [142], that is responsible for multidrug resistence, hence constitutes a serious issue for cancer therapy. Both cells apoptosis and activation of blood platelets involve exposure of PS lipids on the outer monolayer of the plasma membrane, controlled by scramblases, which are constitutively active or are regulated, e.g., by Ca^{2+} [134–136].

Table 2.1 Half-times of spontaneous flip-flop ($t_{1/2}$) measured for various lipids. Different methods were used: Fluorescence with labeled lipids (Fluo), Excimer Fluorescence (EF), Electron Paramagnetic Resonance (EPR), Nuclear Magnetic Resonance (NMR), Shape Change (SC) analysis, Small-Angle Neutron Scattering (SANS), Sum-Frequency Vibrational Spectroscopy (SFVS). The amount of cholesterol in mixed bilayers is expressed in mol %

Lipid	Host	*T* (°C)	$t_{1/2}$	Technique	Preparation	Ref.
TEMPO-DPPC[a]	EPC	30	6.5 h	EPR	SUV	[72]
		40	46–178 min			
C8PC	POPC	37[b]	87 h	EF	SUV	[58]
C10PC			173 h			
C12PC			347 h			
C8PE			6 h			
C12PE			10 h			
C8PG			69 h			
C12PG			69 h			
C8PA			29 h			
C12PA			30 h			
C5-DMB-Cer[c]	POPC	22	$1.3 \cdot 10^3$ s	Fluo	LUV	[10]
C5-DMB-DAG[d]			70 ms			
C5-DMB-SM[e]			$1.2 \cdot 10^4$ s			
C5-DMB-PC[f]			$2.4 \cdot 10^4$ s			
C6-NBD-PC[g]	DPPC	41	9 ± 2 min	Fluo	SUV	[65]
		50	50 ± 20 min		SUV	
	DMPC	23	1.6 ± 0.2 min		SUV	
			1.9 ± 0.7 min		LUV	
C_6Cer[h]	EPC	20	0.6 ± 0.3 min[i]	SC	GUV	[83]
C_{10}Cer[h]			0.8 ± 0.4 min[i]			
C_{12}Cer[h]			1.2 ± 0.5 min[i]			
DMPC	DMPC	37	350 min	SANS	LUV	[98]
	DMPC, 20%C		1300 min			
	DMPC, 40%C		>7000 min			
POPA	POPA		450 min			
POPC	POPC		>1000 h			
DPPC	DPPC	20	>4555 h	NMR	LUV	[87]
		40	75 ± 6 h			
		50	147 ± 9 h			

(continued)

Table 2.1 (continued)

Lipid	Host	T (°C)	$t_{1/2}$	Technique	Preparation	Ref.
DMPC	DMPC	4.2	226 ± 5 min	SFVS	SLB	[81]
		20.4	1.30 ± 0.003 min			
DPPC	DPPC	27.7	146 ± 1 min			
		36.6	9.20 ± 0.07 min			
DSPC	DSPC	41.7	312 ± 2 min			
		51.3	25.9 ± 1 min			

[a] 1,2-dipalmitoyl-*sn*-glycero-3-phospho(tempo)choline
[b] Estrapolated from data at various temperatures
[c] N-[5-(5,7-dimethyl BODIPY)-1-pentanoyl]-D-*erythro*-sphingosine
[d] 1,2 (palmitoyl-5,7-dimethyl BODIPY-1-pentanoyl)-diacylglycerol (10 mol %)
[e] N-[5-(5,7-dimethyl BODIPY)-1-pentanoyl]-sphingosylphosphorylcholine (D-*erythro*-isomer)
[f] 1,2 (palmitoyl-5,7-dimethyl BODIPY-1-pentanoyl)-phosphatidylcholine
[g] 1-acyl-2-(6-(7-nitrobenz-2-oxa-1,3-diazol-4-yl)amino)hexanoyl-*sn*-glycero-3-phosphocholine
[h] C_n-ceramide
[i] This is one half of the $t_{1/2}$ value reported in Ref. [83]

Table 2.2 Half-times of spontaneous flip-flop ($t_{1/2}$) measured for cholesterol (C) in different hosts. Different techniques were used, among which Nuclear Magnetic Resonance (NMR), Small-Angle Neutron Scattering (SANS), Sum-Frequency Vibrational Spectroscopy (SFVS)

Host	%C	T (°C)	$t_{1/2}$	Technique	Preparation	Ref.
DPPC	47	37	~6 days	[^{3}H]sterol exchange	SUV	[113]
DMPC	48.5	37	3.25 h	[^{3}H]sterol exchange	SUV	[114]
DPPC			3 h			
DSPC			2.5 h			
DOPC			1.5 h			
EPC	50	37	<1.5 h	[^{3}H]sterol exchange	SUV	[9]
EPC	41		≤1 m	Cholesterol oxidase		
EPC		37	4.5–19 h[a]	[^{3}H]sterol exchange	LUV	[120]
SOPC		37	<1–2 min	[^{3}H]sterol exchange[b]	LUV	[79]
HRC		37	<1 s	[^{3}H]sterol exchange[b]		[133]
POPC, POPA (1:1)		37	≤10 ms	NMR	LUV	[23]
POPC	35	50	200 min	SANS	SUV	[47]
DSPC	15	23	<10 min	SFVS	SLB	[82]
	30	5	<10 min			

[a] This is one half of the $t_{1/2}$ value reported in Ref. [120]
[b] Using methyl-β-cyclodextrin (MBCD)

Spontaneous flip-flop rates vary depending on the head group structure and acyl chain length [58, 130, 131], on the composition and on the physical state of the membrane. A look at the values in Table 2.1 shows that, in the case of pure lipid systems, the measured rates are considerably lower below the gel to liquid crystal phase transition temperature (T_m), i.e., in the gel phase, than above it, i.e., in the liquid crystalline phase. A dramatic increase of the flip-flop rate was observed close to the phase transition [36, 65]. This can be explained considering that around the phase transition strong density fluctuations occur, which cause a drastic increase of the gel to liquid crystal interfaces. These may facilitate not only spontaneous flip-flop, but also the permeability of small molecules [32], as well as the action of interfacially active enzymes such as phospholipases A2 [96]. An increase of the translocation rate by up two orders of magnitude was observed also under the action of mechanical forces [117, 137, 158], which would be consistent with the hypothesis that flip-flop occurs via the formation of defects, i.e., temporary localized disordering of the lipid bilayer structure [118, 151].

There are factors that accelerate flip-flop, and among these the presence of oxidized phospholipids, as was observed already in the beginning of the '80s [127]. The reason for this acceleration was first attributed to the formation of non-bilayer structures [11] and, later on, to altered conformational dynamics of the oxidatively modified acyl chains [147]. Also, it was shown that spontaneous flip-flop of phospholipids is influenced by external agents, used as detergents [104] or carriers for incorporating hydrophobic molecules into cell membranes [141].

It was found that also membrane spanning peptides may accelerate lipid flip-flop [76], but the presence of cholesterol may inhibit this effect [70]. Keeping in mind that, relative to the plasma membrane, the early secretory pathway membrane of the ER is highly depleted of cholesterol – hence less thick and rigid, and more permeable than the plasma membrane – Kol et al. [70] suggested that rapid phospholipid translocation in biogenic membranes is "mediated via membrane-spanning segments of a subset of proteins, characterized by a small number of transmembrane helices". This implies that spontaneous flip-flop in the plasma membrane of eukaryotes is irrelevant for maintaining lipid asymmetry, and also that in the ER flip-flop does not require dedicated flippases. That any transmembrane proteins should accelerate flip-flop was also hypothesized for the cytoplasmic membranes of prokaryotes [69].

Cholesterol is ubiquitous in the membranes of eukaryotes [144] and may have a synergic effect on flip-flop. Remarkably, just 1 mol % cholesterol is sufficient to inhibit the flipping activity, toward newly synthesized phospholipids, of the biogenic membrane flippase in reconstituted proteoliposomes [51, 115]. Also, it was found that the addition of 30 mol % cholesterol to DMPC or DPPC liposomes suppresses phospholipid flip-flop [65]. This may be attributed to the increase in conformational order of the acyl chains, as it is known that, at that concentration, cholesterol induces the formation of the liquid-ordered phase [62].

2.3 Experimental Methods

Transbilayer diffusion of lipids cannot be directly monitored, therefore indirect detection methods have to be devised. Different assays have been proposed to this purpose, which have to fulfill specific criteria [38]. In general a non-equilibrium asymmetric distribution of lipids is generated, and then its evolution towards a uniform distribution is monitored.

Experimental studies have been done using model membranes and biomembranes. In the former case, flip-flop is investigated under controlled conditions, without the presence of the active processes that occur in natural biosystems. Model systems are prepared in the form of liposomes or planar supported lipid bilayers (SLB). Liposomes are generally distinguished according to their size: small unilamellar vesicles (SUV), ranging in size from 20 to 100 nm, large unilamellar vesicles (LUV) with a diameter of hundreds of nanometers and giant unilamellar vesicles (GUV), whose diameter can reach several tens of micrometers. SUVs are suitable to investigate the effects of curvature, whereas GUVs are often used to measure mechanical and rheological properties, because their changes in shape can be directly visualized under the microscope. SLBs offer the advantage that they can be studied by surface-sensitive techniques.

2.3.1 Labeled Lipid Analogues

The first insights on the transbilayer dynamics of lipids were obtained using compounds labeled with specific probes, either attached to the head group or to an acyl chain, which could be detected by a suitable technique. In the pioneering experiment by Kornberg and McConnell [72], egg phosphatidylcholine (EPC) vesicles doped with a paramagnetic analogue of phosphatydilcholine were investigated by electron paramagnetic resonance (EPR) spectroscopy. The analogue carried a 6-member nitroxide ring on the phospholipid head group, in place of one of the methyls of the quaternary ammonium (TEMPO-PC, see Fig. 2.2). The flip-flop rate was inferred from the time needed to recover the original spectrum after a reducing agent,

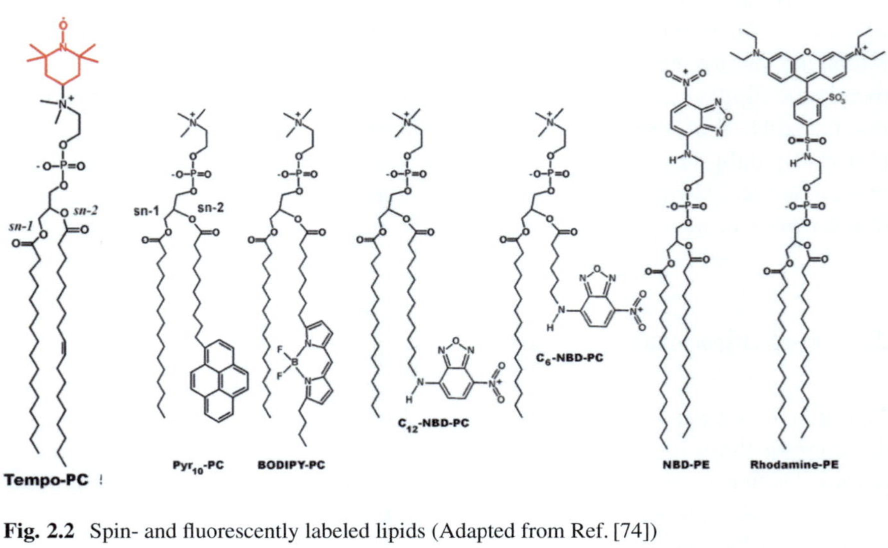

Fig. 2.2 Spin- and fluorescently labeled lipids (Adapted from Ref. [74])

which could access only the external leaflet, was added to solutions containing the vesicles. Likewise, spin-labeled aminophospholipids were employed to measure the translocation rate in HRC membranes [95, 125]. Besides head group labeled phospholipids, also analogues bearing a nitroxide radical attached to an acyl chain were employed. In this case, the so called back-exchange assay was used to monitor the translocation process: lipid analogues having a long *sn*-1 chain and a short *sn*-2 chain were removed from the external leaflet of the membrane, either by liposomes or by fatty acid free bovine serum albumin (BSA), and their amount was then determined by EPR. It is worth remarking that the observation of the fast redistribution of analogues of aminophospholipids (PE, PS) in HRCs was the first indication of the existence of an energy-dependent transport of specific phospholipids in the plasma membrane [125].

Also fluorescent probes, bearing a bulky fluorophore like NBD, fluorescein, pyrene and rhodamine (see Fig. 2.2), have been used to determine flip-flop rates. The methods to determine the transbilayer distribution of fluorophores are analogous to those used for spin-labeled lipids. One is extraction from the outer leaflet (back-exchange) of the analogues, whose amount is then measured by fluorescence spectroscopy. Alternatively, in the case of fluorescent probes that remain exposed to solvent, a water soluble, non-penetrating reducing agent, such as ascorbate or dithionite, can be used. This destroys the probes located in the outer leaflet, so that only the fluorophores that have flipped into the inner monolayer contribute to the fluorescence spectrum. A different methodology is based on the use of pyrene labels, whose fluorescence spectrum is characterized by two signals, one arising from excited monomers and the other from excited dimers (excimers), with a relative intensity that depends on the fluorophore concentration. Phospholipid analogues bearing a pyrene moiety in the *sn*-2 chain are incorporated into vesicles (donors), which are mixed with analogous undoped vesicles (acceptors). Then, the redistribution of fluorophores between the outer and the inner monolayer of the acceptor is followed by monitoring the change of fluorescence emission ratio. In Ref. [58] this method was used to study the transmembrane dynamics of pyrene labeled phospholipids with different head groups and acyl chains, in POPC LUVs. A strong dependence on the head group was found, with rates increasing in the order phosphatidylcholine (PC) $<$ phosphatidylglycerol (PG) $<$ phosphatidic acid (PA) $<$ phosphatidylethanolamine (PE). These groups differ in size and charge: PA and PG have a net negative charge, whereas both PC and PE are neutral, but the former is bulkier than the latter. The translocation rates for PE lipids were found to be at least 10 times greater than those for the homologous PC derivatives. A weaker dependence of rates on acyl chain length was observed, with approximately two-fold decrease of the rate upon addition of two methylene groups.

Recently, optical techniques like fluorescence interference contrast (FLIC) microscopy [31] and fluorescence correlation spectroscopy [146], were used to monitor the evolution of the transversal asymmetry in fluorophore distribution in SLBs. Asymmetric bilayers with a desired composition of leaflets were produced using vesicle fusion or monolayer deposition, in the two variants Langmuir-Blodgett (LB) and Langmuir-Blodgett/Schäfer (LB-LS). Interestingly, it was found that a given bilayer can retain full lipid asymmetry if prepared by the former method, but loses part of its asymmetry during deposition of the distal monolayer if the latter method is used. This illustrates the strong influence of experimental conditions on flip-flop, and thus the importance of precisely defining and controlling them.

A main issue in the use of synthetic lipid probes concerns possible specific effects of the label and perturbations (packing, polarity) that they may induce in the bilayer, which raises the question whether their behavior can be representative of that of natural lipids. This has been largely debated in the literature, see for instance Refs. [40] and [1] for recent reviews.

2.3.2 *Isotopically Substituted and Native Lipids*

The problem of bulky labels can be circumvented using isotopic substitution, which has a much less perturbing effect. In most assays donor and acceptor vesicles, which contain isotopically labeled and natural lipids, respectively, are mixed. Then, the transfer of labeled molecules from donors to acceptors is monitored using a suitable technique. The transfer includes both inter-vesicle exchange and flip-flop, and the kinetic parameters of the latter are extracted from modeling of the whole process.

In earlier studies, radioisotopes were used and the two types of vesicles (donors and acceptors) were separated according to their charge [150] or size [151]. More recently, the use of time-resolved small-angle neutron scattering (SANS) [47, 97, 98] was proposed, which exploits the difference in scattering length between deuterated and undeuterated lipids. From experiments with LUVs, flip-flop rates of DMPC comparable with those determined using radioisotopes [150] were obtained, with half-times ranging from 9 h, around room temperature, to 1 h at higher temperature. No transbilayer diffusion was observed for POPC, which has longer acyl chains than DMPC, one of which is unsaturated [97, 98]. Using the same technique, a flip-flop half-time of a few hours was determined for cholesterol in POPC SUVs [47]; this is much longer than the estimates from previous experiments, which ranged from milliseconds to seconds [23, 55, 65, 79].

A variant of SANS, neutron reflectometry (NR) [48], was applied to isotopically asymmetric bilayers prepared by the LB-LS technique, with one leaflet containing only deuterated and the other one only undeuterated lipids. The recorded reflectivity contains a clear signature of the bilayer asymmetry, which is lost in the course of an experiment, as the composition becomes uniform because of transbilayer movements. In NR experiments on DMPC:DSPC bilayers, no flip-flop was detected in the gel phase, whereas in the liquid crystal phase translocation was found to occur on a time-scale of 1–2 s, which is orders of magnitude shorter than the value inferred from SANS experiments [48]. This fast rate was confirmed by further NR measurements on supported bilayers of DMPC, in which an asymmetric population was generated by exposition to a solution of isotopically labeled vesicles [49].

SLBs were also investigated by sum-frequency vibrational spectroscopy (SFVS). This is a coherent, nonlinear vibrational spectroscopy that is sensitive to the symmetry of the system. The decrease in asymmetry of the bilayer is determined by recording the intensity of specific vibrational modes, which allow to distinguish between natural and deuterated lipids. This technique was first applied to DSPC bilayers in the gel phase and the flip-flop rate was found to increase with temperature up to the melting transition; above this, the process became too fast to be measured [80]. Analogous results were obtained for DMPC and DPPC bilayers [81]. The translocation rate was found to decrease with the acyl chain length, with half-times equal to 3 min for DMPC, 14 min for DPPC and 38 min for DSPC, at ∼5 °C below the main phase transition. During the last decade SVFS has been applied to a variety of systems and detailed insight into the temperature dependence of phospholipid flip-flop in the gel phase has been reached. Beside the role of the acyl chain length, also that of the head group was investigated, and for instance it was found that the flip-flop rate in DSPE bilayers is nearly two orders of magnitude higher than in DSPC bilayers at the same temperature [4]. Moreover, SVFS studies showed that that the flip-flop rate increases in the presence of peptides [3, 5] and in the presence of cholesterol, with a dependence on the membrane phase in the latter case [2, 82].

In a very recent study the translocation of chain- and head-deuterated phospholipids between the leaflets of DPPC LUVs, with an initial asymmetric distribution, was monitored by isotope sensitive ^{1}H-NMR spectroscopy in the presence of a paramagnetic reagent, which shifts the resonances for lipids in the outer leaflet [87]. Flip-flop half-times of the order of days to weeks were determined in the liquid crystal phase, whereas no appreciable flip-flop was detected in the gel phase over 250 h. This is in stark contrast to the results of SFVS [1, 80, 81] and NR [48] measurements on SLBs, and confirms the difference between flip-flop in vesicles and in supported bilayers, already pointed out in Ref. [49]. To explain this difference, a defect-mediated acceleration of flip-flop in SLBs was proposed, due to the presence of long-lived, submicron-sized holes, resulting from incomplete surface coverage, which would be the sites of fast lipid translocation.

Most methods for the detection of the change in symmetry in the composition of lipid bilayers make use of spectroscopic techniques; an interesting alternative approach exploits the changes in the shape of vesicles, which are controlled by the bilayer elasticity. In a pioneering work, HRCs were found to exhibit morphological changes upon incorporation of amphiphatic drugs differently distributed in the two leaflets [128]. Likewise, in the early studies of transbilayer motion by EPR it was observed that, immediately after addition of spin-labeled lipid analogues, the HRCs became crenated and the original shape was subsequently recovered, with a kinetics similar to that of transverse diffusion [125]. In Ref. [35] the morphological changes were used to monitor the incorporation and translocation of phosphatidylserines in HRCs. The same method was applied to investigate the transbilayer dynamics of exogenous lipids in EPC GUVs [83, 105]. The shape of GUVs is very sensitive to the area asymmetry, defined as the ratio of the area difference between the monolayers to the area of the bilayer mid-surface; an asymmetry of the order of 0.1% was found

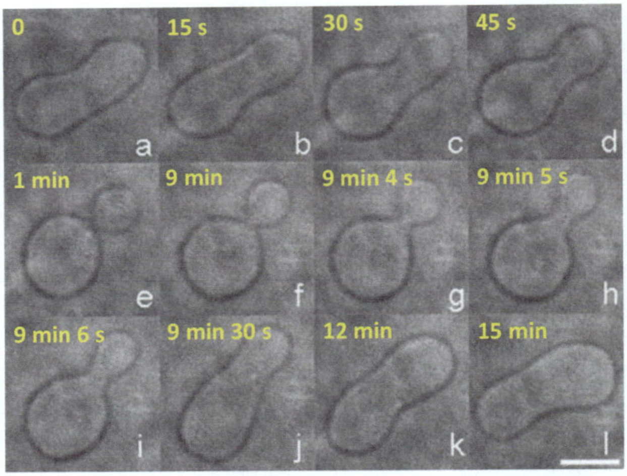

Fig. 2.3 Shape recovery after injection of C6-ceramide into an EPC GUV at T = 20 °C. Images taken at different times after injection. The scale bar corresponds to 10 μm (Adapted from Ref. [83])

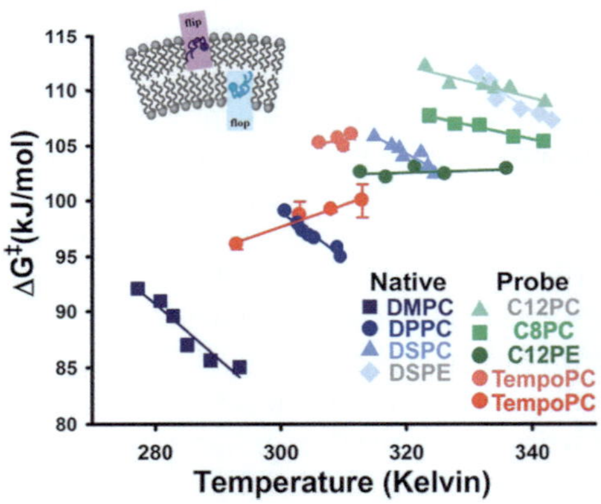

Fig. 2.4 Activation free energy of flip-flop, determined for native lipids in the gel phase by SFVS (*dark* to *light blue*) [5, 6, 81], and for labeled lipids using different techniques: fluorescence for pyrene-labeled PCs in liquid crystalline POPC (*dark* to *light green*) [58]; EPR for TEMPO-DPPC in EPC (*red*) [72]; SFVS for TEMPO-DPPC in gel phase DSPC (*orange*) [1] (From Ref. [1])

to be sufficient to induce a change in shape. The effect is evident in Fig. 2.3, which shows the evolution in time of the shape of a GUV after injection of C6-ceramide.

To close this Section, a comment can be made with the help of Fig. 2.4, which collects data taken from different experiments. Here $\Delta G^{\ddagger}$ is the free energy barrier for the translocation process, and was determined from the temperature dependence of the rate constant k_f (see text under Eq. 2.1), according to $k_f \propto \exp(-\Delta G^{\ddagger}/RT)$, where R is the gas constant [81]. All data refer to zwitterionic lipids, yet they show appreciable differences not only in the magnitude of $\Delta G^{\ddagger}$, but also in its temperature dependence. It can be inferred from the figure that the flip-flop rate decreases with increasing length, and this occurs for both PC lipids and pyrene labeled PC probes. We can also see that the $\Delta G^{\ddagger}$ values for pyrene labeled analogues are higher than for native lipids. However, it is not obvious that this difference can be ascribed to the probe, as the data refer to different systems and conditions: SUV of liquid crystalline POPC for the labeled species, SLBs of pure PC lipids in the gel phase for the unlabeled. Indeed, flip-flop is sensitive to a variety of factors, and this makes it difficult to draw conclusions from the comparison of different experiments. Likewise, it is not obvious that kinetic or mechanistic insights obtained from experiments performed under specific conditions can be directly transferred to biological systems.

2.4 Theoretical and Computational Insights

Theoretical and computational studies aim at providing a molecular description of the energetics and kinetics of flip-flop. During the last 20 years, classical Molecular Dynamics (MD) has been widely employed for this purpose and has given new microscopic insights. Here, before presenting an overview of this method and of its main achievements, the features of flip-flop, described as a roto-translational diffusion process, are outlined.

2.4.1 Flip-Flop as a 2D Diffusion Process

Important features of the molecular mechanism of flip-flop can be captured in the framework of low-dimensional diffusion models, of which there are many successful applications in biophysics [17, 149, 159]. In view of the polarity (in shape and charge distribution) of lipids and of the symmetry of bilayers, flip-flop necessarily involves molecular translation and rotation. Moreover, other degrees of freedom, whether molecular, like conformational angles, or collective, like elastic deformation variables of the bilayer, may play a role in this process. If these additional degrees of freedom are faster or do not give a major contribution, flip-flop can be described as diffusion on a free energy surface defined by the orientational and rotational variables. This assumption could be reasonable for the case of cholesterol, which is relatively rigid, thus conformational variables are not expected to significantly affect its transbilayer movement. Figure 2.5b shows the free energy surface reported in Ref. [107] for cholesterol in a DPPC bilayer, in the liquid crystalline phase at $T = 323$ K. Here the variables are the position of the origin (CM) of the molecular frame along the bilayer normal $\mathbf{N}$ (Z coordinate) and the tilt angle (β) of the long molecular axis (z) with respect to $\mathbf{N}$. CM coincides with the hydrodynamic center of mobility [67] and z is parallel to the $C_{17} \rightarrow C_3$ direction (see Fig. 2.5a). The free energy surface was

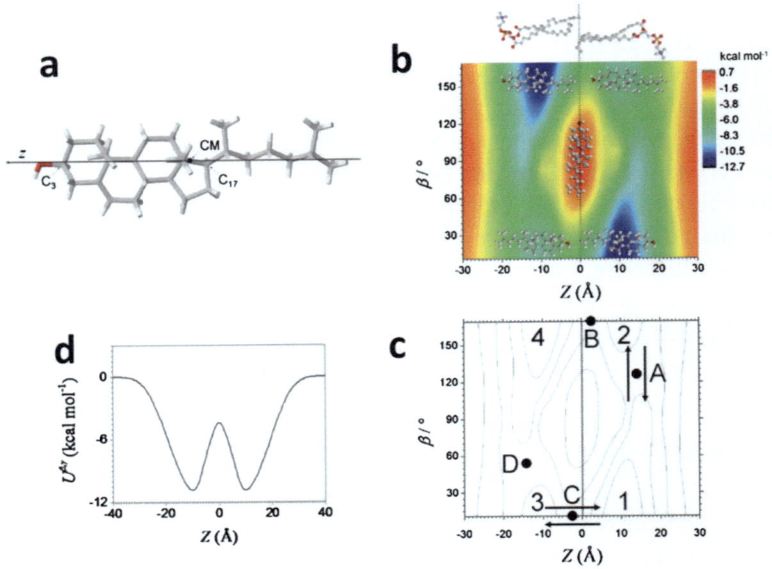

Fig. 2.5 (**a**) Structure of cholesterol and reference frame used to define the molecular position and orientation in the lipid bilayer. The origin of the molecular frame is located in the hydrodynamic center of mobility (*CM*). (**b, c**) Free energy surface calculated for cholesterol in a DPPC bilayer at 323 K, as a function of the *CM* position (*Z*) along the bilayer normal (**N**), and of the angle (β) between **N** and the long molecular axis *z*. At the bilayer mid-plane $Z = 0$. In (**b**), the molecular structures superimposed to the free energy surface show the position and orientation of cholesterol in the minima and in the maximum; also DPPC molecules are shown as a reference In (**c**) the labels **1–4** indicate the position of the free energy minima, whereas **A–D** indicate the position of saddle points. The arrows indicate transitions between the free energy minima across the saddle points. (**d**) Free energy profile calculated as the Boltzmann projection of the 2D free energy surface onto the *Z* coordinate (Adapted from Refs. [107] and [108])

calculated using a model that combines an atomistic description of cholesterol with an implicit representation of the water/membrane environment, whose anisotropy and longitudinal non-uniformity are taken into account through the gradients of density, dielectric permittivity, lateral pressure and acyl chain order parameters [106]. Appropriate values for pure DPPC at 323 K were used to calculate the free energy surface in Fig. 2.5b. The free energy surface exhibits two absolute minima (**1** and **4**), which are equivalent and correspond the canonical upright orientations of cholesterol, on either side of the bilayer. In such positions the molecule has its *z* axis slightly tilted from **N**, with the hydrophobic backbone in the region of highest lipid density and the hydroxyl group at the level of the polar heads. The free energy surface exhibits two additional equivalent relative minima (**2** and **3**), which differ from the stable states because cholesterol has reversed its orientation and points its tail towards water, whereas the hydroxyl group is buried deep in the hydrophobic region. These minima are much higher in energy than the stable states (**1**, **4**), owing to the high cost for moving the polar head away from the interface.

The free energy surface has a maximum in the central region, for configurations in which cholesterol lies in the middle of the bilayer, approximately perpendicular to **N**. The upright position of cholesterol is in agreement with the results of neutron scattering experiments in liquid crystalline DMPC [77]. However the distribution of this molecule was found to change considerably with the lipid composition [86], and in polyunsaturated membranes cholesterol was found to preferentially lie flat in the middle of the bilayer [56, 57].

With reference to the free energy landscape shown in Fig. 2.5b, flip-flop is the process in which cholesterol moves from one of the stable states (**1**, **4**) to the other. This can occur through different pathways, which imply different combinations of rotations and translations [7, 109]. Direct transfer from **1** to **4**, through the free energy maximum, is very unlikely, since this requires crossing of the high energy barrier in the middle of the free energy surface. More likely paths involve the passage through the saddle points between the minima, which are labeled by **A**, **B**, **C**, **D** in Fig. 2.5c. Transitions through **A** and **D** mainly involve the upside-down reorientation of cholesterol, whereas those through **B** and **C** involve translation of the molecule from one leaflet to the other without significantly changing its orientation. Thus, two pathways can be identified, in which reorientation takes place before and after crossing the bilayer midplane, respectively. The process was described using a four-step kinetic model, obtained by reducing the Smoluchowski equation for time evolution of the positional-orientational probability density function to a Master equation for transitions between the free energy minima. This reduction is based on the multidimensional extension of Kramers theory [73, 75, 94], which is justified when the barrier between the minima is sufficiently high. Within this framework, transition rates between the minima are expressed in terms of energetic and frictional parameters: height of the free energy barriers, curvatures of the free energy surface in the minima and in the saddle points, and molecular roto-translational friction tensor. From the time required to reach the equilibrium distribution, starting from one of the absolute free energy minima, a flip-flop rate constant $k_{ff} = 2 \cdot 10^4\ \mathrm{s}^{-1}$ was determined for cholesterol in liquid crystalline DPPC at $T = 323$.

A simpler description of flip-flop is obtained by taking the Z-position as the single reaction coordinate. Figure 2.5d shows the mono-dimensional (1D) free energy profile calculated as the Boltzmann projection of the free energy surface on the positional coordinate. Each point of the curve is obtained by averaging over all molecular orientations, with a proper orientational distribution function. In this way the roto-translational coupling along the flip-flop pathway is neglected, which may be reasonable only if rotations are much faster than translational displacements [28]. Using the 1D description and the potential shown in Fig. 2.5d, a rate constant $k_{ff} = 5.8 \cdot 10^4\ \mathrm{s}^{-1}$ was determined for cholesterol, which is almost three times higher than the value obtained using the 2D model.

A 2D description of flip-flop, with a kinetic analysis of the process according to the multidimensional extension of Kramers theory, was reported also for oleic acid in liquid crystalline POPC at $T = 303$ K [148]. In this case the free energy surface was calculated using umbrella sampling simulations (see Sect. 2.4.2). The flip-flop rate evaluated by a two-dimensional diffusion model was found to be in good agreement with that obtained directly from counting the translocation events in MD simulations, and both were three/five-fold lower than the rate calculated using a 1D description of the process.

The features of the free energy landscape, which determine the flip-flop pathway and its rate, may considerably change with the composition and physical state of the bilayer. Also the concentration of cholesterol is very important, since this has well known ordering and condensing effect [59, 63]. Table 2.3 reports the rate constant k_{ff}, derived from various MD simulations studies (see the next subsection for details on the method). Some of such studies report the number of translocation events (n_f) observed during the simulation time (τ), from which the rate constant can be evaluated as $k_{ff} = 2n_f/N_C\tau$, where N_C is the number of cholesterol molecules in the sample. Other studies report the rate constant for cholesterol transfer from one bilayer leaflet to the other, $k_f = k_{B\leftarrow T} = k_{T\leftarrow B} = k_{ff}/2$ (see Eq. 2.1). In these studies k_f is usually determined from the 1D free energy profile of cholesterol as a function of its position across the lipid bilayer, according to the method proposed in Ref. [12]. We can see that the rate constants reported Table 2.3 cover a wide range. On one side the discrepancies between data have a physical origin, since different systems and conditions were examined by simulations. On the other side, also merely computational aspects have to be considered, in particular regarding the form of the interaction potential used in simulations. Indeed, significant differences have been evidenced between the results obtained using some of the most common empirical potentials, both in the distribution of cholesterol and in its translocation rate [86].

Comparison of the k_{ff} values in Table 2.3 with the half-times reported in Table 2.2 shows that the calculated flip-flop rates are generally faster than the measured values. Actually, in some cases the experimental half-times are upper limits, since the process was too fast to be detected with the methodology employed. Another aspect to be considered is that computer simulations necessarily refer to ideal conditions, which can be well different from those of real systems: the simulated samples are in general pure, symmetric bilayers, whereas biomembranes are asymmetric and contain a variety of components, including proteins. Moreover, mechanism and rate of flip-flop are affected by deformations, curvature and defects, which are often present in experimental systems.

2.4.2 Molecular Dynamics Simulations

Within the classical MD framework a system is represented as a set of particles, which may either correspond to atoms (all-atom, AA) or to groups of atoms (coarse-grain, CG), whose time evolution is determined by numerical integration of Newton's equations of motion. The forces acting on particles are computed from a potential energy function, accounting for inter- and intra-molecular interactions

Table 2.3 Flip-flop rate constants, k_{ff}, of cholesterol (C), from different MD studies. See text for the definition of k_{ff} and its relation with the flip-flop half-time

Host	Components	T (K)	k_{ff} (s^{-1})	Model	Ref.
POPC	38 POPC, 4 C	300	$(0.44 \pm 0.30) \cdot 10^6$ [a]	CG	[90]
SAPC	38 SAPC, 4 C		$(4 \pm 2.4) \cdot 10^6$ [a]		
DAPC	38 DAPC, 4 C		$(20 \pm 5) \cdot 10^6$ [a]		
DTPC	38 DTPC, 4 C		$(6.7 \pm 2.2) \cdot 10^6$ [a]		
DAPC	152 DAPC, 2 C	323	$4.6 \cdot 10^7 - 3.2 \cdot 10^8$ [b]	CG	[12]
SAPC	152 SAPC, 2 C		$6.6 \cdot 10^6 - 3.4 \cdot 10^7$ [b]		
POPC	152 POPC, 2 C		$6.0 \cdot 10^5 - 9.4 \cdot 10^6$ [b]		
DPPC	152 DPPC, 2 C		$2.4 \cdot 10^5 - 3.2 \cdot 10^6$ [b]		
DPPC,C	152 DPPC, 104 C		$14.4 \cdot 10^3 - 5.0 \cdot 10^4$ [b]		
DAPC	64 DAPC, 2 C		$(15.0 \pm 0.4) \cdot 10^7$ [a]		
SAPC	64 SAPC, 2 C		$(13.4 \pm 1.0) \cdot 10^6$ [a]		
POPC	64 POPC, 2 C		$(3.0 \pm 0.6) \cdot 10^6$ [a]		
DPPC	64 DPPC, 2 C		$(14.8 \pm 4.2) \cdot 10^5$ [a]		
DPPC,C	38 DPPC, 26 C		$>2.8 \cdot 10^4$ [a]		
DAPC	72 DAPC, 2 C		$10.4 \cdot 10^5 - 7.4 \cdot 10^6$ [b]	AA	
DPPC	64 DPPC, 2 C		$2.4 \cdot 10^4 - 13.2 \cdot 10^5$ [b]		
DPPC,C	52 DPPC, 12 C		$2.4 \cdot 10^4 - 16.2 \cdot 10^3$ [b]		
DPPC,C	38 DPPC, 26 C		$18.8 - 10.0 \cdot 10^2$ [b]		
DPPC	98 DPPC, 2 C	323.15	$2.4 \cdot 10^3 - 3.2 \cdot 10^4$ [b]	AA	[64]
POPC	98 POPC, 2 C	303.15	$9.6 \cdot 10^2 - 8.0 \cdot 10^3$ [b]		
DAPC	98 DAPC, 2 C		$12.0 \cdot 10^5 - 16.0 \cdot 10^6$ [b]		
DAPC	38 lipids, 4 C	300	$2.54 \cdot 10^7$ [a]	CG	[103]
SAPC			$6.64 \cdot 10^6$ [a]		
POPC			$0.856 \cdot 10^6$ [a]		
DMPC	100 DMPC, 1 C	303	$13.6 \cdot 10^6$ [b]	AA	[86]
DLPC	100 DLPC, 1 C		$5.0 \cdot 10^7$ [b]		
POPC	64 POPC, 1 C	323	$9.0 \cdot 10^4$ [b]	AA	[14]
Raft	22 POPC, 22PSM, 23 C		$8.8 \cdot 10^{-4}$ [b]		
DPPC	360 DPPC, 152 C	323	$6 \cdot 10^4$ [a]	AA	[26]
POPC,PSM,C[b]	64 POPC, 32 PSM, 32 C	310	$8.0 \cdot 10^2 - 6 \cdot 10^4$ [b]	AA	[102]
POPC	127 POPC, 1 C		$1.2 \cdot 10^4 - 9.2 \cdot 10^5$ [b]		
DOPC,DOPS	504 DOPC, 504 DOPS, 196 C	320	$(5.75 \pm 2.73) \cdot 10^6$ [a]	CG	[155]
DSPC,DSPS	504 DSPC, 504 DSPS, 196 C		$(1.27 \pm 0.65) \cdot 10^6$ [a]		
DOPC,DSPC	504 DOPC, 504 DSPS, 196 C		$(3.10 \pm 1.00) \cdot 10^6$ [a]		
DOPS,DSPS	504 DOPS, 504 DSPS, 196 C		$(1.00 \pm 0.53) \cdot 10^6$ [a]		
Plasma	20,000 lipids (60 kinds)	310	$(13.1 \pm 0.02) \cdot 10^6$ [a]	CG	[61]
Raft	828 DPPC, 540 $diC_{18-2}PC$, 576 C	295	$1.0 \cdot 10^7$ [a,c]	CG	[119]

[a]From counting of translocation events

[b]From 1D free energy profile

[c]In liquid crystalline environment; no flip-flop detected in liquid-ordered environment

that are described by an empirical model (force field). The knowledge that can be obtained from a simulation depends crucially on the force field. Major issues related to the application of MD techniques to the study of flip-flop in membranes concern the time scale and the size of the simulated systems. The general scheme for integrating Newton's equations makes use of a finite difference algorithm, based on the partitioning of the trajectory into small steps (time steps). Thus, the computational cost increases with the length of the trajectory, i.e., the number of time steps, and with the size of the sample, i.e., the number of interaction sites, since at each time step the evaluation of forces requires the calculation of the interaction potential between particles. Flip-flop may have characteristic times of the order of several hours; it may occur as a collective process, involving large-scale changes in the membrane, and also be connected with other non equilibrium processes in the system. Recent advances in hardware, software, and algorithms have lead to an impressive increase in the accessible time scales, which can reach the millisecond regime, whereas the size of samples can be as large as several thousands of lipids. This is still far below what would be required for a reasonable sampling of translocation events under realistic conditions. However, suitable methodologies have been developed that, combined with appropriate choices of systems and conditions, make it possible to shed light into molecular aspects of the inter-leaflet translocation of lipids.

2.4.2.1 Minimalist Models

Minimalist models can provide general insights into the physical behavior of lipid bilayers. An example, which can be useful to illustrate their capability and limitations, is provided by the study that is probably the first attempt to use MD to investigate flip-flop [60]. Each lipid was represented as a flexible chain of beads, water was modeled as a single bead, and a combination of Lennard-Jones and soft repulsive potentials was assumed to mimic hydrophobic and hydrophilic interactions between the beads. It was found that flip-flop is an activated process, with a barrier which is mainly determined by the translocation of the hydrophilic head through the hydrophobic region. Thus, the flip-flop rate was shown to increase with the chain length, which is in agreement with experimental findings [58, 65, 130]. On the other hand, the translocation rate was found to decrease with increasing chain stiffness, in apparent contrast to the rapid flip-flop observed in phospholipids bearing polyunsaturated chains [8, 130]. Indeed, the comparison with simulations cannot be taken too strictly, because mapping of a generic model into a real system is not that straightforward. A model similar to that used in Ref. [60] was adopted also in a subsequent study where, exploiting an enhanced sampling technique, several translocation events could be monitored [91]. This work confirmed that flip-flop is an activated process and showed that the lipid translocation is accompanied by some rearrangement of the surroundings. However, the size of the system (76 lipid) could be inadequate for mechanistic insights.

A much bigger system (3321 lipids) was investigated in Ref. [7]. In this case, a millisecond time scale could be reached using Dissipative Particle Dynamics (DPD), with particles interacting via soft potentials, chosen so as to reproduce the hydrodynamic behavior of a fluid [145]. Also in this case lipids were modeled as chains of hydrophilic beads with a hydrophilic head, and it was found that the rate of translocation decreases with increasing chain length. Taking advantage of the smooth free energy landscape, several translocation events could be sampled, and it was shown that flip-flop has the features of a Poisson process. It was also possible to distinguish three different pathways by which, combining rotation and translation, lipids move from one monolayer to the other: "push-in", in which the molecule translates towards the center of the bilayer, then rotates and finally translates towards the equilibrium position in the opposite leaflet; "sliding", where the process occurs via simultaneous rotation and translation; "rotation", where first the molecule rotates and then translates.

An analogous model, within the DPD scheme, was used to mimic active flip-flop [116]. It was found that a finite difference in the lipid number density between the two leaflets, caused by asymmetric flip-flop (i.e., a different average number of up-down and down-up translocations), induces morphological changes in the membrane, in the form of buds or blisters, depending on the flip-flop rate. This is in agreement with theoretical predictions [92] and experimental findings (see Sect. 2.3.2).

In a very recent MD study, an implicit solvent coarse-grain model [112] was used to investigate SLBs. It was found that pore-like structures induced by the density difference between the two monolayers can mediate flip-flop. This could explain the discrepancies between the flip-flop rates measured in vesicles and in SLBs [49, 87].

2.4.2.2 Specific Models for Lipids

Experiments have shown that the rate and the mechanism of flip-flop can strongly depend on the chemical structure of lipids. Therefore it is important to implement chemical details into computational models. This has motivated the development of accurate force fields, which typically are parameterized on the basis of quantum mechanical calculations and experimental data. Indeed, the derivation of a force field able to provide a reliable description of the behavior of lipids in their environment may be difficult, because this is controlled by a subtle balance of entropic and enthalpic effects. Several force fields have been proposed, but further refinement is needed (see [84, 110] for very recent reviews). Beside all-atom models for lipids, also CG versions have been proposed, and among the latter Martini [88, 89] has become very popular.

Specific models for lipids are in principle able to provide detailed molecular insight into the mechanism of flip-flop. However, despite the increase in computational power and the advantages offered by ad hoc CG parameterizations, statistically significant sampling of transbilayer movements by brute force MD simulations remains out of reach, especially in the case of phospholipids. Therefore, also other approaches have been adopted (see [109] for a recent review). For instance, restrained simulations are used to calculate the free energy surface in which the flipping particle is moving, and the kinetic parameters of the process are subsequently determined from the features of this surface. Flip-flop is a process that in principle involves several degrees of freedom, both of the flopping lipid and of its environment, and a kinetic description encompasses the reduction to the (few) slow variables (reaction coordinates) that characterize the evolution of the system on the time scale of the process. Other degrees of freedom, fluctuating on much shorter time scales, can be averaged out without losing relevant information on flip-flop. However the identification of the reaction coordinates is not straightforward and statistical sampling of the free energy surface, beside having a high computational cost, may be affected by significant errors, due to the presence of the so-called hidden barriers, which cause slow relaxation of the degrees of freedom orthogonal to the reaction coordinates [28, 71, 99]. In most studies of lipid flip-flop the strategy proposed in Ref. [138] has been adopted, which envisages the determination of the free energy profile, or potential of mean force (PMF) experienced by a flipping lipid as a function of its position along the bilayer normal. This is generally calculated using the technique known as umbrella sampling.

In AA simulations of phospholipids with charged head groups, few isolated translocation events could be detected under conditions promoting the formation of transient water pores, which facilitated the translocation of ionic moieties across the membrane in a solvated state. This was first evidenced in a self-assembling DPPC vesicle [41], and then translocating lipids were also observed in the presence of perturbing agents like buthanol (DPPC) [42], dimitylsulfoxide (DPPC, DMPC or POPC) [53], antimicrobial peptides [78] and ions (POPC) [66], (DMPC) [53]. To enhance the process, simulations of a double DMPC bilayer in the presence of a significant transmembrane imbalance of cations were performed, and 50 translocation events could be detected in long simulations [52]. As a result of this study, it was proposed that the formation of a hydrophilic pore spanning the entire membrane would be the rate-determining step, followed by fast diffusion of the phospholipid through the pore (on the nanosecond scale). This mechanism is in line with an early observation of similar activation energies (about 80 kJ mol^{-1}) for spontaneous chloride permeation [140] and for flip-flop [72].

The formation of water pores, with the same structure as in the presence of perturbing agents, was also observed in AA umbrella sampling MD simulations, in which the zwitterionic head of a lipid was constrained in the center of the bilayer [138]. This kind of behavior was found for different PC bilayers (DLPC, DPPC, DMPC, POPC, DOPC) [15, 123]. In agreement with experiments [58, 65, 130] and MD simulations with generic models [7, 60], the flip-flop rate was found to increase with decreasing chain length, because of the lower cost to form a pore and the higher stability of large pores in thinner bilayers [15]. This is illustrated in Fig. 2.6, which reports free energy profiles calculated for DLPC and DPPC in pure bilayers [13], showing a typical shape with a barrier in the mid-plane, mainly due to the cost for rearranging neighboring lipids and water to form a pore. The

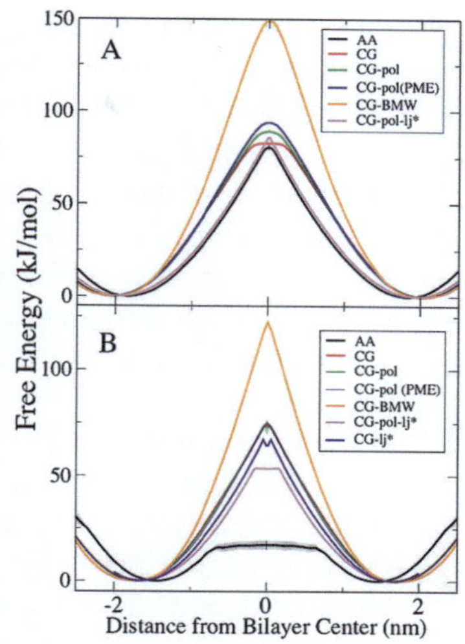

Fig. 2.6 PMFs for phospholipid flip-flop as a function of the distance of the lipid head-group from the bilayer mid-plane. (**a**) DPPC in a DPPC bilayer. (**b**) DLPC in a DLPC bilayer. The different profiles were obtained by umbrella sampling MD simulations, using different force fields (Adapted from Ref. [13])

height of this barrier was found to increase in the presence of cholesterol; actually, in this case the translocation mechanism itself was modified since cholesterol was found to prevent the formation of water pores by increasing the acyl chain order and decreasing the bilayer fluidity [12]. Likewise, no pores were detected in the case of unsaturated lipid bilayers, where flip-flop was found to be much slower than in saturated systems. This result is in line with the slowing down of flip-flop with increasing chain stiffness, which was predicted in Ref. [60] using a minimalist lipid model, but is the opposite of the experimental trend [8, 130]. This discrepancy was ascribed to a poor parametrization for double bonds in the force field used in the simulations [123]. Actually, thorough MD investigations have shown that the formation of pores in lipid bilayers is extremely sensitive to the force field used; therefore very accurate tuning of the parameters is needed to achieve predictive power level [16]. The definition of the force field is particularly delicate in the case of CG models, where a possible issue could be related to the reduced entropy of water [13].

Even though there remain critical aspects, MD simulations have been very useful to provide a detailed molecular picture of various aspects of flip-flop; the main results can be summarized as follows.

(i) Flip-flop of cholesterol and its analogues. Several simulation studies have shown that the transbilayer dynamics of sterol is orders of magnitude faster than that of charged or zwitterionic phospholipids. This is in general agreement with the experimental findings, even though there are discrepancies between the results of different experiments (see Table 2.2) as well as between the results of different simulations (see Table 2.3). The flip-flop rate was found to depend on the sterol structure, with polar substituents playing a special role; for instance, replacement of the hydroxyl group of cholesterol by a carbonyl was found to increase the translocation rate [102, 121], in agreement with the prediction of the 2D diffusion model [107].

(ii) Importance of the size and especially of the charges of the head group. At both the AA [14] and the CG level [61, 103], DAGs and ceramides, which have a neutral and relatively small head group, were reported to undergo relatively fast flip-flop, without the mediation of defects and pores, as for cholesterol. Under the same simulation conditions, the flip-flop rate for DAGs was found to be similar to that of cholesterol, and appreciably higher than that of ceramides. This is in agreement with experiment [10, 46].

(iii) Presence of synthetic and natural peptides. As experiments [3, 4, 68], also simulations indicate that the incorporation of synthetic and natural peptides causes an increase of the flip-flop rate. In particular, it was observed in AA simulations that the synthetic peptides WALP23 and KALP23 cause a decrease of the free energy barrier in DOPE and DOPG bilayers, whereas they do not appreciably affect the free energy barrier in DOPC bilayers [123].

(iv) Physical state of the bilayer. The flip-flop rate was found to depend on the physical state of the bilayer; for instance, slowing down of the process, on moving from a liquid crystalline to a liquid-ordered environment, was reported for cholesterol, DAGs and ceramides [14, 119].

2.5 Conclusions

Nowadays it is well recognized that cell membranes are not simply elastic envelopes, but take an active part in biological processes, e.g., in the transmission of signals. The molecular composition and asymmetry represent important tools to control the mechanical properties of membranes and the fluxes across them. Here flip-flop plays a crucial role, and a clear insight into the molecular aspects of this process is needed to reach a detailed mechanistic understanding of intra- and inter-cellular exchanges. Despite the fact that since the beginning of the '70s flip-flop has been the subject of several investigations, a clear and consistent picture is still missing, and there is an impressive lack of quantitative understanding. This is partly due to the experimental difficulty to monitor the time evolution of the bilayer asymmetry. There are also problems related to the preparation of samples and to the standardization of measurements. Some experiments make use of bulky probes whose capability to reproduce the behavior of native lipids is still debated. Other assays have an intrinsic flaw, due the fact that flip-flop is measured together with other processes, and the separation of the different contributions may be non-straightforward. Then, there may be issues related to the time resolution of experiments, especially in the case of fast translocation. Further difficulty derives from the strong dependence of flip-flop on the molecular composition and physical state of the bilayer, which can affect the rate and probably also the mechanism of the process. So, the information obtained from a given measurement may be limited to specific experimental conditions.

To answer biologically relevant questions and in view of medical applications [101], a deep microscopic insight is needed, because even relatively small changes in the molecular structure of the flipping lipid may be crucial. Just to give an example, it is known that an extra hydroxyl group near the end of its side chain enables 24S-hydroxycholesterol to overcome the blood brain barrier and be excreted from the brain [19], which is essential for the maintenance of cholesterol steady state in the brain. At the present stage we are far from this detailed understanding. However we can imagine that in the next years new knowledge on molecular aspects will be provided by studies on model systems, using techniques that allow to monitor the time evolution of bilayer asymmetry without exogenous probes. Valuable insight will be provided also by large scale computer simulations based on accurate force fields.

It must be said, however, that detailed knowledge on model systems will not necessarily be sufficient to elucidate the role of flip-flop in biological processes. Transmembrane movement is just a step within a network of interconnected processes, which include lipid metabolism and membrane deformations, and occur under non-equilibrium conditions [45]. The capability to perform detailed examination of lipid distribution and leaflet asymmetry in such complex systems is a challenge for experimental techniques. From the point of theory and simulations, a challenge is represented by the integration of microscopic models into a general framework that can account for the coupling between the various processes.

References

1. Allhusen JS, Conboy JC (2017) The ins and out of lipid flip-flop. Acc Chem Res 50:58–65
2. Allhusen JS, Kimball DR, Conboy JC (2017) Structural origins of cholesterol accelerated lipid flip-flop studied by sum-frequency vibrational spectroscopy. J Phys Chem B 120:3157–3168
3. Anglin TC, Liu J, Conboy JC (2007) Facile lipid flip-flop in a phospholipid bilayer induced by gramicidin A measured by sum-frequency vibrational spectroscopy. Biophys J 92:L01–L03
4. Anglin TC, Brown KL, Conboy JC (2009) Phospholipid flip-flop modulated by transmembrane peptides WALP and melittin. J Struct Biol 168:37–52
5. Anglin TC, Conboy JC (2009) Kinetics and thermodynamics of flip-flop in binary phospholipid membranes measured by sum-frequency vibrational spectroscopy. Biochemistry 48:10220–10234
6. Anglin TC, Cooper MP, Li H, Chandler K, Conboy JC (2010) Free energy and entropy of activation for phospholipids flip-flop in planar supported lipid bilayers. J Phys Chem B 114:1903–1914
7. Arai N, Akimoto T, Yamamoto E, Yasui M, Yasuoka K (2014) Poisson property of the occurrence of flip-flops in a model membrane. J Chem Phys 140:064901
8. Armstrong VT, Brzustowicz MR, Wassall SR, Jenski, LJ, Stillwell W (2003) Rapid flip-flop in polyunsaturated (docosahexaenoate) phospholipid membranes. Arch Biochem Biophys 414:78–82
9. Backer JM, Dawidowicz EA (1979) The rapid transmembrane movement of cholesterol in small unilamellar vesicles. Biochim Biophys Acta 551:260–270

10. Bai J, Pagano RE (1997) Measurement of spontaneous transfer and transbilayer movement of BODIPY-labeled lipids in lipid vesicles. Biochemistry 36:8840–8848
11. Barsukov LI, Victorov AV, Vasilenko IA, Evstigneeva RP, Bergelson LD (1980) Investigation of the inside-outside distribution, intermembrane exchange and transbilayer movement of phospholipids in sonicated vesicles by shift reagent NMR. Biochim Biophys Acta 598:153–168.
12. Bennett WFD, MacCallum JL, Hinner MJ, Marrink SJ, Tieleman DP (2009) Molecular view of cholesterol flip-flop and chemical potential in different membrane environments. J Am Chem Soc 131:12714–12720
13. Bennett WFD, Tieleman DP (2011) Water defect and pore formation in atomistic and coarse-grained lipid membranes: pushing the limits of coarse graining. J Chem Theory Comput 7:2981–2988
14. Bennett WFD, Tieleman DP (2012) Molecular simulation of rapid translocation of cholesterol, diacylglycerol, and ceramide in model raft and non raft membrane. J Lipid Res 52:421–429
15. Bennett WFD, Sapay N, Tieleman DP (2014) Atomistic simulations of pore formation and closure in lipid bilayers. Biophys J 106:210–219
16. Bennett WFD, Hong CK, Wang Y, Tieleman DP (2016) Antimicrobial peptide simulations and the influence of force field on the free energy for pore formation in lipid bilayers J Chem Th Comput 12:4524–4533
17. Best RB, Hummer G (2006) Diffusive model of protein folding dynamics with Kramers turnover rule. Phys Rev Lett 96:228104
18. Bevers EM, Williamson PL (2010) Phospholipid scramblase: an update. FEBS Lett 584:2724–2730
19. Bjöerkhem I, Meaney S (2004) Brain cholesterol: long secret life behind a barrier. Arterioscler Thromb Vasc Biol 24:806–815
20. Bretscher MS (1972) Asymmetrical lipid bilayer structure for biological membranes. Nat New Biol 236: 11–12
21. Bretscher MS, Munro S (1993) Cholesterol and the Golgi apparatus. Science 261:1280–1281
22. Brown KL, Conboy JC (2013) Free energy and entropy of activation for phospholipids flip-flop in planar supported lipid bilayers. J Phys Chem B 117:15041–15050
23. Bruckner RJ, Mansy SS, Ricardo A, Mahadevan, Szostak JW (2009) Flip-flop-induced relaxation of bending energy: implications for membrane remodeling. Biophys J 97:3113–3122
24. Carquin M, D'Auria L, Pollet H, Bongarzone ER, Tyteca D (2016) Recent progress on lipid lateral heterogeneity in plasma membranes: from rafts to submicrometric domains. Prog Lipid Res 62:1–24
25. Castegna A, Lauderback CM, Mohammad-Abdul H, Butterfield DH (2004) Modulation of phospholipid asymmetry in synaptosomal membranes by the lipid peroxidation products, 4-hydroxynonenal and acrolein: implications for Alzheimer's disease. Brain Res 1004: 193–197
26. Choubey A, Kalia RK, Malmstadt N, Nakano A, Vashishta P (2013) Cholesterol translocation in a phospholipid membrane. Biophys J 104:2429–2436
27. Coleman JA, Vestergaard AL, Molday RS, Vilsen B, Andersen JP (2012) Critical role of a transmembrane lysine in aminophospholipid transport by mammalian photoreceptor P4-ATPase ATP8A2. Proc Natl Acad Sci USA 109:1449–1454
28. Comer J, Schulten K, Chipot C (2014) Diffusive models of membrane permeation with explicit orientational freedom. J Chem Theory Comput 10:2710–2718
29. Connor J, Schroit AJ (1987) Determination of lipid asymmetry in human red blood cells by resonance energy transfer. Biochemistry 26:5099–5105
30. Contreras F-X, Sánchez-Magraner L, Alonso A, Gõni FM (2010) Transbilayer (flip-flop) lipid motion and lipid scrambling in membranes. FEBS Lett 584:1779–1786
31. Crane JM, Kiessling V, Tamm LK (2005) Measuring lipid asymmetry in planar supported bilayers by fluorescence interference contrast microscopy. Langmuir 21:1377–1388

32. Cruzeiro-Hansson L, Mouritsen OG (1988) Passive ion permeability of lipid membranes modelled via lipid-domain interfacial area. Biochim Biophys Acta 944:63–72
33. Cullis PR, de Kruijff B (1973) ^{31}P-NMR studies of unsonicated aqueous dispersions of neutral and acidic phospholipids. Effects of phase transitions, pH and divalent cations on the motion in the phosphate region of the polar headgroup. Biochim Biophys Acta 507:207–218
34. Daleke DL (2008) Regulation of phospholipid asymmetry in the erythrocyte membrane. Curr Opin Hematol 15:191–195
35. Daleke DL, Huestis WH (1985) Incorporation and translocation of aminophospholipids in human erythrocytes. Biochemistry 24:5406–5416
36. De Kruijff B, Van Zoelen EJJ (1978) Effect of the phase transition on the transbilayer movement of dimyristoylphosphatidylcholine in unilamellar vesicles. Biochim Biophys Acta 511:105–115
37. Devaux PF (1991) Static and dynamic lipid asymmetry in cell membranes. Biochemistry 30:1163–1173
38. Devaux PF, Herrmann A (eds) (2011) Transmembrane dynamics of lipids. Wiley, Hoboken
39. Devaux PF, Morris R (2004) Transmembrane asymmetry and the lateral domains in biological membranes. Traffic 5:241–246
40. Devaux PF, Fellmann P, Herve P (2002) Investigation on lipid asymmetry using lipid probes. Comparison between spin-labeled lipids and fluorescent lipids. Chem Phys Lipids 116:115–134
41. de Vries AH, Mark AE, Marrink SJ (2004) Molecular dynamics simulation of the spontaneous formation of a small DPPC vesicle in water in atomistic detail. J Am Chem Soc 126:4488–4489
42. Dickey AN, Faller R (2007) How alcohol chain-length and concentration modulate hydrogen bond formation in a lipid bilayer. Biophys J 92:2366–2376
43. Dzikovski B, Freed JH (2008) Membrane fluidity. In: Begley TP (ed) Wiley encyclopedia of chemical biology. Wiley
44. Fahy E, Subramaniam S, Murphy RC, Nishijima M, Raetz CR, Shimizu T, Spener F, van Meer G, Wakelam MJ, Dennis EA (2009) Update of the LIPID MAPS comprehensive classification system for lipids. J Lipid Res 50:S9–S14
45. Frickenhaus S, Heinrich R (1999) Kinetic and thermodynamic aspects of lipid translocation in biological membranes. Biophys J 76:1293–1309
46. Ganong BR, Bell RM (1997) Transmembrane movement of phosphatidylglycerol and diacylglycerol sulfhydryl analogues. Biochemistry 23:4977–4983
47. Garg S, Porcar L, Woodca AC, Butler PD, Perez-Salas U (2011) Noninvasive neutron scattering measurements reveal slower cholesterol transport in model lipid membranes Biophys J 101:370–377
48. Gerelli Y, Porcar L, Fragneto G (2012) Lipid rearrangement in DSPC/DMPC bilayers: a neutron reflectometry study. Langmuir Langmuir 28:15922–15928
49. Gerelli Y, Porcar L, Lombardi L, Fragneto G (2013) Lipid exchange and flip-flop in solid supported bilayers. Langmuir 29:12762–12769
50. Goren MA, Morizumi T, Menon I, Joseph JS, Dittman JS, Cherezov V, Stevens RC, Ernst OP, Menon AK (2014) Constitutive phospholipid scramblase activity of a G protein-coupled receptor. Nat Commun 5:5115
51. Gummadi SN, Kumar KS (2005) The mistery of phospholipid flip-flop in biogenic membranes. Cell Mol Biol Lett 10:101–121
52. Gurtovenko AA, Vattulainen I (2007) Molecular mechanism for lipid flip-flops. J Phys Chem B 111:13554–13559
53. Gurtovenko AA, Onike OI, Anwar J (2008) Chemically induced phospholipid translocation across biological membranes. Langmuir 24:9656–9660
54. Gurtovenko AA, Anwar J, Vattulainen I (2010) Defect-mediated trafficking across cell membranes: insights from in *silico* modeling. Chem rev 2010:6077–6103
55. Hamilton JA (2003) Fast flip-flop of cholesterol and fatty acids in membranes: implications for membrane transport proteins. Curr Opin Lipidol 14:263–271

56. Harroun TA, Katsaras J, Wassall SR (2006) Cholesterol hydroxyl group Is found to reside in the center of a polyunsaturated lipid membrane. Biochemistry 45:1227–1233.
57. Harroun TA, Katsaras J, Wassall SR (2008) Cholesterol is found to reside in the center of a polyunsaturated lipid membrane. Biochemistry 47:7090–7096
58. Homan R, Pownall HJ (1988) Transbilayer diffusion of phospholipids: dependence on headgroup structure and acyl chain length. Biochim Biophys Acta 938:155–166
59. Hung W-C, Lee M-T, Chen F-Y, Huang HW (2007) The condensing effect of cholesterol in lipid bilayers. Biophys J 92:3960–3967
60. Imparato A, Shillcock JC, Lipowsky R (2003) Lateral and trans- verse diffusion in two-bilayer component membrane. Eur Phys J E 11:21–28
61. Ingólfsson HI, Melo MN, van Eerden FJ, Arnarez C, Lopez CA, Wassenaar TA, Periole X, de Vries AH, Tieleman DP, Marrink SJ (2014) Lipid organization of the plasma membrane. J Am Chem Soc 136:14554–14559
62. Ipsen JH, Karlström G, Wennerström H, Mouritsen OG, Zuckermann MJ (1987) Phase equilibria in the phosphatidylcholine-cholesterol system. Biochim Biophys Acta 905:162–1672
63. Ipsen JH, Mouritsen OG, Bloom M (1990) Relationships between lipid membrane area, hydrophobic thickness, and acyl-chain orientational order. The effects of cholesterol. Biophys J 57:405–412
64. Jo S, Rui J, Lim JB, Klauda JB, Im W (2010) Cholesterol flip-flop: insights from free energy simulation studies. J Phys Chem B 114:13342–13348
65. John K, Schreiber S, Kubelt J, Herrmann A, Müller P (2002) Transbilayer movement of phospholipids at the main phase transition of lipid membranes: implications for rapid flip-flop in biological membranes. Biophys J 83:3315–3323
66. Kandasamy S, Larson R (2006) Cation and anion transport through hydrophilic pores in lipid bilayers. J Chem Phys 125:074901
67. Kim S, Karrila SJ (2005) Microhydrodynamics. Principles and Selected Applications. Dover, Mineola
68. Kol MA, de Kroon AIPM, Rijkers DTS, Killian JA, de Kruijff B (2001) Membrane-spanning peptides induce phospholipid flop: a model for phospholipid translocation across the inner membrane of E. coli. Biochemistry 40:10500–10506
69. Kol MA, van Dalen A, de Kroon AIPM, de Kruijff B (2003) Translocation of phospholipids is facilitated by a subset of membrane-spanning proteins of the bacterial cytoplasmic membrane. J Biol Chem 278:24586–24593
70. Kol MA, de Kroon AIPM, Killian JA, de Kruijff B (2004) Transbilayer movement of phospholipids in biogenic membranes. Biochemistry 43:2673–2681
71. Kopelevich DI (2013) One-dimensional potential of mean force underestimates activation barrier for transport across flexible lipid membranes J Chem Phys 139:134906
72. Kornberg RD, McConnell HM (1971) Inside-outside transitions of phospholipids in vesicle membranes. Biochemistry 10:1111–1120
73. Kramers HA (1940) Brownian motion in a field of force and the diffusion model of chemical reactions. Physica 7:284–304
74. Kyrychenko A (2015) Using fluorescence for studies of biological membranes: a review Methods Appl Fluoresc 3:042003
75. Langer JS (1969) Statistical theory of the decay of metastable states. Ann Phys 54:258–275
76. LeBarron J, London E (2016) Effect of lipid composition and amino acid sequence upon transmembrane peptide-accelerated lipid transleafleat. Biochim Biophys Acta 1858:1812–1820
77. Leonard A, Escrive C, Laguerre M, Pebay-Peyroula E, Neri W, Pott T, Katsaras J, Dufourc EJ (2001) Location of cholesterol in DMPC membranes. A comparative study by neutron diffraction and molecular mechanics simulation. Langmuir 17:2019–2030
78. Leontiadou H, Mark AE, Marrink SJ (2006) Antimicrobial peptides in action. J Am Chem Soc 128:12156–12161

79. Leventis R, Silvius JR (2001) Use of cyclodextrins to monitor transbilayer movement and differential lipid affinities of cholesterol. Biophys J 81:2257–2267
80. Liu J, Conboy JC (2004) Direct measurement of the transbilayer movement of phospholipids by sum-frequency vibrational spectroscopy. J Am Chem Soc 126:8376–8377
81. Liu J, Conboy JC (2005) 1,2-Diacyl-phosphatidylcholine flip-flop measured directly by sum-frequency vibrational spectroscopy. Biophys J 89:2522–2532
82. Liu J, Brown KL, Conboy JC (2013) The effect of cholesterol on the intrinsic rate of lipid flip-flop as measured by sum-frequency vibrational spectroscopy. Faraday Discuss 161:45–61
83. López-Montero I, Rodriguez N, Cribier S, Pohl A, Vélez M, Devaux PF (2005) Rapid transbilayer movement of ceramides in phospholipid vesicles and in human erythrocytes J Biol Chem 280:25811–25819
84. Lyubartsev AP, Rabinovich AL (2016) Force field development for lipid membrane simulations. Biochim Biophys Acta 1858:2483–2497
85. Ma S, Li H, Tian K, Ye S, Lou Y (2014) In situ and real-time SFG measurements revealing organization and transport of cholesterol analogue 6-ketocholestanol in a cell membrane. J Phys Chem Lett 5:419–424
86. Marquardt D, Heberle FA, Greathouse DV, Koeppe RE, Standaert RF, Van Oosten BJ, Harroun TA, Kinnun JJ, Williams JA, Wassall SR, Katsaras J (2016) Lipid bilayer thickness determines cholesterol's location in model membranes Soft Matter 12:9417–9428
87. Marquardt D, Heberle FA, Miti T, Eicher B, London E, Katsaras J, Pabst G (2017) ^{1}H NMR shows slow phospholipid flip-flop in gel and fluid bilayers Langmuir 33:3731–3741
88. Marrink SJ, Tieleman DP (2013) Perspective on the Martini model. Chem Soc Rev 42:6801–6822
89. Marrink SJ, Risselada HJ, Yefimov S, Tieleman DP, De Vries AH (2007) The MARTINI force field: coarse grained model for biomolecular simulations. J Phys Chem B 111:7812–7824
90. Marrink SJ, de Vries AH, Harroun TA, Katsaras J, Wassall SR (2008) Cholesterol shows preference for the interior of polyunsaturated lipid membranes. J Am Chem Soc 130:10–11
91. Martí J, Csajka FS (2004) Transition path sampling study of flip-flop transitions in model lipid bilayer membranes. Phys Rev E 69:061918
92. Miao L, Seifert U, Wortis M, Dobereiner HG (1994) Budding transitions of fluid-bilayer vesicles: the effect of area-difference elasticity. Phys Rev E 49:5389–1994
93. Montigny C, Lyons J, Champeil P, Nissen P, Lenoir G (2016) On the molecular mechanism of flippase- and scramblase-mediated phospholipid transport. Biochem Biophys Acta 1861: 767–783
94. Moro GJ, Ferrarini A, Polimeno A, Nordio PL (1989) Models of conformational dynamics. In: Dorfmüller Th (ed) Reactive and flexible molecules in liquids. Kluwer Academic Publishers, Dordrecht, pp 107–139
95. Morrot G, Hervé P, Zachowski A, Fellmann P, Devaux PF (1989) Aminophospholipid translocase of human erythrocyte: phospholipid substrate specificity and effect of cholesterol. Biochemistry 28:3456–3462
96. Mouritsen OG, Jørgensen K (1992) Dynamic lipid-bilayer heterogeneity: a mesoscopic vehicle for membrane function? BioEssays 14:129–136
97. Nakano M, Fukuda M, Kudo T, Endo H, Handa T (2007) Determination of interbilayer and transbilayer lipid transfers by time-resolved small-angle neutron scattering. Phys Rev Lett 98:238101
98. Nakano M, Fukuda M, Kudo T, Matsuzaki N, Azuma T, Sekine K, Endo H, Handa T (2009) Flip-flop of phospholipids in vesicles: kinetic analysis with time-resolved small-angle neutron scattering. J Phys Chem B 113:6745–6748
99. Neale C, Pomeés R (2016) Sampling errors in free energy simulations of small molecules in lipid bilayers Biochim Biophys Acta 1858:2539–2548
100. Nicolson GL (2014) The fluid-mosaic model of membrane structure: still relevant to understanding the structure, function and dynamics of biological membranes after more than 40 years. Biochim Biophys Acta 1838:1451–1466

101. Nicolson GL, Ash ME (2014) Lipid replacement therapy: a natural medicine approach to replacing damaged lipids in cellular membranes and organelles and restoring function. Biochim Biophys Acta 1838:1657–1679
102. Neuvonen M, Manna M, Mokkila S, Javanainen M, Rog T, Liu Z, Bittman R, Vattulainen I, Ikonen E (2014) Enzymatic oxidation of cholesterol: properties and functional effects of cholestenone in cell membranes. PLoS ONE 9:e103743
103. Ogushi F, Ishitsuka R, Kobayashi T, Sugita Y (2012) Rapid flip-flop motions of diacylglycerol and ceramide in phospholipid bilayers. Chem Phys Lett 522:96–102
104. Pantaler E, Kamp D, Haest CW (2000) Acceleration of phospholipid flip-flop in the erythrocyte membrane by detergents differing in polar head group and alkyl chain length. Biochim Biophys Acta 1509:397–408
105. Papadopulos A, Vehring S, López-Montero I, Kutschenko L, Stöckl M, Devaux PF, Kozlov M, Pomorski T, Herrmann A (2007) Flippase activity detected with unlabeled lipids by shape changes of giant unilamellar vesicles. J Biol Chem 282:15559–15568
106. Parisio G, Ferrarini A (2010) Solute partitioning into lipid bilayers: an implicit model for nonuniform and ordered environment. J Chem Theory Comput 6:2267–2280
107. Parisio G, Sperotto MM, Ferrarini A (2012) Flip-flop of steroids in phospholipid bilayers: effects of the chemical structure on transbilayer diffusion. J Am Chem Soc 134:12198–12208
108. Parisio G, Stocchero M, Ferrarini A (2013) Passive membrane permeability: beyond the standard solubility-diffusion model J Chem Theory Comput 9:5236–5246
109. Parisio G, Ferrarini A, Sperotto MM (2016) Model studies of lipid flip-flop in membranes. Int J Adv Eng Sci Appl Math 8:134–146
110. Poger D, Caron B, Mark AE (2016) Validating lipid force fields against experimental data: progress, challenges and perspectives. Biochim Biophys Acta 1858:1556–1565
111. Pomorski GP, Menon AK (2016) Lipid somersaults: uncovering the mechanisms of protein-mediated lipid flipping. Prog Lip Res 64:69–84
112. Pousoroush A, Sperotto MM, Laradji M (2017) Phase behavior of supported lipid bilayers: A systematic study by coarse-grained molecular dynamics simulations. J. Chem. Phys. 146:154902
113. Poznansky MJ, Lange Y (1976) Transbilayer movement of cholesterol in dipalmitoyllecithin-cholesterol vesicles. Nature 259:420–421
114. Poznansky MJ, Lange Y (1978) Transbilayer movement of cholesterol in phospholipid vesicles under equilibrium and non-equilibrium conditions. Biochim Biophys Acta 506:256–264
115. Rajasekharan A, Gummadi SN (2012) Inhibition of biogenic membrane flippase activity in reconstituted ER proteoliposomes in the presence of low cholesterol levels. Cell Mol Biol Lett 17:136–152
116. Ramachandran S, Kumar PBS, Laradji M (2008) Lipid flip-flop driven mechanical and morphological changes in model membranes. J Chem Phys 129:125104
117. Raphael RM, Waugh RE (1996) Accelerated interleaflet transport of phosphatidylcholine molecules in membranes under deformation. Biophys J 71:1374–1388
118. Raphael RM, Waugh RE, Svetina S, Zeks B (2001) Fractional occurrence of defects in membranes and mechanically driven interleaflet phospholipid transport Phys Rev E 64:051913
119. Risselada HJ, Marrink SJ (2008) The molecular face of lipid rafts in model membranes. Proc Natl Acad Sci USA 105:17367–17372
120. Rodrigueza WV, Wheeler JJ, Klimuk SK, Kitson CN, Hope MJ (1995) Transbilayer movement and net flux of cholesterol and cholesterol sulfate between liposomal membranes Biochemistry 34:6208–6217
121. Róg T, Stimson LM, Pasenkiewicz-Gierula M, Vattulainen I, Karttunen M (2008) Replacing the cholesterol hydroxyl group with the ketone group facilitates sterol flip-flop and promotes membrane fluidity. J Phys Chem B 112:1946–1952
122. Sanyal S, Menon AK (2009) Flipping lipids: why an' what's the reason for? ACS Chem Biol 4:895–909

123. Sapay N, Bennett WFD, Tieleman DP (2009) Thermodynamics of flip-flop and desorption for a systematic series of phosphatidylcholine lipids. Soft Matter 5:3295–3302
124. Schaffer JE (2002) Fatty acid transport: the road taken. Am J Physiol Endocrinol Metab 282:E239–E346
125. Seigneuret M, Devaux PF (1984) ATP-dependent asymmetric distribution of spin-labeled phospholipids in the erythrocyte membrane: relation to shape changes. Proc Natl Acad Sci USA 81:3751–3755
126. Sharom FJ (2011) Flipping and flopping- lipids on the move. IUBMB Life 63:736–746
127. Shaw JM, Thompson TE (1982) Effect of phospholipid oxidation products on transbilayer movement of phospholipids in single lamellar vesicles. Biochemistry 21:920–927
128. Sheetz MP, Singer SJ (1974) Biological membranes as bilayer couples. A molecular mechanism of drug-erythrocyte interactions. Proc Natl Acad Sci USA 71:4457–4461
129. Simon K, Sampaio JL (2011) Membrane organisation and lipid rafts. Cold Spring Harb Perspect Biol 3:a004697
130. Son M, London E (2013) The dependence of lipid asymmetry upon phosphatidylcholine acyl chain structure. J Lip Res 54:223–231
131. Son M, London E (2013) The dependence of lipid asymmetry upon polar head group structure. J Lip Res 54:3385–3393
132. Steck TL, Lange Y (2012) How slow is the transbilayer diffusion (flip-flop) of cholesterol? Biophys J 102:945–946
133. Steck TL, Ye J, Lange Y (2002) Probing red cell membrane cholesterol movement with cyclodextrin. Biophys J 83:2118–2125
134. Suzuki J, Umeda M, Sims PJ, Nagata S (2010) Calcium-dependent phospholipid scrambling by TMEM16F. Nature 468:834–838
135. Suzuki J, Denning DP, Imanishi E, Horvitz HR, Nagata S (2013) Xk-Related protein 8 and CED-8 promote phosphatidylserine exposure in apoptotic cells. Science 341:403–406
136. Suzuki J, Fujii T, Imao T, Ishihara K, Kuba H, Nagata S (2013) Calcium-dependent phospholipid scramblase activity of TMEM16 protein family members. J Biol Chem 288:13305–13316
137. Svetina S, Zeks B, Waugh RE, Raphael RM (1998) Theoretical analysis of the effect of the transbilayer movement of phospholipid molecules on the dynamic behavior of a microtube pulled out of an aspirated vesicle. Eur Biophys J 27:197–209
138. Tielemann DP, Marrink S-J (2006) Lipids out of equilibrium: energetics of desorption and pore mediated flip-flop. J Am Chem Soc 128:12462–12467
139. Tocanne J-F, Dupou-Cézanne L, Lopez A (1994) Lateral diffusion of lipids in model and natural membranes. Progr Lipid Res 33:203–237
140. Toyoshima Y, Thompson TE (1975) Chloride flux in bilayer membranes: chloride permeability in aqueous dispersions of single-walled, bilayer vesicles. Biochemistry 14:1525–1531
141. Urbina P, Alonso A, Contreras FX, Goñi FM, López DJ, Montes LR, Sot J (2006) Alkanes are not innocuous vehicles for hydrophobic reagents in membrane studies. Chem Phys Lipids 139:107–114
142. van Helvoort A, Smith AJ, Sprong H, Fritzsche I, Schinkel AH, Borst P, van Meer P (1996) MDR1 Pglycoprotein is a lipid translocase of broad specificity, while MDR3 P-glycoprotein specifically translocates phosphatidylcholine. Cell 87:507–517
143. van Meer G, de Kroon AIPM (2011) Lipid map of the mammalian cell. J Cell Sci 124:5–8
144. van Meer G, Voelker DR, Feigenson GW (2008) Membrane lipids: where they are and how they behave. Nat Rev Mol Cell Biol 9:112–124
145. Venturoli M, Sperotto MM, Kranenburg M, Smit B (2006) Mesoscopic models of biological membranes. Phys Rep 437:1–54
146. Visco I, Chiantia S, Schwille P (2014) Asymmetric supported lipid bilayer formation via methyl-β- cyclodextrin mediated lipid exchange: influence of asymmetry on lipid dynamics and phase behavior Langmuir 30:7475–7484
147. Volinsky R, Cwiklik L, Jurkiewicz P, Hof M, Jungwirth, Kinnunen PKJ (2011) Oxidized phosphatidylcholines facilitate phospholipid flip-flop in liposomes. Biophys J 101:1376–1384

148. Wei C, Pohorille A (2014) Flip-flop of oleic acid in a phospholipid membrane: rate and mechanism. J Phys Chem B 118:12919–12926
149. Wilson MW, Pohorille A (1996) Mechanism of unassisted ion transport across membrane bilayers. J Am Chem Soc 118:6580–6587
150. Wimley WC, Thompson TE (1990) Exchange and flip-flop of dimyristoylphosphatidylcholine in liquid-crystalline, gel, and two-component, two-phase large unilamellar vesicles. Biochemistry 29:1296–1303
151. Wimley WC, Thompson TE (1991) Transbilayer and interbilayer phospholipid exchange in dimyristoylphosphatidylcholine/dimyristoyl- phosphatidylethanolamine large unilamellar vesicles. Biochemistry 30:1702–1709
152. Wood WG (1990) Asymmetric distribution of a fluorescent sterol in synaptic plasma membranes: effect of chronic ethanol consumption. Biochim Biophys Acta 1025:243–246
153. Wood WG, Igbavboa U, Müller ME, Eckert GP (2011) Cholesterol asymmetry in synaptic plasma membranes. J Neurochem 116:684–689
154. Yamon Y, Broccardo C, Chambenoit O, Luciani M-F, Toti F, Chaslin S, Freyssinet J-M, Devaux PF, Niesh J, Marguet D, Chimini G (2000) ABC1 promotes engulfment of apoptotic cells and transbilayer redistribution of phosphatidylserine. Nat Cell Biol 2:399–406
155. Yesylevskyy SO, Demchenko AP (2015) Cholesterol behavior in asymmetric lipid bilayers: insights form molecular dynamics simulations. Methods Mol Biol 1232:291–306
156. Zachowski A (1993) Phospholipids in animal eukaryotic membranes: transverse asymmetry and movement. Biochem J 294:1–14
157. Zachowski A, Devaux PF (1990) Transmembrane movement of lipids. Experientia 46:644–656
158. Zhelev DV (1996) Exchange of monooleoylphosphatidylcholine with single egg phosphatidylcholine vesicle membranes. Biophys J 71:257–273
159. Zhou H-X (2010) Rate theories for biologists. Q Rev Biophys 43:219–293

Chapter 3
Membrane Lipid-Protein Interactions

Michael F. Brown, Udeep Chawla, and Suchithranga M.D.C. Perera

Abstract In this review we describe how the properties of cellular membranes govern protein structure and activity. Lipids can modulate protein functional states through general bilayer properties, or by specific binding and acting as allosteric regulators. Hydrophobic matching by solvation of the protein surface entails short-range interactions of the lipids, and cellular water affects bilayer structure through hydrating the lipid polar head groups and protein hydrophilic domains. Biomembranes have important analogies to supercritical fluids leading to raft-like nanostructures with cholesterol. Additional long-range interactions of the lipids and proteins involve the curvature stress field, where a flexible surface model (FSM) describes how collective properties of the lipids affect the conformational energetics of membrane proteins. Curvature elasticity and hydrophobicity of native lipid mixtures play key roles in functional proteolipid couplings and give insights into protein activation mechanisms in cellular membranes.

Keywords Cholesterol • Critical fluids • Flexible surface model • Fluid-mosaic model • G-protein–coupled receptor • Hydrophobic matching • Lipid bilayer • Membrane curvature • Rafts • Rhodopsin • Spontaneous curvature

3.1 Introduction

Understanding how membrane lipids interact with proteins can markedly impact our knowledge of cellular function at the intersection of biology and physics [14, 33, 51, 56, 79, 87]. The amphiphilic nature of membrane proteins clearly distinguishes them from the globular and fibrous proteins. Examples increasingly show how the activities of G-protein–coupled receptors (GPCRs) [12, 29, 70, 89], ion channels

M.F. Brown (✉)
Department of Chemistry and Biochemistry, University of Arizona, Tucson, AZ, 85721, USA

Department of Physics, University of Arizona, Tucson, AZ, 85721, USA
e-mail: mfbrown@u.arizona.edu

U. Chawla • S.M.D.C. Perera
Department of Chemistry and Biochemistry, University of Arizona, Tucson, AZ, 85721, USA

R.M. Epand, J.-M. Ruysschaert (eds.), *The Biophysics of Cell Membranes*,
Springer Series in Biophysics 19, DOI 10.1007/978-981-10-6244-5_3

[3, 77, 79], and transporters [8] are regulated by their interactions with water and membrane lipids [15, 23, 34]. Although X-ray crystal structures capture functional states of membrane proteins [32, 38], they mainly provide a static depiction, and do not fully account for reactions that involve the lipid and water components of liquid-crystalline bilayers.

Molecular spectroscopy, on the other hand, provides both structural and dynamical information and is not restricted to crystalline solids. Because they are components of supramolecular assemblies, studies of proteins in lipid bilayers require approaches that are highly synergistic with both X-ray crystallography and cryoelectron microscopy. The aim of this review is to highlight experimental and theoretical avenues for investigating how the membrane lipids influence proteins to produce the characteristic functions of life itself. Our goal is to build on previous accounts of lipid-protein interactions [13, 15, 56, 57, 79, 99] by emphasizing how material properties of the lipids correspond to the forces acting on proteins. Curvature elasticity, critical fluctuations, and hydrophobic matching are discussed in relation to proteolipid membrane couplings and cellular function.

3.2 Lipid-Protein Interactions in Biomembrane Function

Various lines of thinking apply to understanding biomembrane function. For example, the standard fluid mosaic model (FMM) [87] states that membranes are two-dimensional solutions of amphiphilic proteins dissolved within a fluid lipid bilayer. The lipids act as the solvent for membrane proteins, giving the flexibility needed for conformational changes to occur. In this view, the lipid bilayer acts as a permeability barrier to ions and polar molecules, allowing for membrane fluidity, and presenting different faces to the extracellular medium and the cytoplasm (Fig. 3.1). The alternative is that the membrane lipids are more actively implicated with protein-linked functions [8, 14, 64, 79, 85, 99]. Specific influences of lipids on membrane protein functions are summarized by several excellent reviews [56, 57]. How chemically nonspecific bilayer properties modulate the functions of membrane proteins [14, 15] is where the current thinking departs from previous approaches.

The new biophysics involves considering whether the lipids exert their influences through material properties of the bilayer, or by molecularly specific interactions [16, 51, 57, 64]. E.g., specific lipids could directly bind to integral membrane proteins, and act as allosteric modulators of GPCRs, ion channels, pumps, or transporters [23, 29]. Biophysical properties of the membrane lipid bilayer [14, 53, 79] can likewise be implicated. Recently, attention has been focused by the new structures being solved for various integral membrane proteins [32, 38]. The structural knowledge enhances our understanding of how lipids interact with membrane proteins and influence their functions [57]. Even so, the X-ray or cryo-electron microscopy structural snapshots need to be combined with spectroscopic and functional studies to yield a full understanding of cellular reaction mechanisms occurring at the membrane level.

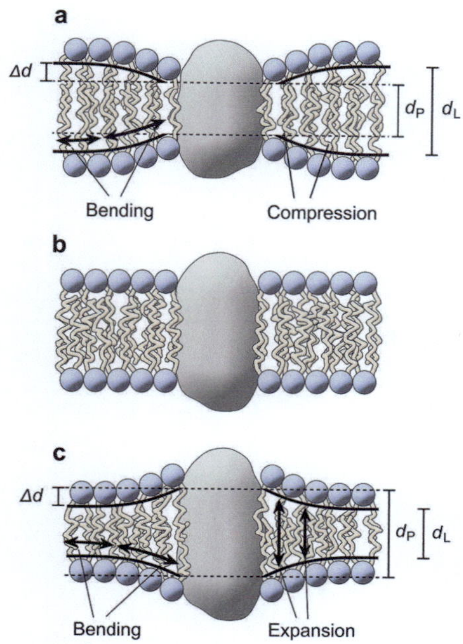

Fig. 3.1 Membrane proteins are solvated by hydrophobic matching to the lipids and entail curvature deformation of the bilayer. Solid-like objects denote proteins embedded within the bilayer; lipid head groups are indicated as circles; and noodle-like strings depict the acyl chains. The hydrophobic thickness at the proteolipid interface (d_P) can be (**a**) smaller than, (**b**) the same, or (**c**) larger than the average bilayer thickness (d_L). Elastic bilayer distortion includes bending plus compression or stretching (expansion) of the hydrocarbon chains (Figure adapted with permission from Ref. [17])

3.3 Short- and Long-Range Proteolipid Couplings in Biomembranes

Increasingly, experimental results are uncovering the influences of the membrane lipid bilayer on the activities of integral membrane proteins [56, 57]. Properties due to the tightly regulated lipid compositions are known to affect protein function in various cellular membranes [7, 40, 49, 50, 56, 59, 68, 74, 97, 99]. E.g., a balance of lamellar- and nonlamellar forming lipids in *A. laidlawii* [59] and *E. Coli* [71] is important for cellular growth, suggesting the spontaneous (intrinsic) curvature is a key factor in membrane function. For rhodopsin, a direct effect of membrane lipids on the conformational energetics of an integral membrane protein was discovered for the first time [7, 14, 97]. The pH-dependent equilibrium between the metarhodopsin states upon light activation is clearly modulated by the membrane lipid composition (Fig. 3.2a,b). The native head groups and acyl chains enable native-like activation of rhodopsin to occur by absorption of light; yet neither alone is sufficient (Fig. 3.2b) [61]. These results suggest that the standard FMM is due for revision.

With rhodopsin as an example, lipid substitution (replacement) experiments have led to the following conclusions. (1) The membrane lipid composition can perturb the formation of active rhodopsin. (2) The metarhodopsin equilibrium is reversible, and can be shifted forward or backward, depending on the thermodynamic state variables (*T*, *P*, osmotic stress). (3) Altering the lipid head groups can compensate for changes in the acyl chains, where various acyl chain substitutions are possible. (4) Proximity to a lamellar–nonlamellar lipid phase boundary favors rhodopsin

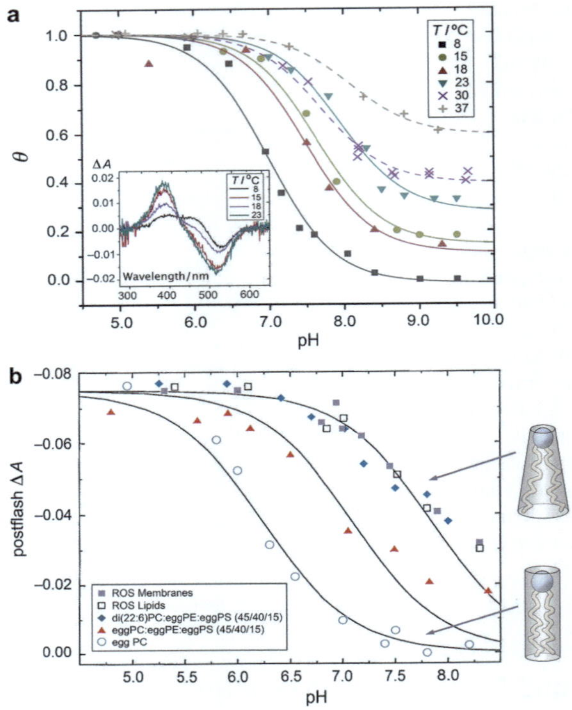

Fig. 3.2 Membrane lipids affect light activation of rhodopsin due to general bilayer-mediated properties. (**a**) Active rhodopsin (MII state, fraction θ) in equilibrium with inactive rhodopsin (MI state) versus pH. Note the alkaline endpoint does not reach zero at higher temperatures showing an ensemble of states. Inset: UV–visible difference spectra (light minus dark) show temperature dependence of MI–MII equilibrium. (**b**) Postflash absorbance change ($\lambda = 478$ nm) (active MII) graphed against pH at $T = 28$ °C. The apparent pK_A and alkaline endpoint depend on the phospholipid head groups and acyl chain composition (shown in figure). Shifting the metarhodopsin equilibrium is connected with the average lipid shape or alternatively the intrinsic (spontaneous) curvature of the lipids (Figure modified with permission from Ref. [15])

activation, implying that membrane curvature forces are involved. (5) Lastly, the metarhodopsin equilibrium depends on the lipid to rhodopsin ratio. Chemically nonspecific bilayer properties are causal factors that affect rhodopsin function [12]. Lipid head groups can replace acyl chains, and there is an influence of the lipid/protein molar ratio on activation.

3.3.1 Short-Range Boundary Lipid-Protein Interactions

Significant possibilities of functional protein-lipid interactions can exist for membrane lipid bilayers (Fig. 3.1). Uncovering how short-range solvation of integral membrane proteins occurs due to a single shell of annular or boundary lipids has benefited from spin-label EPR spectroscopy [42, 64]. Attention has been recently focused on lipid structural specificity [56, 57] by the X-ray crystal structures of integral membrane proteins. Membrane proteins often contain shallow protrusions or surface grooves that can be occupied by the fatty acyl groups of the boundary lipids, allowing them to be densely packed within the membrane. Resolved lipid molecules in the X-ray crystal structures of membrane proteins tend to be localized between the transmembrane (TM) helices, or found in a partial band about the protein hydrophobic surface [57]. Further insights into lipid-protein associations have recently come from native mass spectrometry [65]. Taken together, the X-ray, spin-label EPR, and native mass spectrometry results affirm how the flexible lipids can act as a solvent for membrane proteins.

3.3.2 Solvation of the Hydrophobic Protein Surface by Membrane Lipids

For amphiphilic membrane proteins, it is of great interest to understand how the polar and nonpolar regions are solvated by lipids and water (Fig. 3.1). One should recall that the hydrophobic effect is short range, and mainly involves the molecularly thin interface between hydrocarbon and water [16]. In X-ray crystal structures, the perturbation of water is found to involve mainly the first hydration shell, and not to reach beyond the second shell of water. Likewise, according to spin-label EPR spectroscopy [42, 64], perturbation of the lipids extends mainly to the first solvation shell. Lipids next to the protein are deformed by protrusions or clefts between the TM helices, and exchange rapidly with bulk lipids by lateral diffusion. There is low selectivity for the head group composition, and the composition of the boundary (annular) lipids resembles the main bilayer [56]. E.g. rhodopsin has low selectivity for polar head groups, and other proteins show a marginal selectivity for anionic phospholipids [64].

3.3.3 Long-Range Collective Lipid-Protein Interactions in Cellular Membranes

As a rule, the influences of the protein solvation by the lipids and by water are greatest for the shell of molecules surrounding the protein [42, 64]. Even so, an arresting counterexample [11] entails the influence of the lipid/protein molar ratio on light activation of rhodopsin. Increasing the number of lipids (mole fraction) yields greater stabilization of the active state [12]. Set against the molecular size, the distance scale of the forces is large—a fundamental caveat of elasticity theory. In terms of the material properties of the lipid bilayer, the work of deforming the protein shape within the membrane comes into play [14, 15]. Both short and longer-range proteolipid interactions are evidently important. Below, we consider the various models put forth to describe the lipid-protein interactions at the atomistic and the mesoscopic level, falling between the molecular dimensions and the bulk lipid bilayer.

3.4 Solvation of Membrane Proteins: Hydrophobic Matching to Lipids

The most obvious property to account for nonspecific lipid-protein interactions entails their hydrophobic matching at the intramembranous proteolipid boundary. According to existing models [3, 35, 51, 57, 64, 72], the physical constraints imposed upon an integral membrane protein involve a coupling to the bilayer hydrocarbon thickness (Fig. 3.1). Typically it is considered that the lipid chains adapt to the protein intramembranous surface, which acts like a rigid body [51]. Alternatively, a fluid mechanical (two-way) coupling of lipid and protein deformations might occur. Examples (Fig. 3.1) include a protein whose hydrophobic thickness is less than the bilayer ($d_P < d_L$), is equal to the lipid thickness ($d_P = d_L$), or exceeds the bilayer thickness ($d_P > d_L$). The free energy cost goes as $|d_P - d_L|^2$ in a harmonic approximation, where d_P denotes the intramembranous protein hydrophobic thickness, and d_L is the lipid hydrophobic thickness [72] (see Fig. 3.1). Stretching or compressing the lipid chains shrinks or expands the cross-sectional area per lipid molecule at the aqueous interface. The area per lipid next to the protein is less than for the bulk lipids if the protein hydrophobic thickness exceeds the unperturbed bilayer. By contrast, the hydrophobic matching of a protein to a thick membrane compresses the fatty acyl chains of lipids, yielding a greater area per lipid at the aqueous interface.

3.4.1 Short-Range Solvation Versus Collective Proteolipid Couplings

Despite that hydrophobic matching is simple and intuitively appealing, it focuses mainly on short-range interactions. It may overlook longer-range collective interactions among the lipids [19] and proteins [12, 16]. Among the basic differences between membrane proteins and globular proteins is that the forces are more isotropic for the latter. By contrast, the forces acting on membrane proteins are anisotropic; they exist in the stress field of the lipid bilayer. Because the forces are not averaged to zero over the protein surface, long-range elastic interactions come into play [14]. As first shown [11, 12], experimentally it is found that increasing the lipid to protein molar ratio yields a greater population of light-activated rhodopsin (MII state) [12, 75, 88], The influence of the lipid/protein ratio implies a persistence length for the perturbation extending away from the proteolipid boundary, suggesting a role of curvature forces in the elastic membrane deformation [45].

3.5 Lipid Miscibility and Raft-Like Mixtures in Biomembranes

However, first let us turn to the subject of lipid rafts in mammalian cellular membranes. The topic has been thoroughly discussed [30, 85, 93], and here only a brief synopsis will be given. Cellular membranes are very diverse in their lipid and protein compositions, and raft-like microdomains are proposed to exist in the 10–100 nm size range due to collective intermolecular interactions [1]. In the plasma membranes of mammalian cells, rafts are associated with cholesterol and high-melting lipids like sphingomyelin, which can separate from lower-melting lipids in the bilayer [41, 95]. They may be involved with protein segregation and signaling, membrane trafficking, budding of viruses, and other key cellular processes.

3.5.1 Miscibility of Cholesterol in Mammalian Cellular Plasma Membranes

Models for raft formation often rely on lipid mixtures with cholesterol that segregate into liquid-ordered (lo) and liquid-disordered (ld) regions [95]. When cholesterol and high-melting lipids like sphingomyelin or PE occur together with lower-melting lipids like PC, nanoscale heterogeneities or phase separations of lo regions from ld regions can result [1, 41, 60]. For raft-like (lo) nano- or microstructures, evidently the lipid-protein interactions are comparable to the lipid-lipid interactions [85]. Because of nonideal mixing with cholesterol, the effects of proteins on raft-like lipid mixtures can differ from their effects on the lipid solid-ordered to liquid-disordered (so–ld) phase transition. The lipid-protein coupling resembles interactions among the lipids themselves, either in the ld state, in lipid rafts, or when liquid-liquid immiscibility is present. Broadening of the order–disorder transition due to cholesterol can facilitate interactions of lo or ld regions with proteins, and thereby modulate their activities [1, 85].

3.5.2 Lamellar to Nonlamellar Phase Transitions and Cholesterol in Lipid Mixtures

In mammalian cells, cholesterol is predominantly located in the plasma membrane, and moreover lipids are present with a tendency to form nonlamellar phases. The major influence of cholesterol is a condensation of the bilayer plus an umbrella effect, whereby it is located beneath the polar head groups, acting as a spacer between the lipid molecules [20]. Furthermore, cholesterol affects the membrane curvature free energy, because the local membrane thickening correlates with an increased bilayer bending modulus, accompanied by a weakly negative spontaneous curvature (see below). Cholesterol strongly favors the planar bilayer state, and because like dissolves like [69], it will segregate from lipids with a more negative spontaneous curvature. Such membrane curvature-driven interactions have been previously discussed for proteins within regard to general bilayer-mediated lipid influences on function.

Accordingly, the order-disorder phase transitions or the presence of lipid rafts are moderately affected by the proteolipid coupling, consistent with the standard FMM. Evidently that is not the case for the lamellar to nonlamellar phase transition of lipids with fluid, liquid-crystalline chains (see below). The lamellar to nonlamellar phase transition can be more strongly perturbed by cholesterol or proteins [31]. The strong proteolipid coupling is a marked shift away from the FMM [87], and might account for separations of nanosocale raft-like heterogeneities in cellular plasma membranes. For example, membrane curvature forces might stabilize raft-like lipid clusters, where greater local bilayer thickness due to cholesterol is compensated by curvature of the surrounding membrane. Strong out-of-plane curvature forces giving rise to the lamellar to nonlamellar phase transition might be connected to protein stability in nansocale raft-like heterogeneities with cholesterol. Another possibility is that the physical state of the lipids might resemble critical fluids, involving a proximity to a miscibility critical point in the temperature-composition phase diagram [46], as considered below.

3.6 Critical Behavior in Cellular Membranes

The possible influences of lateral phase separation and/or critical behavior have long been discussed with regard to collective lipid-lipid and lipid-protein interactions. For lipid rafts [1, 85], these concepts have been recast by supposing the membrane lipid composition is near a miscibility critical point [46]. For example, large susceptibilities are exhibited by supercritical systems, meaning that small inputs can produce sudden or large changes. Shifts can occur from one state to another with more sensitivity than would otherwise be possible. In analogy with supercritical solvents, compositional fluctuations could explain lipid influences on protein function in mammalian cellular membranes.

3.6.1 Miscibility Critical Points of Lipid Mixtures and Biomembranes

Previously, it has been suggested that the cellular lipid composition is regulated so the order-disorder phase transition is optimized versus the growth temperature [26].

For biomembranes, the collective interactions can involve clusters of the protein and lipid molecules due to nonideal mixing. Local regions of the proteolipid assembly can differ in the bilayer thickness or area per lipid molecule [54, 78]. What is more, the lipid compositions of plasma membranes of living cells may be controlled to be near a critical point at the growth conditions, e.g., as shown by fluorescence microscopy [46, 94]. In the region of a critical point, nanoscale collective interactions among the lipid and protein components could affect membrane function. Homoeostasis of criticality could apply to cellular plasma membranes with high levels of cholesterol, whereby living cells adjust their proximity to a miscibility critical point due changes in growth temperature or nutrient conditions. Large-scale fluctuations of the system may be enhanced, in which small perturbations yield a transition from one functional state to another [94, 95].

3.6.2 Lipid Bilayers and Biomembranes Treated as Critical Fluids

For model lipid mixtures with cholesterol, collective fluctuations are detected by fluorescence microscopy and by NMR spectroscopy at nearly critical temperatures [95], with correlation lengths in the 10-nm to μm range. Ternary lipid mixtures with a critical composition may traverse a miscibility critical point with changing temperature, where the correlation length for the compositional fluctuations diverges. Besides model ternary lipid mixtures, such critical behavior has been proposed to occur for the plasma membranes of living cells [94]. Upon cooling, plasma membrane vesicles are observed to phase separate into coexisting liquid-like phases [94] (Fig. 3.3). The correlation length depends on the proximity to the critical point in the temperature-composition phase diagram, meaning that a large range of cluster sizes can occur (10-nm to μm range) (Fig. 3.3).

In living cells, such critical behavior is associated with the presence of functional raft-like lipid domains. For example, mammalian cellular plasma membranes may be tuned to criticality, and might exploit properties of the supercritical solvent to detect and respond to environmental stimuli. Moreover critical Casimir forces could affect or mediate long-range interactions among the protein molecules over 5–10 nm length scales and beyond, potentially involving raft-like heterogeneity in cellular membranes. Yet, there are counterexamples of membranes low in cholesterol or sphingomyelin [15, 59], where appreciable influences of lipid-protein interactions exist. Because cholesterol is mainly found in mammalian cellular plasma membranes, the universality of raft-like phenomena or supercritical behavior might be somewhat overstated.

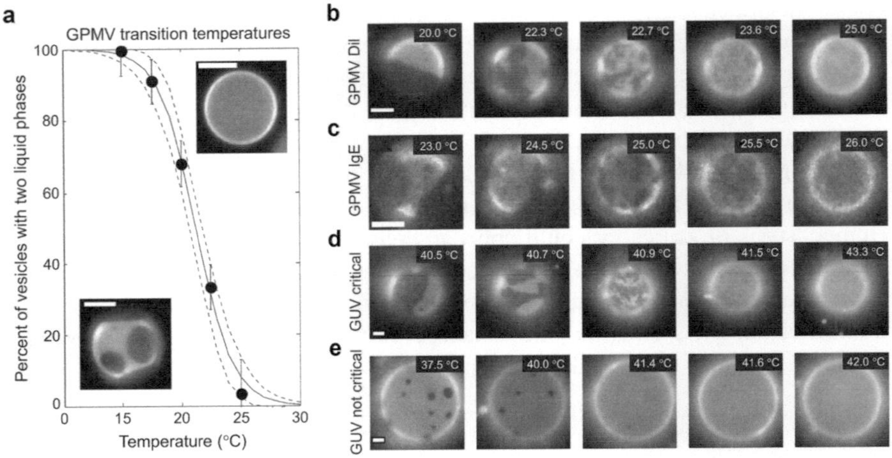

Fig. 3.3 Critical fluctuations are proposed to occur in giant plasma membrane vesicles derived from living rat leukemia (mast) cells [94]. (**a**) Percentage of lipids from mammalian cells in coexisting liquid phases versus temperature. (**b**)–(**d**) Fluorescent micrographs for individual giant membrane vesicles. Domain-boundary fluctuations occur at lower temperatures and lateral compositional fluctuations at higher temperatures. (**e**) Bottom row shows control lipid mixture (Figure adapted with permission from Ref. [94]. Copyright (2008) American Chemical Society)

3.7 Lipid Membranes as Generators of Curvature Stress

An alternative to considering hydrophobic mismatch or lipid criticality is to treat a continuous membrane lipid film [3, 13, 14, 21, 79]. Both plant and animal membranes contain lipids with a tendency to form nonlamellar phases [31, 59, 97]. For membrane shape deformations or fluctuations due to proteolipid couplings, applying the differential geometry of curves and surfaces [47] leads to the flexible surface model (FSM) [10, 14]. In this approach, the free energy of elastic curvature deformation of the membrane lipid bilayer affects the stability and appearance of protein functional states that modulate key biological functions, such as transport, ion conduction, and signaling.

3.7.1 Bilayer Force Field and Profile of Lateral Pressures

For the lipids (Fig. 3.4), within the head group region, attractive and repulsive interactions at the polar–nonpolar interface affect the area per molecule [62, 73, 78]. The lateral pressure profile has three main contributions (Fig. 3.4a, b) [22, 63, 84]. (1) The attractive pressure ($F_{L/W}$) (negative) corresponds to the surface tension ($\gamma_{L/W}$) of the hydrophobic acyl groups with water [84]. The lipid/water interfacial tension $\gamma_{L/W}$ manifests the hydrophobic effect, and minimizes the interfacial area.

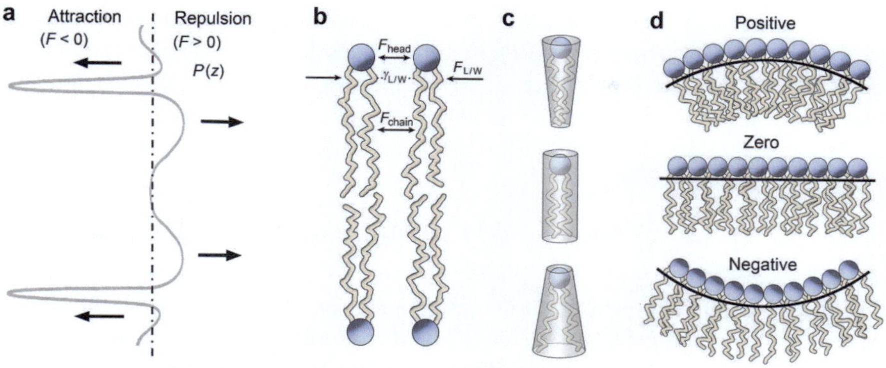

Fig. 3.4 Phospholipid self-assembly entails a balance of forces due to the polar head groups and nonpolar hydrocarbon chains. (**a**) Schematic plot of lateral pressure along the bilayer normal versus bilayer depth. (**b**) The attractive force (pressure) at the lipid/water interface ($F_{L/W}$) from the hydrophobic effect balances the repulsive force (pressure) of the head groups (F_{head}) and acyl chains (F_{chain}). (**c**) Molecular packing parameter for lipids with different head groups and acyl chains (shown within their geometrical shapes). (**d**) The spontaneous (intrinsic) curvature of a lipid monolayer due to the lateral force imbalance: positive (towards hydrocarbon), zero (planar), or negative (towards water) (Figure adapted with permission from Ref. [16]. Copyright (2012) American Chemical Society)

Additional attractive interactions include head group dipole and hydrogen-bonding forces, that act together with the long-range van der Waals force involving the acyl chains of the two monolayers [78]. (2) In the head group region, the lateral pressure (F_{head}) arises from steric, hydration, and electrostatic effects; though typically repulsive, it is can include an attractive part from hydrogen-bonding interactions. (3) Within the chain region, the repulsive lateral pressure (F_{chain}) is due to thermal bond rotational isomerizations. Together, the repulsive head group pressure and the acyl chain pressure compensate the attractive lipid/water interfacial tension, so the resultant lateral pressure is zero at equilibrium. For a particular head group size, the area per lipid restricts the acyl packing [78], yielding the observed microstructures. Competition of attractive and repulsive forces governs the self-assembly and polymorphism of the lipids, including their interactions with proteins and peptides.

The molecular packing parameter is due to the balance of opposing forces acting upon the lipid polar head groups and the nonpolar acyl chains [48, 84, 91]. Amphiphiles with a greater head group size versus the chains, such as lysophospholipids, single-chain detergents, or gangliosides, favor packing into a *conical* molecular shape on average (Fig. 3.4c, top). They form micelles or normal hexagonal H_{I} phases, and are analogous to an oil-in-water dispersion [84]. Lipids with intermediate size head groups, for instance PC whose head group is methylated versus PE, pack on average with a *cylindrical* molecular shape, and form a planar lipid bilayer (Fig. 3.4c, middle). Finally, lipids with small head groups as opposed the chains, such as PE, pack into an *inverted conical* molecular shape on average

(Fig. 3.4c, bottom). They form the reverse hexagonal $H_{\rm II}$ phase, which corresponds to a water-in-oil dispersion [84, 91]. Optimum packing of the lipids thus depends on the lateral pressure profile (Fig. 3.4a), and corresponds directly to the curvature energy [59, 76].

3.7.2 Intrinsic (Spontaneous) Curvature of a Lipid Monolayer

The reader should appreciate that the lateral pressure profile is invisible; it cannot be measured experimentally. By contrast, the spontaneous curvature H_0 of an individual lipid monolayer is accessible for nonlamellar phases of membrane lipids exposed to dual solvent stress [81]. The difference of the optimum cross-sectional areas of the lipid head groups versus the chains gives a bending moment for the lipid monolayer. The spontaneous (intrinsic) monolayer curvature (H_0) can be positive (towards hydrocarbon), zero, or negative (towards water) (Fig. 3.4d). It becomes more negative as temperature rises, or hydration is less, giving a sequence of nanostructures in the phase diagram [16]. Additional phases with curvature include microemulsions and bicontinuous cubic phases. The gyroid (G), Schwartz diamond (D), and primitive (P) minimal surfaces (where the mean curvature is everywhere zero) are connected via the Bonnet transformation [47]. Here the lipid or surfactant film is wrapped or draped onto an infinite periodic minimal surface, yielding a labyrinth of aqueous channels (so-called plumber's nightmare).

Now when the optimum head group separation exceeds the chains, there is a tendency to curl towards hydrocarbon (Fig. 3.4d, top). The head groups have their greatest aqueous exposure, as in the case of single-chain surfactants (e.g., lysolipids), glycolipids, and gangliosides. The positive spontaneous (intrinsic) curvature (H_0) is expressed by formation of small micelles or the type-1 hexagonal ($H_{\rm I}$) phase (or wormlike micelles), with the head groups outside and the chains inside the aggregate. By contrast, lipids with smaller head groups or larger chains, as for double-chain phospholipids, involve less exposure to water. Hence they favor a more condensed membrane surface, with a smaller interfacial area per lipid. If the optimal head group separation matches the chains, there is just a minor tendency of a monolayer to curl, as in the case of PCs. The spontaneous curvature H_0 is approximately zero and the planar lipid bilayer is stabilized, as in the fluid-mosaic model (Fig. 3.4d, middle). Lastly, the lipids with smaller head groups are even more dehydrated. Because the optimum polar head group separation is less than the chains, the lipid monolayer bends towards water (Fig. 3.4d, bottom), e.g., as in unsaturated and polyunsaturated PEs with a negative spontaneous curvature. As a result, the type-2 hexagonal ($H_{\rm II}$) (or cubic) phases are formed, where the head groups are inside and the chains outside the lipid nanostructure.

3.7.3 Curvature in Lipid-Protein Interactions

At this point the reader should note that the above curvatures are not implicit. They correspond to bending of a pivotal (neutral) plane, a mathematical surface located beneath the monolayer aqueous interface, where the lateral area is constant [81]. What is striking is not so much the topology of the nanostructures, but rather the monolayer interfacial curvature (Fig. 3.4d). Lipids that adopt nonlamellar phases, such as the H_{II} phase, have a negative spontaneous curvature H_0 (toward water). In a planar bilayer, there is a deviation of the geometric mean curvature H (which is zero) from the H_0 spontaneous curvature. The two monolayers are held together by the hydrophobic effect and packing forces. For an individual monolayer, the curvature mismatch is due to the chain packing interactions with the other monolayer. Despite the fact that a bilayer is flat on average, the two monolayers can still have an intrinsic tendency to curl [14, 43, 84].

3.8 Lipid-Protein Interactions and Curvature Forces

The implication of curvature elastic stress in membrane protein conformational changes was initially proposed based on experimental studies of rhodopsin [39, 97]. Lipid substitution experiments plus monitoring of protein transitions provides a connection of lipid properties to functional states [17]. The lipid effects are chemically nonspecific, indicating that biophysical properties are involved, although chemically specific interactions may also be significant [56, 57]. Experimental tests of the validity of the biophysical concepts have been carried out using UV-visible and FTIR spectroscopy [12, 88–90].

3.8.1 Elastic Curvature Stress of Proteolipid Membranes

One possibility is to conduct numerical membrane simulations [21], or alternatively a simpler continuum approximation can be implemented [55, 66]. As an illustration, let us return to the lateral pressure profile [22, 51, 63, 92]. Its influence on lipid-protein interactions amounts to investigating the curvature elastic force field of the membrane [14, 63]. So-called frustration of the spontaneous curvature (H_0) by the chain packing energy of the bilayer yields elastic deformation of the monolayer leaflets [14]. Because the bilayer thickness is much smaller than its area, the Helfrich approach of introducing differential geometry and curvature elasticity is applicable at the cellular membrane level [21, 45, 79]. Still, the extension to the mesoscopic regime of lipid-protein interactions—intermediate between the molecular size and the macroscopic dimensions—needs some additional development.

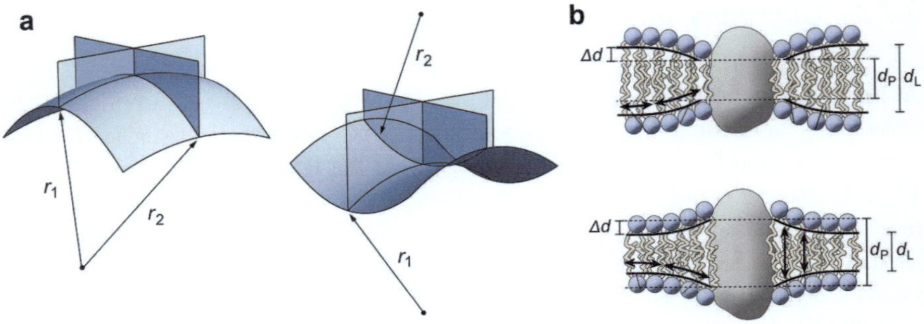

Fig. 3.5 Application of differential geometry explains lipid influences on membrane protein function. (**a**) Membrane distortion is described by two principal curvatures (given by the intersection of normal planes with the surface). The principal curvatures can be the same or opposite in sign as depicted. (Figure courtesy of J. Kinnun.) (**b**) Flexible surface model (FSM) for membrane lipid-protein interactions. Top: hydrophobic thickness of protein is less than lipid bilayer giving compression of the acyl groups. Bottom: protein hydrophobic thickness exceeds the bilayer yielding stretching (expansion) of lipid chains. Competition of monolayer bending with the proteolipid solvation energy governs the functional protein states (Figure modified with permission from Ref. [17])

3.8.2 Bending Free Energy of a Flexible Surface

Now in a flexible surface model (FSM), curvature deformation at the proteolipid boundary competes with hydrophobic solvation of the protein by the lipids (Fig. 3.5). Following Helfrich & Servuss [45], the curvature free energy for deforming a soft (flexible) surface is given by,

$$G_c = \oint_S g_c dA = \oint_S \left[\frac{1}{2}\kappa \left(\frac{1}{r_1} + \frac{1}{r_2} \right)^2 + \frac{\overline{\kappa}}{r_1 r_2} \right] dA. \tag{3.1}$$

Here r_1 and r_2 are the two principal radii of curvature for a particular topology (geometrical shape) (Fig. 3.5a). The mean-curvature bending modulus is denoted by κ and $\overline{\kappa}$ is the modulus of Gaussian curvature [47]. Starting from Eq. (3.1), the free energy density can be expanded in $c_1 = 1/r_1$ and $c_2 = 1/r_2$, which are the two principal curvatures (Fig. 3.5a). By expressing the free energy in terms of the mean curvature $H = (1/2)(1/r_1 + 1/r_2)$, and introducing H_0 as the spontaneous curvature (where the bending energy is minimized), one can separate out a Gaussian curvature term $K = 1/r_1 r_2$ yielding further simplification. The resulting free energy density per unit area depends explicitly on the curvatures and the corresponding elastic moduli. For an individual monolayer of a lipid bilayer, it reads [45]

$$g_c = \kappa (H - H_0)^2 + \overline{\kappa} K, \tag{3.2}$$

where the curvature free energy is obtained by integrating over the surface. Eq. (3.2) states that the displacement of the mean curvature from its spontaneous curvature together with the Gaussian curvature is related to the curvature energy. When the curvature free energy is not at a minimum, the system is considered as frustrated.

3.8.3 Competition of Curvature and Hydrophobic Forces

For membrane systems, additional free energy terms must counterbalance the curvature free energy, e.g., due to the proteolipid solvation energy [7, 10]. Let us next consider two states of a membrane protein (designated 1 and 2) in a mixture of two lipid types (designated *i* and *j*). The hydrophobic thickness is related to the protein hydrophobic surface with areas A_1 and A_2 and with lipid mean curvatures H_1 and H_2 in the two states (Fig. 3.5b). The spontaneous curvature is then given by $H_0 = H_0^i X_i + H_0^j X_j$ where in the mixture X_i and X_j are the mole fractions of the two lipids. For simplicity, we assume that lipid *i* favors the planar bilayer (zero H_0), and lipid *j* favors the nonlamellar (reverse hexagonal) state ($H_0 < 0$), yielding $H_0 \approx H_0^j X_j$ for the spontaneous curvature. The standard free energy change ΔG^{o} per protein molecule is then,

$$\Delta G^{\mathrm{o}}/N_{\mathrm{A}} = -\kappa \left(A_{\mathrm{L}} H_2 H_0^j\right) N_{\mathrm{L}} X_j + \gamma \Delta A, \tag{3.3}$$

$$= -\kappa \left(A_{\mathrm{L}} H_2 N_{\mathrm{L}}\right) H_0 + \gamma \Delta A, \tag{3.4}$$

where $H_1 \approx 0$ and to linear order $H_2 << 1$. The spontaneous curvature H_0^j is due to the nonlamellar-favoring lipids, and $\Delta A \equiv A_2 - A_1$ is the change in protein hydrophobic surface area. In addition γ is the proteolipid surface tension, which is approximately the same for both lipid types; A_{L} is the cross-sectional area per lipid [78], N_{L} is the number of lipids per protein, and all other symbols have their standard meanings. For simplicity, neither the Gaussian (saddle splay) curvature nor the protein free energy is included explicitly. Note the value of H_0 is approximately the inverse radius of curvature (R_{W}) of the H_{II}-phase nanotubes for nonlamellar lipids under dual solvent stress [81].

In Eqs. (3.3) and (3.4) the first term is negative, yielding a driving force for the conformational change. The curvature frustration is counterbalanced by the second positive term, due to the work of expanding the proteolipid hydrophobic surface area (volume). According to the FSM, the free energy change is linear in the mole fraction (X_j) of the nonlamellar-forming lipid in the mixture. In addition, the free energy change depends on the lipid/protein molar ratio (N_{L}), as first shown for rhodopsin [12]. The equilibrium constant K for the protein transition thus depends on both the spontaneous curvature and proteolipid interfacial tension, as given by Eq. (3.4), where $K = \exp(-\Delta G^{\mathrm{o}}/RT)$. The FSM is a simple robust theory with very little mathematical detail that can obscure the biophysical significance.

3.8.4 Magnitude of the Curvature Free Energy

To summarize at this point, the power of the curvature forces in membrane deformation is illustrated by considering the free energy density given by Eq. (3.2). For a lipid bilayer, we can adopt representative values of $2\kappa = 4 \times 10^{-19}$ J with a mean monolayer curvature of 1/(40 Å) [37, 44], an area per lipid of 70 Å^2 [78], and 100 lipids/protein molecule. Applying Eq. (3.2) to a planar membrane then gives an appreciable curvature free energy of $\approx$ 0.5 MJ/mol protein. What is more, only $\approx$10 kJ/mol protein is needed to shift the equilibrium mainly from reactants to products. Sufficient free energy is thus contained within the bilayer stress field to drive the membrane protein conformational changes.

3.9 Curvature Elastic Stress in Membrane Protein Function

The flexible surface model (FSM) has been proposed as an option to the standard fluid-mosaic model that includes the strong proteolipid couplings [10, 12, 14, 39]. It remains under active discussion to explain functional lipid-protein interactions [16]. Here a material science picture is introduced at the mesoscopic length scale, in between the macroscopic membrane and the molecular size. For integral membrane proteins, the long-range elastic bilayer deformation energy and the short-range proteolipid solvation energy influence the general bilayer-mediated interactions. The absence of molecular specifics is both a strength and weakness of the simple continuum theory, which can be refined once the underlying bilayer forces are identified.

3.9.1 Flexible Surface Model for Membrane Lipid-Protein Interactions

Experimentally, the effects of nonlamellar-forming lipids on membrane protein function [6, 15, 27, 31, 34, 49, 50, 74, 82, 96, 98] point to an influence of curvature elastic forces [5, 10, 12, 14, 28, 40, 55, 66, 97]. There are two relevant surfaces in a continuum view: the lipid/water interface and the lipid/protein interface [40]. A key prediction of the FSM (Fig. 3.5b) is that nonlamellar-forming lipids modulate the energetics of integral membrane proteins, as experimentally first shown for rhodopsin [14, 97], and subsequently for other proteins such as mechanosensitive ion channels [77]. Lipid interactions with the intramembranous protein surface (solvation energy) and with water (hydrophobic effect) yield small differences in large opposing forces that affect the lipid-protein interactions [12].

According to the FSM, the functional lipid–protein interactions are explained by elastic deformation of the individual monolayers of the membrane film [12, 14].

If the protein shape is altered (Fig. 3.5b), then the free energy of the system of interacting protein, lipids, and water is modified [10]. In Eqs. (3.3) and (3.4) changing the monolayer curvature H away from its spontaneous (intrinsic) curvature H_0 gives a contribution to the free energy that is frustrated (balanced) by the lipid-protein solvation energy. If the spontaneous (intrinsic) curvature of a lipid monolayer differs from the curvature at the proteolipid boundary, then a lipid-mediated force exists that affects the energetics of the membrane protein shape (conformational) changes. The spontaneous curvature (H_0) amounts to bending a neutral (pivotal) plane, where the lipid cross-sectional area remains constant (Fig. 3.5b). Hydrophobic coupling entails the local expansion or compression of the bilayer next to the proteolipid interface ($d_P \equiv$ thickness adjacent to proteolipid interface; $d_L \equiv$ thickness of unperturbed bilayer).

For sake of illustration, let us assume a membrane protein with an equilibrium between two states, differing in its protrusion from the lipid bilayer (e.g., the inactive MI and active MII states of rhodopsin) [83]. Then work is done in the conformational change, and decreasing the curvature frustration due to lipids with a negative spontaneous (intrinsic) curvature (like PE) (Fig. 3.5b) can stabilize the more protruded (thicker) state of the protein. With rhodopsin as an example, lipids with approximately zero spontaneous curvature H_0 shift the activation equilibrium toward the inactive MI state; while lipids with negative H_0 adjust it toward the active MII state [40, 97]. Additional molecular support for the FSM [14] comes from X-ray crystallography [25] and site-directed spin labeling studies [2], which show an extension of TM helix H5 plus a change in tilt of helix H6 upon rhodopsin activation.

3.9.2 *Free Energy Landscape and Functional States of Membrane Proteins*

Coming back to rhodopsin as a sensor of curvature stress [31, 97], according to Eq. (3.3) the lipid/protein ratio affects the formation of the active MII state [12, 75, 88], as first shown by Botelho et al. [11]. We now consider the second FSM prediction: in addition Eqs. (3.3) and (3.4) imply the free energy change is linear in the mole fraction (X_j) of the nonlamellar-forming lipids in the mixture [10, 39]. Here we recall that unsaturated PEs have negative H_0 (curvature toward water) whereas $H_0 \approx 0$ for PCs (low tendency to curve) [91]. For rhodopsin plots of ln$K$ for the MI–MII equilibrium (Fig. 3.6a) in a series of DOPE/DOPC recombinant membranes versus the effective spontaneous curvature show an approximately linear dependence [10], where $K = \exp(-\Delta G^{\circ}/RT)$. Further support comes from studies of recombinants of rhodopsin with head-group methylated PEs and polyunsaturated SDPE lipids [89] (Fig. 3.6b). Once again, an approximately linear dependence is obtained when ln$K$ is plotted against the radius of spontaneous curvature ($R_0 \equiv 1/H_0$). For all the PE-containing membranes, a positive linear slope is found, consistent with a negative curvature (H_2) (towards water) at the

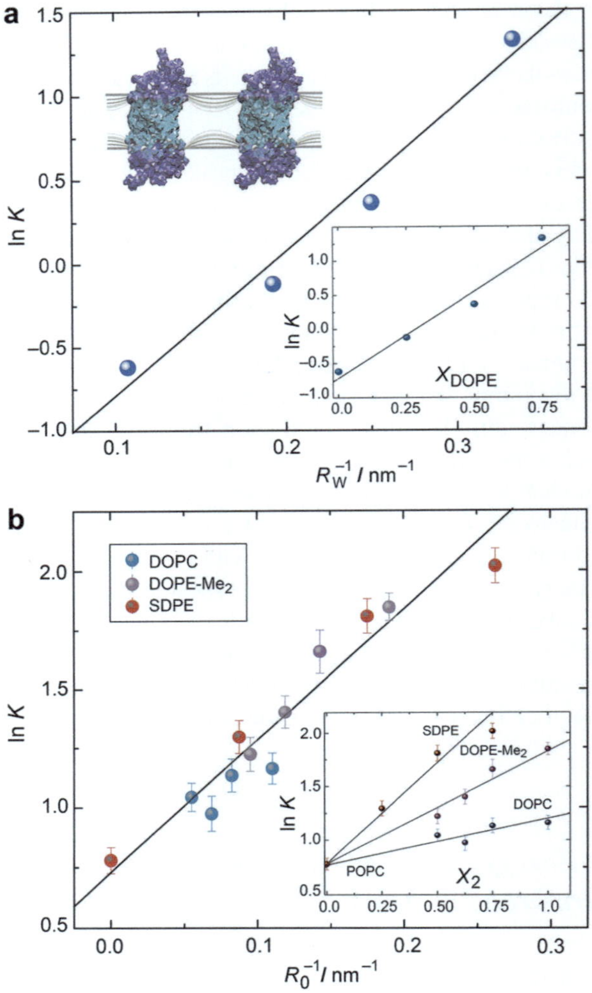

Fig. 3.6 Phosphoethanolamine (PE) lipids govern rhodopsin light activation in membranes due to curvature elastic deformation. (**a**) Plot of ln$K$ (at $T = 28$ °C) for rhodopsin in DOPE/DOPC membranes against spontaneous curvature (Eq. 3.4) given by inverse water-core radius (R_W) of H_{II}-phase cylinders. Inset: corresponding plots versus mole fraction (X_{DOPE}) of DOPE lipids. (**b**) Graph of ln$K$ (at $T = 37$ °C) for rhodopsin against spontaneous curvature defined by inverse radius ($H_0 \equiv 1/R_0$) of H_{II}-phase nanotubes for head-group methylated PEs. Inset: plots versus mole fraction of methylated PE lipids and polyunsaturated SDPE lipids. The results collapse to a universal relation as a function of spontaneous curvature (H_0) of the lipid mixtures (Rhodopsin data are from Refs. [10, 89] and lipid data are from Refs. [37, 44, 52]. Figure modified with permission from Ref. [18])

proteolipid boundary [10, 14, 39]. What is more, the above inferences are supported by plasmon-waveguide resonance (PWR) studies [83], and by the latest X-ray structures of active rhodopsin [25, 32]. Similar treatments of mechanosensitive channels [77, 79] also involve coupling of the protein shape and conformation to mechanical curvature forces.

3.10 Soft Matter and the Role of Membrane Water

Although X-ray crystallography, cryoelectron microscopy, and NMR spectroscopy are widely used for investigating membrane proteins, elucidating the role of water plus membrane lipids (soft matter) needs complementary strategies. For rhodopsin, molecular dynamics (MD) simulations show that bulk water molecules flood the protein interior during light activation [58]. Recent experimental osmotic stress studies have uncovered how changes in its hydrated volume within the lipid bilayer are coupled to signaling [24]. In proteolipid membranes, osmotic stress involves the effects of water on both the proteins and the lipids. For the lipid effect, osmotic stress (dehydration) increases the (absolute) monolayer spontaneous membrane curvature [16], which stabilizes the active MII state [15]. Dehydration of the lipid bilayer also increases its thickness [62], and likewise favors the active MII formation [15]. The osmotic lipid effect is opposite to the protein effect: experimental studies show that osmotically stressing rhodopsin (dehydration) stabilizes the preactive MI state upon light absorption. Using thermodynamic relations approximately 60–80 water molecules flood the protein core. Because of the large influx of bulk water [58], the protein effect is dominant. The balance can be tipped from one state to another, depending on the level of hydration of rhodopsin within the lipid membrane.

The flexible surface model (FSM) includes the soft matter of the membrane, explaining the influences of both the lipids and water. For rhodopsin, water does not just passively enter the binding pocket upon opening of the protein. It directly couples to the metarhodopsin equilibrium that occurs upon light illumination. Most surprising, the apparent number of water molecules can have a positive value, due to a water influx upon rhodopsin activation, or a negative value, involving displacement (efflux) of water by transducin-derived peptides. In this way wet-dry cycling of the binding cleft of rhodopsin for its cognate G-protein transducin explains the high fidelity, rapid visual signaling by an sponge-like alternating mechanism [24]. Properties of the soft membrane matter (lipids and water) play an important role, whereby the energetic balance modulates the activity in visual signaling.

3.11 Membrane Proteins as Inducers and Sensors of Curvature Stress

According to the flexible surface model, curvature elasticity involves long-range proteolipid interactions (extending over >1 molecular diameter), and is fundamental to membrane structure, assembly, and function. Collective lipid and protein interactions can extend across multiple length scales, and emerge from the atomistic-level forces. The coupling of the lipids and proteins provides various ways that cells or organelles can change their shape, e.g., as in the case of endocytotic vesicles or caveolae, or budding of membrane-encapsulated viruses like HIV or influenza. The different means whereby proteins can interact with membrane lipids to produce curvature stress are highly active topics of biophysical investigations at present [4, 86].

3.11.1 Proteins That Induce and Sense Membrane Curvature

Increasingly it is becoming understood how curvature-inducing or curvature-sensing proteins can dramatically affect lipid structural changes, enabling them to partake in trafficking, membrane fusion or fission, or/and cell division. Recently, cellular shape transformations have led to a focus on proteins known as BAR (Bin/amphiphysin/Rvs) domains (Fig. 3.7a) [36, 67, 86]. The BAR domain superfamily includes amphiphysin and endophilin, which may participate in clathrin-mediated endocytosis. Transitions in membrane shape or topology and remodeling of lipid microstructures [4, 67, 86] entail an important role for multiscale lipid-protein interactions. Because the size of the membrane is much greater than the thickness, a continuum approach [47] that is founded on differential geometry is applicable. Here we have further elaborated with regard to nanoscale lipid-protein interactions how such a continuum view is complementary to molecular approaches.

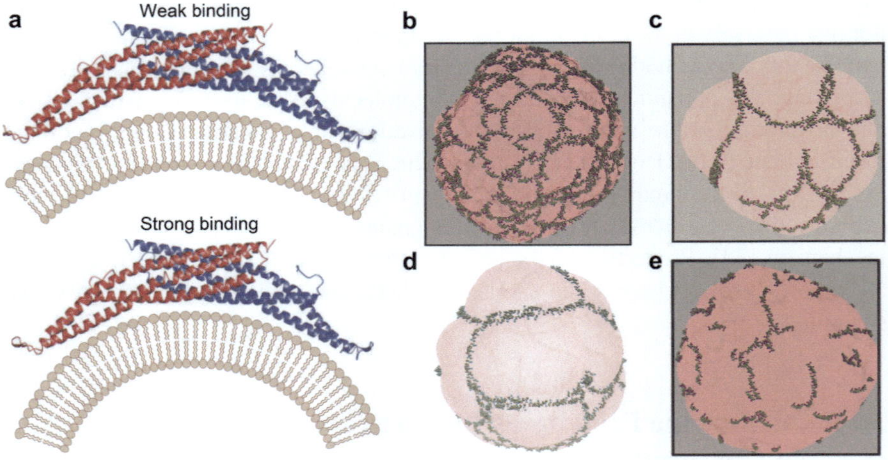

Fig. 3.7 Membrane remodeling in cellular biology entails lipid-protein interactions. (**a**) Depiction of how BAR (Bin/amphiphysin/Rvs) domains bind to lipid membranes as dimers through their concave surface [67]. The amphiphysin BAR domain interacts either with low-curvature (above) or high-curvature (below) lipid membranes. (**b**) Multiscale coarse-grain simulation of vesicle remodeling by N-BAR domains [86]. (**c**)–(**e**) The N-BAR proteins organize into string-like meshworks leading to buds, tubules, or vesicles (Figure modified with permission from Refs. [67 and 86])

3.11.2 Bilayer Remodeling and Membrane Shape Transitions

At the protein level, the insertion of amphipathic helices into bilayers takes on added importance by providing molecular insights into membrane bending. Amphiphysin and endophilin are among the proteins of the BAR domain superfamily that are involved in eliciting the membrane curvature [36]. The BAR domains include an N-terminal amphipathic α-helix (N-BAR domains) that acts as a wedge, causing local buckling of the bilayer polar-apolar interface (Fig. 3.7a). In addition the F-BAR proteins are involved with membrane curvature induction, and the I-BAR proteins (inverse-BAR) stabilize tubulations [36]. Cryoelectron microscopic reconstructions have uncovered how the BAR proteins are organized on membrane tubules. Notably, coarse-grain molecular dynamics (MD) models of N-BAR domains interacting with flat membranes and vesicles yield further structural insights (Fig. 3.7b) [86]. These findings show that binding of N-BAR domain proteins to lipid membranes entails their linear association or polymerization into webs or networks (Fig. 3.7c–e). Curvature stress imposed by N-BAR proteins yields endocytotic vesicular deformations, in which proteins assemble at the base of the emerging membrane buds [86]. An increasing number of diseases connected with BAR protein dysfunction suggests new therapeutic strategies in molecular and translational medicine [36].

3.12 Summary

The viewpoint of lipids as liquid-crystalline soft matter occupies an important place in cellular membrane biophysics. Identifying how dysfunction and dysregulation happen at the cellular membrane level can contribute to our understanding of diseases such as cancer, cardiovascular disease, and visual disorders like *retinitis pigmentosa*. Elastic remodeling within the membrane is related to conformational transitions of proteins [14, 16] that affect folding, stability, and organelle or cellular shape changes. The spontaneous (intrinsic) curvature H_0 is a material property that accounts not only for the membrane lipid nanostructures, but also the functions of lipid-embedded proteins such as GPCRs, ion channels, pumps, and transporters. Cellular processes that occur within the stress field of the membrane [14] include budding of viruses like influenza, dengue, or HIV; membrane biogenesis and trafficking; and membrane protein folding mechanisms [9, 80]. Understanding how membrane lipids control protein states within the bilayer motivates the new biophysics at the crossroads of structure, function, and mechanism.

Acknowledgments We thank B. Kobilka, G. Lindblom, K. Palczewski, A. Parsegian, T. Huber, G. Voth, and H. Wennerström for discussions. Thanks are due to J. Kinnun for assisting with the figures, and to past and present members of our research group for contributing to this work. Research from our laboratory is supported by the U.S. National Institutes of Health.

References

1. Ackerman DG, Feigenson GW (2015) J Phys Chem B 119:4240–4250
2. Altenbach C, Kusnetzow AK, Ernst OP, Hofmann KP, Hubbell WL (2008) Proc Natl Acad Sci U S A 105:7439–7444
3. Andersen OS, Koeppe RE II (2007) Annu Rev Biophys Biomolec Struct 36:107–130
4. Antonny B (2011) Annu Rev Biochem 80:101–123
5. Aranda-Espinoza H, Berman A, Dan P, Pincus P, Safran S (1996) Biophys J 71:648–656
6. Attard GS, Templer RH, Smith WS, Hunt AN, Jackowski S (2000) Proc Natl Acad Sci U S A 97:9032–9036
7. Baldwin PA, Hubbell WL (1985) Biochemistry 24:2633–2639
8. Bogdanov M, Heacock P, Guan Z, Dowhan W (2010) Proc Natl Acad Sci U S A 107:15057–15062
9. Booth PJ, Curnow P (2009) Curr Opin Struct Biol 19:8–13
10. Botelho AV, Gibson NJ, Wang Y, Thurmond RL, Brown MF (2002) Biochemistry 41:6354–6368
11. Botelho AV, Huber T, Brown MF (2003) Biophys J 84:55A
12. Botelho AV, Huber T, Sakmar TP, Brown MF (2006) Biophys J 91:4464–4477
13. Brown FLH (2008) Ann Rev Phys Chem 59:685–712
14. Brown MF (1994) Chem Phys Lipids 73:159–180
15. Brown MF (1997) In: Epand RM (ed) Current topics in membranes, vol 44. Academic Press, San Diego, pp 285–356
16. Brown MF (2012a) Biochemistry 51:9782–9795, and references therein
17. Brown MF (2012b) In: Vaidehi N, Klein-Seetharaman J (eds) Membrane protein structure: methods and protocols, methods in molecular biology, vol 914. Springer, LLC, New York, pp 127–153
18. Brown MF (2017) Annu Rev Biophys 46:379–410, and references therein
19. Brown MF, Ribeiro AA, Williams GD (1983) Proc Natl Acad Sci U S A 80:4325–4329
20. Brown MF, Seelig J (1978) Biochemistry 17:381–384
21. Campelo F, Arnarez C, Marrink SJ, Kozlov MM (2014) Adv Colloid Interf Sci 208:25–33
22. Cantor RS (1999) Biophys J 76:2625–2639
23. Chawla U, Jiang Y, Zheng W, Kuang L, Perera SMDC, Pitman MC, Brown MF, Liang H (2016a) Angew Chem 128:598–602
24. Chawla U, Perera SMDC, Struts AV, Pitman MC, Brown MF (2016b) Biophys J 110:83a
25. Choe H-W, Kim YJ, Park JH, Morizumi T, Pai EF, Krauβ N, Hofmann KP, Scheerer P, Ernst OP (2011) Nature 471:651–655
26. Cronan JE (2003) Annu Rev Microbiol 57:203–224
27. Dahlqvist A, Nordström S, Karlsson OP, Mannock DA, McElhaney RN, Wieslander Å (1995) Biochemistry 34:13381–13389
28. Dan N, Pincus P, Safran SA (1993) Langmuir 9:2768–2771
29. Dawaliby R, Trubbia C, Delporte C, Masureel M, Van Antwerpen P, Kobilka BK, Govaerts C (2016) Nat Chem Biol 12:35–39
30. Day CA, Kenworthy AK (2009) Biochim Biophys Acta 1788:245–253
31. Deese AJ, Dratz EA, Brown MF (1981) FEBS Lett 124:93–99
32. Deupi X, Edwards P, Singhal A, Nickle B, Oprian D, Schertler G, Standfuss J (2012) Proc Natl Acad Sci U S A 109:119–124
33. Engelman DM (2005) Nature 438:578–580
34. Epand RM (1998) Biochim Biophys Acta 1376:353–368
35. Fattal DR, Ben-Shaul A (1993) Biophys J 65:1795–1809
36. Frost A, Unger VM, De Camilli P (2009) Cell 137:191–196
37. Fuller N, Rand RP (2001) Biophys J 81:243–254
38. Gao Y, Cao E, Julius D, Cheng Y (2016) Nature 534:347–351
39. Gibson NJ, Brown MF (1991) Photochem Photobiol 54:985–992

40. Gibson NJ, Brown MF (1993) Biochemistry 32:2438–2454
41. Goñi F, Alonso A, Bagatolli LA, Brown RE, Marsh D, Prieto M, Thewalt JL (2008) Biochim Biophys Acta 1781:665–684
42. Griffith OH, Jost PC (1976) In: Berliner LJ (ed) Spin labeling. Theory and applications. Academic press, New York, pp 453–523
43. Gruner SM (1989) J Phys Chem 93:7562–7570
44. Hamai C, Yang T, Kataoka S, Cremer PS, Musser SM (2006) Biophys J 90:1241–1248
45. Helfrich W, Servuss R-M (1984) Nuovo Cim 3:137–151
46. Honerkamp-Smith AR, Veatch SL, Keller SL (2009) Biochim Biophys Acta 1788:53–63
47. Hyde ST, Andersson S, Larsson K, Blum Z, Landh T, Lidin S, Ninham BW (1997) The language of shape. The role of curvature in condensed matter: physics, chemistry and biology. Elsevier, Amsterdam
48. Israelachvili JN (2011) Intermolecular and surface forces, 3rd edn. Academic Press, San Diego
49. Jensen JW, Schutzbach JS (1984) Biochemistry 23:1115–1119
50. Jensen JW, Schutzbach JS (1988) Biochemistry 27:6315–6320
51. Jensen MØ, Mouritsen OG (2004) Biochim Biophys Acta 1666:205–226
52. Keller SL, Bezrukov SM, Gruner SM, Tate MW, Vodyanoy I, Parsegian VA (1993) Biophys J 65:23–27
53. Khelashvili G, Blecua Carrillo Albornoz P, Johner N, Mondal S, Caffrey M, Weinstein H (2012) J Am Chem Soc 134:15858–15868
54. Kinnun JJ, Mallikarjunaiah KJ, Petrache HI, Brown MF (2015) Biochim Biophys Acta 1848:246–259
55. Kralchevsky PA, Paunov VN, Denkov ND, Nagayama K (1995) J Chem Soc Faraday Trans 91:3415–3432
56. Lee AG (2004) Biochim Biophys Acta 1666:62–87
57. Lee AG (2011) Trends Biochem Sci 36:493–500
58. Leioatts N, Mertz BM, Martínez-Mayorga K, Romo TD, Pitman MC, Feller SE, Grossfield A, Brown MF (2014) Biochemistry 53:376–385
59. Lindblom G, Brentel I, Sjölund M, Wikander G, Wieslander Å (1986) Biochemistry 25:7502–7510
60. London E (2002) Curr Opin Struct Biol 12:480–486
61. Mahalingam M, Martínez-Mayorga K, Brown MF, Vogel R (2008) Proc Natl Acad Sci U S A 105:17795–17800
62. Mallikarjunaiah KJ, Leftin A, Kinnun JJ, Justice MJ, Rogozea AL, Petrache HI, Brown MF (2011) Biophys J 100:98–107
63. Marsh D (2007) Biophys J 93:3884–3899
64. Marsh D (2008) Biochim Biophys Acta 1778:1545–1575
65. Marty MT, Hoi KK, Gault J, Robinson CV (2016) Angew Chem Int Ed 55:550–554
66. May S, Ben-Shaul A (1999) Biophys J 76:751–767
67. McMahon HT, Gallop JL (2005) Nature 438:590–596
68. Mitchell DC, Straume M, Miller JL, Litman BJ (1990) Biochemistry 29:9143–9149
69. Molugu TR, Brown MF (2016) Chem Phys Lipids 199:39–51
70. Mondal S, Khelashvili G, Shan J, Andersen OS, Weinstein H (2011) Biophys J 101:2092–2101
71. Morein S, Andersson A-S, Rilfors L, Lindblom G (1996) J Biol Chem 271:6801–6809
72. Mouritsen OG, Bloom M (1993) Annu Rev Biophys Biomol Struct 22:145–171
73. Nagle JF, Tristram-Nagle S (2000) Biochim Biophys Acta 1469:159–195
74. Navarro J, Toivio-Kinnucan M, Racker E (1984) Biochemistry 23:130–135
75. Niu S-L, Mitchell DC (2005) Biophys J 89:1833–1840
76. Österberg F, Rilfors L, Wieslander Å, Lindblom G, Gruner SM (1995) Biochim Biophys Acta 1257:18–24
77. Perozo E, Kloda A, Cortes DM, Martinac B (2002) Nat Struct Biol 9:696–703
78. Petrache HI, Dodd SW, Brown MF (2000) Biophys J 79:3172–3192
79. Phillips R, Ursell T, Wiggins P, Sens P (2009) Nature 459:379–385
80. Popot J-L, Engelman DM (2000) Annu Rev Biochem 69:881–922

81. Rand RP, Fuller NL, Gruner SM, Parsegian VA (1990) Biochemistry 29:76–87
82. Rietveld AG, Koorengevel MC, de Kruijff B (1995) EMBO J 14:5506–5513
83. Salamon Z, Wang Y, Brown MF, Macleod HA, Tollin G (1994) Biochemistry 33:13706–13711
84. Seddon JM (1990) Biochim Biophys Acta 1031:1–69
85. Simons K, Gerl MJ (2010) Nat Rev Mol Cell Biol 11:688–699
86. Simunovic M, Srivastava A, Voth GA (2013) Proc Natl Acad Sci U S A 110:20396–20401
87. Singer SJ, Nicolson GL (1972) Science 175:720–731
88. Soubias O, Teague WE Jr, Hines KG, Gawrisch K (2015) Biophys J 108:1125–1132
89. Soubias O, Teague WE Jr, Hines KG, Mitchell DC, Gawrisch K (2010) Biophys J 99:817–824
90. Teague WE Jr, Soubias O, Petrache H, Fuller N, Hines KG, Rand RP, Gawrisch K (2013) Faraday Discuss 161:383–395
91. Thurmond RL, Lindblom G, Brown MF (1993) Biochemistry 32:5394–5410
92. van den Brink-van der Laan E, Killian JA, de Kruijff B (2004) Biochim Biophys Acta 1666:275–288
93. van Meer G, Voelker DR, Feigenson GW (2008) Nat Rev Mol Cell Biol 9:112–124
94. Veatch SL, Cicuta P, Sengupta P, Honerkamp-Smith A, Holowka D, Baird B (2008) ACS Chem Biol 3:287–293
95. Veatch SL, Soubias O, Keller SL, Gawrisch K (2007) Proc Natl Acad Sci U S A 104:17650–17655
96. Vikström S, Li L, Karlsson OP, Wieslander Å (1999) Biochemistry 38:5511–5520
97. Wiedmann TS, Pates RD, Beach JM, Salmon A, Brown MF (1988) Biochemistry 27:6469–6474
98. Zidovetzki R (1997) In: Epand RM (ed) Current topics in membranes, vol 44. Academic Press, San Diego, pp 255–283
99. Zimmerberg J, Gawrisch K (2006) Nat Chem Biol 2:564–567

Chapter 4
Principles of Mechanosensing at the Membrane Interface

Navid Bavi, Yury A. Nikolaev, Omid Bavi, Pietro Ridone, Adam D. Martinac, Yoshitaka Nakayama, Charles D. Cox, and Boris Martinac

Abstract Mechanotransduction is a general term for all physiological processes through which living cells sense and respond to external and/or internal mechanical stimuli. These stimuli are converted into electrochemical intracellular signals via various mechanosensory transducers eliciting specific cellular responses. Among the many molecular mechanosensors found in living cells, mechanosensitive (MS) ion channels form a group of the fastest signaling molecules essential for cellular mechanotransduction. In this chapter, we discuss the basic principles of ion channel mechanosensitivity and highlight the importance of the surrounding lipid bilayer, cytoskeleton and extracellular matrix. We also discuss how these facets of channel mechanosensitivity may be reduced to changes of the transbilayer pressure profile and MS channel conformations that mutually affect each other according to the 'force-from-lipids' paradigm.

Keywords Mechanotransduction • Mechanosensitive channels • MscL • Piezo1 • Force-from-lipids • Transbilayer pressure profile • Force-from-filament • Liposomes • Patch clamp • MD simulations

---

N. Bavi • P. Ridone • C.D. Cox • B. Martinac (✉)
Victor Chang Cardiac Research Institute, Lowy Packer Building, Darlinghurst, NSW, 2010, Australia

St Vincent's Clinical School, University of New South Wales, Darlinghurst, NSW, 2010, Australia
e-mail: B.Martinac@victorchang.edu.au

Y.A. Nikolaev
Victor Chang Cardiac Research Institute, Lowy Packer Building, Darlinghurst, NSW, 2010, Australia

School of Biomedical Sciences and Pharmacy, University of Newcastle, Newcastle, NSW, 2308, Australia

O. Bavi
Department of Physics, University of Tehran, Tehran 1439955961, Iran

A.D. Martinac • Y. Nakayama
St Vincent's Clinical School, University of New South Wales, Darlinghurst, NSW, 2010, Australia

R.M. Epand, J.-M. Ruysschaert (eds.), *The Biophysics of Cell Membranes*, Springer Series in Biophysics 19, DOI 10.1007/978-981-10-6244-5_4

4.1 Mechanotransduction

Mechanotransduction refers to the physiological processes through which living cells sense and respond to mechanical stimuli by converting them into electrochemical signals that in turn elicit specific cellular responses [1]. In bacteria and plants, mechanotransduction processes are involved in osmoregulation, and gravitropism. In mammals, they include cell migration, anoikis and differentiation as well as whole array of sensory systems [1–11].

4.2 Mechanical Force Pathways in Cellular Signaling

In order for an efficient mechanotransduction process to occur, the living organisms possess cellular structures able to detect mechanical stimuli called mechanosensors. Thus, mechanosensors are molecular reporters within cellular structures that link external mechanical stimuli at the cell surface to intracellular signalling events and downstream effectors through mechanosensory proteins (Fig. 4.1). There are three main known mechanosensors in eukaryotes: (i) those formed by the wrinkling and invagination of plasma membrane (e.g., caveolae, microvilli, and possibly cilia), (ii) intra- or extracellular elements such as the actin cytoskeleton, glycocalyx, a cadherin rich cell-cell junction and (iii) mechanosensitive (MS) ion channels. A combination of the above elements can also be involved in a mechanotransduction process such

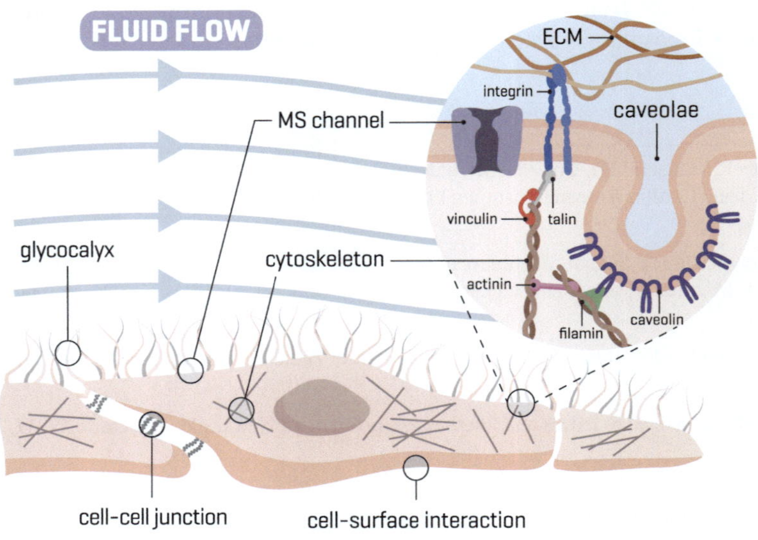

Fig. 4.1 Diagram of mechanical stimulation on vascular endothelial cells. Various types of force transmission pathways in endothelial cells including membrane deformation, extracellular matrix (ECM) and cytoskeletal elements, cell-cell junctions, cell-surface interaction and MS channels

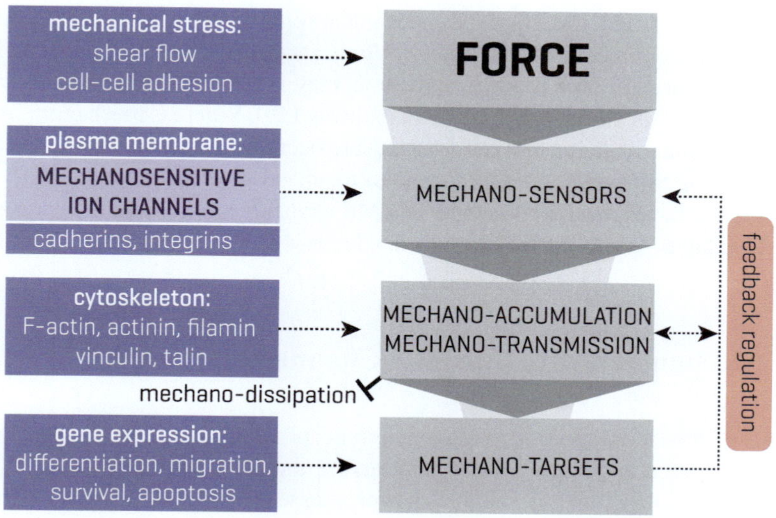

Fig. 4.2 Pathways linking mechanical force at the cell surface to intracellular signalling and downstream effectors. *Left*: Examples of processes and proteins involved in each step. *Right*: External force is sensed by mechanosensors in the plasma membrane, which is linked to extra- and intracellular adaptors that transmit mechanical signals to targets inside the cell. These pathways may exhibit positive/negative feedback regulation to all these steps. These mechanotransduction processes result in modulations of protein expression and cellular functions, including cell migration, proliferation and differentiation (The diagram has been modified from Pruitt et al. [14])

as dynamic coupling of the inner ear stereociliary (hair) bundle and MS channel complex which underlies hearing [12, 13].

Cells are exposed to various types of mechanical stimuli, which can be categorised as (i) extracellular: such as shear forces stemming from fluid flow, tensile/traction forces from osmotic forces or through the extracellular matrix (ECM) complex, (ii) intercellular: through cell-cell junctions or mechanosensitive ion channels, (iii) intracellular: contractile forces generated by cytoskeleton (actomyosin contraction, microtubule etc). Mechanosensor-driven pathways often involve positive/negative feedback regulation in response to different cell-type specific stimuli (Fig. 4.2) [14]. Cells can exhibit different behaviors such as stress stiffening (reinforcement), fluidization and rejuvenation through cytoskeletal reorganization in response to different mechanical stimuli (both magnitude and type) and thus transmission/amplification or dissipation of mechanical forces [15]. This could be a mechanism for several cellular processes such as cell migration or structural remodeling in arteries in response to high blood pressure [16, 17]. Note that the basic flowchart shown in Fig. 4.2 does not show some of the auxiliary and secondary mechanosensing elements and features such as protein-protein interactions or membrane polarization.

Among the aforementioned mechanosensors, MS ion channels are the fastest in living cells by responding to mechanical stimuli on a less than millisecond time scale [5, 18]. They can transduce the mechanical force directly into electrochemical signals coupling the cellular machinery within a signaling pathway. For example, Ca^{2+} influx through MS channel activation may be coupled to gene regulation through a calcium/calmodulin-dependent pathway [19]. Whereas mechanotransduction through other cellular mechanotransducers occur over longer time scales (e.g., $\sim$1 s), such as Src activation in response to localized mechanical stress [20]. That is why as the fastest mechanotransducers, MS channels are ubiquitously present in rapid mechanotransduction loci such as in somatosensory neurons.

4.3 Mechanosensitive (MS) Ion Channels

MS ion channels are pore-forming membrane proteins that gate in response to mechanical force and allow ions to flow across the cell membrane. Over the last three decades, interest in MS channels has progressed from being seen as an artifact of patch-clamp recording [21] to becoming a central player in our understanding of a number of physiological processes. These channels have been shown to play a key role in osmoregulation (in plants and bacteria), hearing, touch sensation, proprioception, information processing in the human brain as well as cell volume and blood flow regulation in mammals [5, 6, 22–25].

Given the ubiquity of mechanical force acting at the cell membrane interface, the function of many integral membrane proteins can be modulated by the application of mechanical force [26–28], not just MS channels. For example, membrane tension has also been known to modulate various voltage-, ligand-, or Ca^{2+}-gated channels [29, 30]. To demonstrate the degree of sensitivity to mechanical force, a continuum landscape can be assumed on which all channels reside from highly sensitive (e.g., TREK-1) to almost completely insensitive (TWIK-1) (Fig. 4.3) [5, 31].

MS channels can be gated in different ways by mechanical stimuli, which can be divided into three physically different stress types, stretch-compression, shear and bending [35, 36]. For example, one channel may be more sensitive to membrane stretch compared to curvature and *vice versa,* this is the case for *E. coli* (Ec)MscL versus *M. tuberculosis* (Mt)MscL [37]. Consequently, what is the definition of a mechanosensitive channel? The presently used criteria for defining channel mechanosensitivity are as follows [9, 38]:

(i) *Localization:* The channel must be broadly expressed in the relevant specialized mechanosensory cells and localized at the correct position within the cell (e.g., Piezo2 in Merkel cells).
(ii) *Function:* The channel must be the primary mechanosensor that elicits the initial electrical response of the sensory cell. It is best to be able to modulate channel function (e.g., activation and inhibition) by introducing genetic mutations.
(iii) *'Mimicry':* When the putative MS channel is expressed in heterologous expression systems, in culture or reconstituted in lipid bilayers, the channel current

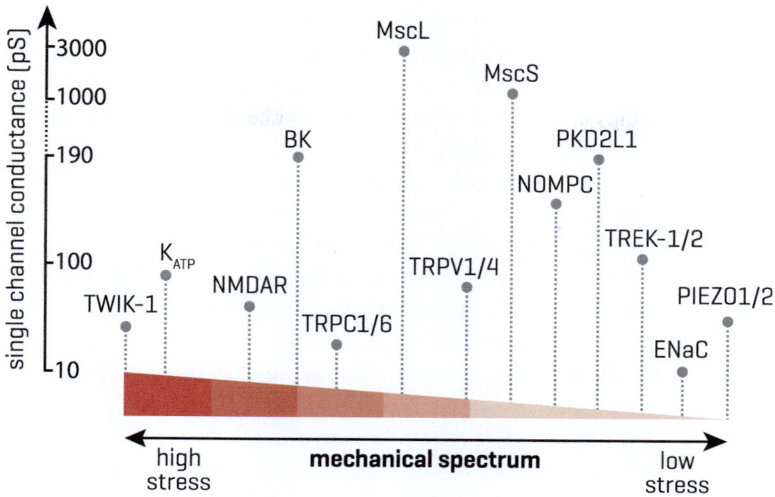

Fig. 4.3 Ion channels sensitivity to mechanical force probed by the patch-clamp technique [32]. All ion channels can be placed on a mechanical continuum from highly sensitive (e.g., Piezo1) to those that are almost insensitive to mechanical force (e.g., TWIK-1). They cover a broad conductance spectrum ranging from tens of pS (e.g., ENaC) to 3 nS (e.g., MscL). Note that several factors such as presence or absence of extracellular or intracellular network and experimental paradigm (i.e., stimulus type) may shift channels along the spectrum (see [32–34]; modified from Cox et al. 2016 [32])

should largely recapitulate the properties of the native channel current. Such properties include conductance, kinetics, activation by agonists, inhibition by antagonists, and in particular ion selectivity. It is possible that some of the channel characteristics such as activation threshold are modulated by the change in the physico-chemical properties of the cell membrane.

(iv) *Mechanosensitivity:* mechanical force (alone) should activate the channel under the similar conditions used to determine 'mimicry'.

Note that very few channels fulfil all these criteria.

To date, a large number of MS channels from organisms of diverse phylogenetic origins have been identified [39]. In prokaryotes, there are two main families, MscL-like and MscS-like channels. The mechanosensitive ion channel of large conductance MscL and mechanosensitive ion channel of small conductance MscS of *E. coli* are the prototypes of two very large and diverse families of MS channels. MscL-like channels are mostly found in bacterial and archaeal cells, although representatives have been identified in the genomes of some fungi (e.g., Neurospora), fungus-like organisms (Phytophthora) and Mycoplasma [40]. The MscS-like family members are found in cell-walled eukaryotes such as fungi, algae, and plants [5]. In *E. coli* there are seven MS channels (one MscL-like and six MscS-like channels) which can be distinguished based on their conductance, selectivity and sensitivity to membrane tension. The three main conductances encountered are

MscM (M for mini), MscS (S for small), and MscL (L for large) [1, 41]. MscM (a member of MscS-like family) has a conductance of ~200–370 pS and is cation selective [1, 41–43]; MscS conductance is ~1 nS and weakly anion selective; MscL conductance is ~3 nS and nonselective. MscS-like channels are activated by lower membrane tensions than MscL channels (Fig. 4.3).

To date, four major membrane protein families of MS channels have been identified in animals. These are transient receptor potential (TRP), 2 pore domain K^+ (K_{2P}), the epithelial Na^+ (DEG/ENaC) and Piezo channels [5, 9, 32, 44–49]. In mammals, MS channels are involved in touch and sound sensations as well as blood pressure regulation. Hence, defects in MS channels are associated with several pathologies including both hereditary (e.g., skeletal dysplasia and stomatocytosis) and multifactorial human diseases (e.g. cardiac hypertrophy and arrhythmias) [50–53].

The structures of MS channels are very diverse and there is little to no structural identity among different families of MS channels hence they differentially interact with their surrounding (i.e. lipids, extracellular matrix and cytoskeleton). Among all known MS channels, structural information regarding the molecular architecture is available for *M. tuberculosis* MscL (PDB: 2OAR), *E. coli* MscS (2OAU), TRAAK (3UM7), TREK-2 (4XDJ) and Piezo1 (3JAC) which have been resolved by either X-ray crystallography or cryo-electron microscopy (cryo-EM). It is noted that structures of several MscL and MscS homologs have been resolved. MscL forms a homopentameric channel, while MscS and TRAAK form homoheptameric and homodimeric channels, respectively. In addition, an increasing number of studies have suggested that TRPV1 (2PNN) and TRPV2 (5AN8) are involved in osmosensation [46, 54, 55].

4.4 Gating Paradigms of Mechanosensitive Channels

Despite knowing the structure and function of many MS channels, the exact details of their gating mechanism(s) are unknown. However, it is widely assumed that MS channels obey one or both of two main gating paradigms: (i) *force-from-lipids* and/or (ii) *force-from-filament*.

4.4.1 Force-from-Lipids

The *force-from-lipid principle* was established about three decades ago from the early studies of the MS channels from *E. coli* [56–58]. The force-from-lipids concept (bilayer model) [1, 29, 56, 59, 60] purports that force is directly transferred through the lipid bilayer onto the embedded MS channel (Fig. 4.4a) [56, 61, 62]. Prokaryotic channels MscL and MscS reconstituted into lipid bilayers retained their

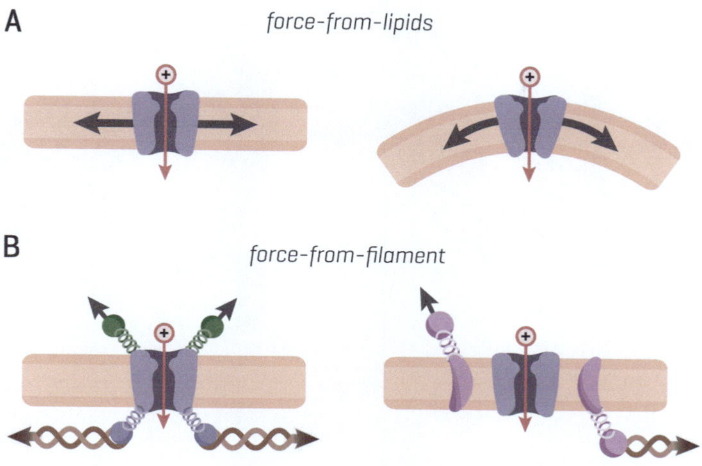

Fig. 4.4 Gating mechanisms of MS channels. (**a**) force-from lipids (bilayer) paradigm. (**b**) force-from-filament (tether or swing) mechanism [32] (*black arrows* indicate force direction)

mechanosensitivity similar to their activity in the native bacterial membrane [61, 63]. This is the key proof for the force-from-lipids being sufficient for activation and modulation of MS channels.

Since prokaryotic cells largely lack an animal-cell-like cytoskeleton [58], it is not surprising that all prokaryotic MS channels are believed to obey the force-from-lipids principle. This however, does not preclude the possibility of interaction of bacterial MS channels with their cytoplasmic components, such as interaction of MscS with cell division protein FtsZ [64]. Moreover, eukaryotic MS channels may not follow this paradigm just because they have cytoskeletal and/or extracellular components [38]. In fact inspired by the pioneering studies on prokaryotic MS channels [63, 65, 66], this paradigm has been rigorously extended to some eukaryotic MS ion channels, most notably TREK-1, TRAAK [31, 67–69] and Piezo1 [70, 71]. Through reconstitution of purified channel proteins into lipid bilayers, it has been shown that like bacterial MS channels, Piezo1 and MS members of the K_{2P} channel family are also inherently mechanosensitive (i.e., gated by mechanical force according to the force-from-lipids principle). Significantly, this implies evolutionary conservation of the basic physical principles of mechanotransduction in living cells from bacteria to humans.

This is particularly important given that Piezo channels have been linked to a number of physiological processes. Piezo1 plays a central role in vascular development [73, 74] and remodeling [75], erythrocyte volume homeostasis [76], endothelial ‘quorum sensing’ [77, 78], urothelium signaling [79], blood pressure control [80], lymphatic vessel development [81], neural stem cell differentiation [82] and axonal pathfinding [83]. Central to these functions is the ability of the MS channel to transduce mechanical force into cellular signals. How the channel

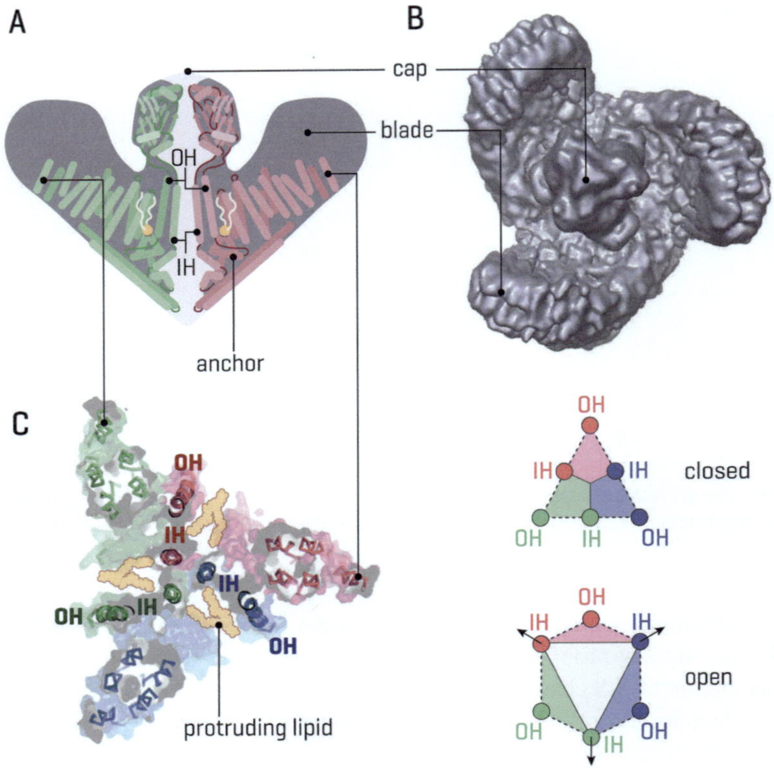

Fig. 4.5 Mammalian mechanosensitive Piezo1 architecture and a putative membrane-mediated gating mechanism. (**a**) Schematic of the side view of Piezo1 structure. (**b**) *Top view* of Cryo-EM structure of mouse Piezo1 as shown in *shaded grey* surface (PDB: 3JAC) [72]. (**c**) View from the *top* of the human Piezo1 (homology model based on mouse Piezo1) shows the interlocked arrangement of its 3 subunits at the level of the hydrophobic core of the lipid bilayer. An increase in lateral bilayer tension is thought to result in a clockwise or counter-clockwise deflection of the 'Blade' domains around the 'Anchor' and outer helix (OH) domains. This movement ultimately results in the displacement of the inner helices (IH) away from the center of the pore to allow ion conduction, as shown in the diagram. This hypothesis aims to explain the intrinsic mechanosensitivity of the channel different from the Blade-deflection model proposed by Ge et al. (2015) [72]

does this at the molecular level is an open question. In the recently published medium resolution cryo-EM structure of the mouse Piezo1 homolog [72], the authors speculated that the large blade domains must play a major role in the gating cycle of the channel (Fig. 4.5). These large structures are suggested to bend towards the plasma membrane in response to shear flow and trigger the cascade of conformational changes necessary for pore opening. Recent work by Syeda et al. in reconstituted droplet interface bilayers [70] and experiments on membrane blebs [71] demonstrated how membrane tension alone is sufficient to open the channel. Whether shear stress and membrane tension instigate different conformational pathways to channel activation is unknown. However, it seems likely that there is a concerted lateral deflection of the transmembrane helices around the central pore in an iris-like fashion, similar to the bacterial channel MscL. Indeed, regions in this area are likely to come into close proximity with lipids (Fig. 4.5c), including the anchor domain which cross-links with cholesterol [84], a known modulator of Piezo1 function [85].

Piezo1 is one of the most sensitive eukaryotic mechanotransducers identified and can be activated by tensions much less than 1 mN/m in a variety of cellular systems (Fig. 4.3). The channel is also characterized by fast inactivation, which might involve the downward movement of the CED domain to block the pore, in a mechanism reminiscent of the classical ball and chain model of N-type inactivation [86, 87]. The loss of inactivation in cell-attached patches means there may also be a component linked to force propagation and cell mechanics. Alterations in the inactivation rate of the channel underlie the pathology of many hereditary Piezo-linked diseases.

Despite the fact that lipids are sufficient to gate MS channels such as TREK-1, TRAAK and Piezo1, their sensitivity is largely influenced by the cytoskeleton and extracellular matrix [88, 89]. This can take the form of shielding the channel from the applied stress ('mechanoprotection' [75, 90]) or sensitizing the channel by focusing the force directly to the channel's locality. In studies of cell mechanics, it is assumed that the membrane shares a tiny fraction of any applied mechanical load and that the cytoskeleton dissipates the resulting stress. Defining what component is transmitted to the membrane is not a trivial task but is essential if we are to understand whether membrane tension is a physiologically relevant gating stimulus.

It is indeed obvious 'that one size does not fit all' in mechanobiology. For example, while Piezo1 can be robustly activated in cell-attached patches by membrane tension this is not the case for Piezo2, which perhaps intimates a larger reliance on tethers. Despite these differences both Piezo1 and Piezo2 are physiologically essential with global knockouts resulting in lethality in mice. In addition, a number of conditional knockouts are also lethal (Piezo1 in vascular endothelium [74], Piezo2 in sensory neurons that innervate the lung [91]).

The interaction between the membrane and structural scaffold proteins indicates that it is not a straight forward task to examine the gating mechanism of MS channels. A case in point is TRPV4, which is involved in a number of mechanosensory processes such as volume regulation, osmosensing and possibly nociception [5, 50, 92–94]. Despite being definitively linked to mechanotransduction, how mechanical force activates this channel is still unclear. Although recordings of TRPV4 currents from oocytes indicate a direct link with the lipid bilayer [45, 95], other work suggests more reliance on tethers [96].

None of the above exclusively precludes the possibility of these MS channels being activated by tether-like molecules it simply shows the evolutionary conservation of this gating modality (force-from-lipids) from prokaryotes to eukaryotes. It is likely that MS channels adopt more than one gating mechanism depending on their physiological role. Furthermore, energetic calculations of the gating free energy shows that the two mechanisms are not mutually exclusive [97], albeit one mechanism may be more efficient than another in particular circumstances, depending on the channel type and their specific physiological roles.

4.4.2 Force-from-Filament

The second gating paradigm is the *force-from-filament* (also called tethered model or gating by spring model), whereby force is transmitted to a channel via an auxiliary structural element (tether) from the intra- or extracellular environment (Fig. 4.4b) [1, 9, 38].

The archetypal example is the MS channel complex responsible for auditory transduction, which is believed to follow the force-from-filament principle [98–102]. Cadherin 23 and protocadherin 15 are the main components of the "tip-links" which connect the stereocilia, and they couple hair bundle movement to MS channel activation during hearing or head position [12, 103]. However, despite increasing efforts, the precise pore-forming subunit of the channel complex remains unknown [32–34, 104]. For more information on hearing mechanotransduction and their relevance to force-from-filament principle, readers are referred to the following articles [13, 33, 103, 105, 106].

So far only NompC, a member of the TRPN channel family, which is involved in various mechanosensitive responses in *Drosophila* including gentle touch in larvae [107, 108], has been shown to follow the force-from-filament gating model. NompC is present at the dendritic tips of mechanosensory neurons in the *Drosophila* (halteres), where it converts forces transmitted through the deformation of the overlying cuticle into neuronal depolarization. This channel possesses 29 ankyrin repeats in its N-terminus, which directly associate with microtubules. It has been shown that this channel can be activated by mechanical force pulling (imitation of gentle touch) on the ankyrin repeats acting as a tether [47]. Although pulling force is directly transmitted from the tether to the protein, it does not necessarily exclude the potential role of membrane lipids in NompC gating.

Similar to TRPN, it has been shown that the activation of TRPV1 channel occurs largely according to the force-from-filament principle via its ankyrin repeats directly interacting with the microtubules [44]. However, recently a cryo-EM structure of TRPV1 has been revealed in a native bilayer using lipid nanodiscs [109]. Using a toxin activator from the Chinese bird spider, it has been shown that the toxin dislodges a lipid from a protein pocket enabling a structural transition to the open state. They speculated that entropy driven removal of lipids from this region by heat may also underlie TRPV1 heat induced gating. This highlights the role of the lipid environment in modulating the gating of some "filament-gated" TRPs.

Furthermore, members of the DEG/ENaC family (i.e., MEC-4/MEC-10), which convey gentle touch in *C. elegans* and regulatory volume increases in brain cells [9, 110–113], are believed to be activated by force-from-filament mechanism [39]. Nevertheless, it has been clearly shown that the presence of cholesterol is relevant for the function of these channels, indicating the importance of the lipid bilayer in force transmission to these channels [39, 114]. Similarly, the role of the bilayer has been investigated for TRPA1, another putative "filament-gated" channel expressed in peripheral sensory neurons [115]. Gating of TRPA1 can be robustly modulated by asymmetric insertion of amphipathic molecules into bilayer leaflets (similar to studies on bacterial MS channels [56, 116]). Consequently, the role of the lipid bilayer even in mechanosensing by "tethered channels" cannot be ignored.

4.5 Mechanical Force Alters MS Channels Conformational Equilibrium

MS channels exist in a conformational equilibrium where different states are populated according to their relative energies. For a system in equilibrium in a canonical ensemble, the probability of the system being in state with energy G is proportional to exp. $(-G/k_BT)$ [117]. Consider a channel in equilibrium between, for example, closed A and open B conformations, named nA and nB respectively. Then,

$$\frac{n_B}{n_A} = \exp\left(-\frac{\Delta G}{k_B T}\right) \tag{4.1}$$

ΔG is the free energy difference between the states, where k_B is Boltzmann's constant and T is absolute temperature. External mechanical force shifts the equilibrium among pre-existing states by the amount of work done on the system W. This leads to a new equilibrium state,

$$\frac{n_B}{n_A} = \exp\left(\frac{-\Delta G + W}{k_B T}\right) \tag{4.2}$$

Numerous forms of forces may act on MS channels with their specific membrane environments and their structure dictating their sensitivity to these forces [32]. The mechanical work W can be transferred to a channel by the lipid bilayer, intra- or extra cellular structures or a combination of both. The mechanical stimuli can result in three distinct types of stress, which are tension/compression, shear and bending (Fig. 4.6).

In thermodynamics, the free energy of a system, or work done on a system, is expressed in terms of pairs of conjugate variables such as membrane tension and areal expansion (Eq. 4.3) or force and displacement (Eq. 4.4). In fact, all thermodynamic potentials are expressed in terms of conjugate pairs. If the bilayer is stretched uniformly, then

$$W = -\tau \Delta A^{A \to B} \tag{4.3}$$

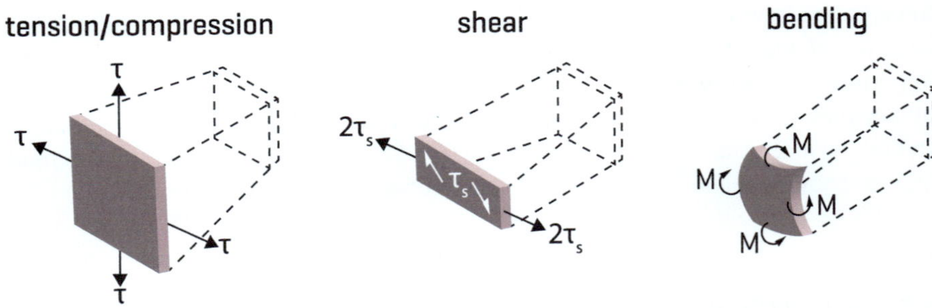

Fig. 4.6 Active physical forces on a membrane patch. Stress is defined as force per unit area and has three basic forms of tension (stretch) τ, shear τ_s, and bending M, respectively. These quantities allow characterization of the mechanical response of materials. Channels may deform differently under tensile/compressive, shear and bending forces

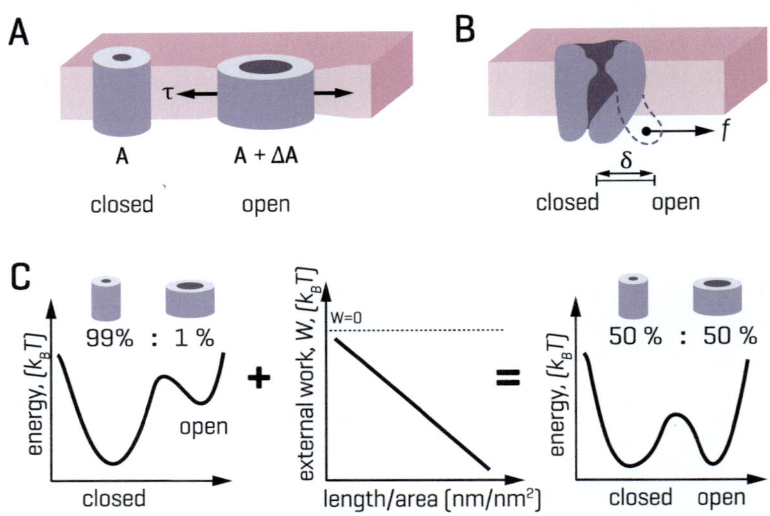

Fig. 4.7 Mechanical force affects channel conformational equilibrium [14]. Schematic illustrating MS channel gating due to the external work done by (**a**) force-from-lipids or (**b**) force-from-filament [120, 121]. (**c**) External work (done by membrane tension) overcomes the energy barrier between the closed and open state of the channel, thus increasing the channel's open probability

with τ being the membrane tension and $\Delta A^{A\rightarrow B}$ being area change of the protein in response to membrane tension (Fig. 4.7a). How specific membrane deformations are linked to MS channel activity is described in Sect. 4.7.

If the force f is directly conveyed from an intra or extra cellular structure to the MS channel, and causes a conformational distance δ (Fig. 4.7b) then,

$$W = -f\delta^{A\rightarrow B} \tag{4.4}$$

The latter model is called the ‘swing model’, ‘force-from-filament’ or ‘gating by linear force’. External force in form of a moment or shear force may also activate the channel. In this case, the external work for a linear bending moment is:

$$\boldsymbol{W} = -\boldsymbol{M}\boldsymbol{\theta}^{\boldsymbol{A}\rightarrow\boldsymbol{B}} \tag{4.5}$$

where M is bending moment and θ could be, for example, the degree of out-of-bilayer plane rotation of a structurally crucial transmembrane helix (e.g., pore lining helix). Finally, when the stress is caused by a pure linear shear (torsional) force S, then

$$\boldsymbol{W} = -\boldsymbol{S}\boldsymbol{\varphi}^{\boldsymbol{A}\rightarrow\boldsymbol{B}} \tag{4.6}$$

where φ is the degree of rotation of a structurally crucial transmembrane helix within the bilayer plane and S is the shear force.

Based on Eq. 4.2, the likelihood of seeing channel activity increases if mechanical force applied (i.e. work is done on the system, W) is sufficient to overcome a certain energy threshold (i.e. gating energy barrier) (Fig. 4.7c). Examples of gating energy barriers for MS channels are; TREK-1 is ~5 k_BT [118] and MscL is ~20 k_BT [1, 119]. Given that one k_BT is equal to ~4.2 pN·nm at physiological temperature, this means pN forces acting over nm distances are sufficient to meaningfully shift the equilibrium between closed and open conformations of an MS channel.

4.6 Importance of Cell Membrane Mechanics

Cell membranes have been experiencing mechanical force throughout evolution [122]. Thus, they have been reinforced by a cell wall (e.g. in bacteria, fungi, algae and plants) or by a cytoskeleton and extracellular matrix (in animal and human cells) to better withstand and cope with environmental stresses [123].

Studying the mechanics of the cell membrane is necessary for understanding: (i) how cells are protected by their membrane, (ii) how cells interact with each other and/or move (cell migration), and (iii) how membrane and proteins (e.g. MS channels) interact in various biological processes.

In addition, assessing cellular elasticity and viscosity provides useful information for comparative characterization between different membrane-mediated processes [15]. This is given that genetic mutations and pathogens that disrupt the cytoskeletal architecture can result in changes in cell mechanical properties such as elasticity, adhesiveness, and viscosity [15, 124].

The choice of experimental and computational tools to measure cell and bilayer mechanical properties requires consideration of the length scale, timescale and the magnitude of forces (or intrinsic mechanical properties of the system) [125]. It has been indicated that some of the cellular rheological behaviors (stress-strain distribution) are empirically similar to the rheology of soft materials such as foams, emulsions, pastes, and slurries [15, 126]. Given the central role of lipids in mechanosensation [32, 127, 128] a detailed understanding of lipid bilayer rheometry is essential [129]. The key question here is how do mechanical stimuli tune the biochemical and biophysical properties of lipids and protein molecules or *vice versa*? This question is of critical importance due to the increasing evidence of the role that MS channels play in health and disease [5, 122].

4.7 Mechanical Coupling Between the Cell Membrane and MS Channel

The function of MS channels can be modulated by a host of mechanical properties of lipids [29]. This modulation may depend on specific chemical interactions between proteins and individual molecules in the bilayer (at the lipid-protein interface). This includes affinity and/or avidity of certain lipid molecules to the lipid-protein interface which has direct influence on the structure and function of the membrane proteins [130]. Different proteins, however, demonstrate different degrees of selectivity in their binding to lipids. This has recently been studied using an elegant approach called ion mobility mass-spectrometry [130]. In that study, the lipid binding selectivity and strength to MscL channel, aquaporin Z (AqpZ), and the ammonia transporter (AmtB) have been investigated. They showed that the degree of selectivity for specific lipid types for these proteins followed the order AmtB > AqpZ > MscL. MscL binds lipids non-selectively, without strong affinity for a particular headgroup or chain length, but binding of any lipid imparts considerable stability. Although the *E. coli* MscL responds almost non-selectively to lipid composition, its homolog MtMscL has been shown to have a slight preference for phosphatidylinositol (PI) binding and a remarkably lower activation threshold in the presence of PI [130, 131]. Given that all the MS channels are embedded in the lipid medium, it is possible that this difference in their local lipid environment is among the reasons for the existing difference between their activation thresholds [37, 131]. AqpZ and AmtB on the other hand have been shown to be highly selective for cardiolipin and phosphatidylglycerol, respectively [130].

Furthermore, the lipid bilayer affects MS channels via its macroscopic biophysical properties such as area and bending elastic moduli, intrinsic monolayer curvature and thickness. From this perspective, variation in acyl chain length, degree and position of chain unsaturation, head group repulsion, incorporation of amphipathic molecules and interactions of co-surfactants all result in a redistribution of the lateral pressure profile of the lipid bilayer [132–135]. Therefore, specific lipids may affect both nano- and macroscopic rheological factors that likely modulate MS channel function, as it has already been shown for MscL and MscS co-reconstituted in liposomes of different lipid composition [136].

4.7.1 Transbilayer Pressure Profile, Surface Tension, Membrane Tension

Lipid bilayers, as a unique medium for membrane proteins, contain water molecules which strongly influence their structural and functional properties, as well as the embedded membrane proteins. Due to the "hydrophobic effect", the amphipathic nature of the lipids drives membrane self-assembly by minimizing the surface exposure of the tails to water. This feature of the lipid bilayer comes with a set of properties including its strong anisotropic internal stress, called "transbilayer pressure profile", which is susceptible to physical and chemical stimuli [132]. The pressure profile may be simplified into three zones: i) a region with positive lateral pressure caused by the repulsion among the hydrophilic headgroups, ii) a surface tension area (negative pressure) at the level of the glycerol backbone, which prevents water exposure of the hydrophobic tails and iii) an area characterized by positive pressure due to the entropic (steric) repulsion between the lipid tails (Fig. 4.8a). In the bilayer, the chain fluctuations are strongly suppressed by the neighboring chains with respect to a free lipid molecule. The amplitude of chain fluctuations in the bilayer under natural conditions (Fig. 4.8a) is attenuated by one order of magnitude with respect to a free chain [137, 138]. Suppression of chain fluctuations leads to a decrease of conformational entropy and, in turn, results in the enhancement of pressure.

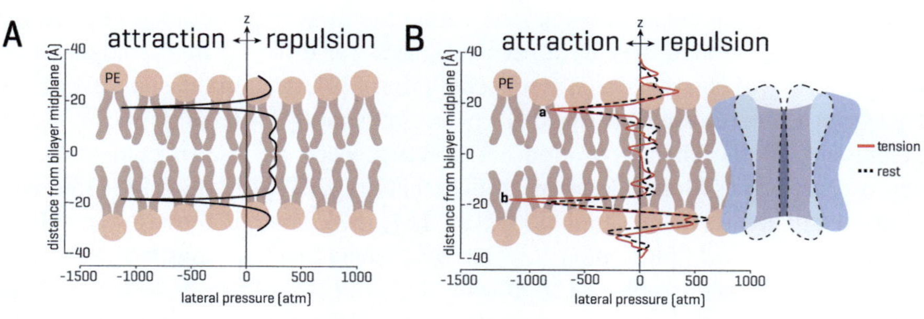

Fig. 4.8 The transbilayer pressure profile. The transbilayer pressure profile is largely inhomogeneous across the bilayer thickness. This stress heterogeneity in bilayers originates from the amphipathic nature of the lipid molecules and the presence of water without MscL. (**a**) An idealized symmetrical transbilayer pressure profile showing characteristic negative peaks at the water-lipid interface and repulsive positive peaks in the headgroup and tail region. z indicates the bilayer thickness direction. (**b**) Transbilayer pressure profiles from molecular dynamics simulation of a POPE bilayer in presence of *E. coli* MscL (*black dashed* line is the lateral pressure profile at rest, and *red* is in the presence of applied tension). Note how in the presence of the protein the pressure profile in a symmetrical lipid bilayer has become noticeably asymmetric. Peak **a** and peak **b** represent the rise in the pressure profile at lipid solvent interface (for more details see Bavi, N. et al. [141]; modified from Cox et al. [32])

To avoid water entry into the hydrophobic tail area, an acute lateral "surface tension" develops at the water-lipid interior interface. A typical surface tension in each monolayer at rest is estimated to be 50 mN/m [139]. This way, a balance is established between the tension generated at the water-lipid interface and the repelling pressure between the tails in "free bilayers". A typical value for the peak pressure inside the lipid bilayer is about 300 atm whereas around the head groups it can be up to 1000 atm [139, 140]. This stress heterogeneity along the bilayer thickness alters the conformation of membrane proteins [28]. External membrane stretch can alter force vectors on embedded proteins and cause their conformation to change [26, 139, 141, 142].

Membrane proteins and MS channels in particular can reciprocally redistribute the transbilayer pressure profile [133, 140, 143]. The information on the MS channel-lipid interactions has largely been acquired using MD simulations [139, 141, 144, 145]. Figure 4.8b illustrates the changes in the pressure profile of a 1-palmitoyl-2-oleoyl-sn-glycero-3-phosphoethanolamine (POPE) lipid bilayer in the presence of *E. coli* MscL, with and without increase in surface tension [141]. Due to presence of MscL in the bilayer, there is a noticeable asymmetry in the lateral pressure profile and a reduction of the pressure peaks in the tails. As shown in Fig. 4.8b, when the bilayer is stressed (surface tension = 75 mN/m), the pressure profile of the lipid in the presence of MscL shows most change at the lipid-water interface (Peak A and Peak B). A comparison between a stressed and a non-stressed bilayer (surface tension$\cong$50 mN/m) in conjunction with the area under the pressure profile, one can estimate the equivalent "membrane tension" of $\sim$22 mN/m required for the activation of MscL, for example [141].

Overall, stress/pressure distributions within the bilayer are difficult to measure experimentally, though they have been estimated for different lipid compositions using a variety of computational approaches [133, 139]. In addition, some estimates have been achieved using NMR and X-ray crystallography [146, 147]. The mainstay of pressure profile characterization has however, been the use of computational approaches such as Monte Carlo, mean-field theory (MFT) [132, 134, 148–150] and molecular dynamics (MD) simulations [139, 151, 152]. This has allowed researchers to probe the effect of lipid composition on MS channel function using both atomistic [139, 153, 154], and continuum approaches [27, 129, 155–160]. Nevertheless, one should be careful about the common methodological problems in pressure profile calculations for lipid bilayers using these methods [135, 161].

4.8 Methods to Study Mechanosensitive Ion Channels

Over the past three decades, various theoretical and experimental approaches have been employed to study structure and function of ion channels. The main theoretical (MD and continuum approaches) and new experimental techniques developed for MS channel characterization are discussed in 4.8.1 and 4.8.2.

4.8.1 Computational Approaches

MD simulations are computer based estimations of interactions between molecules. These interactions are calculated by solving the Newtonian equations of motion. The interacting forces and masses are defined in a force field and used to calculate the new acceleration vector of a molecule or atom, typically for intervals of 2 picoseconds. This often requires simplifications to be imposed on the models; nevertheless results from MD simulations are often in good agreement with experimental data. For more information about the equations used and the approximations made, the NAMD or GROMACS manuals provide the essential background information [162–164]. Furthermore, several recent review articles provide excellent instructions on how to setup and run MD simulations of membrane protein systems [165–168].

Most computational studies have been focused on examining the membrane protein, leaving the membrane as an addendum [169]. As such, with MD simulations of bacterial and mammalian MS channels, it is common to use a membrane model consisting of only 1-palmitoyl-2-oleoyl-sn-glycero-3-phosphocholine (POPC) or POPE [139, 141, 154, 165, 166, 170] and in some cases in combination with 1-palmitoyl-2-oleoyl-sn-glycero-3-phosphoglycerol (POPG) lipids [37, 127]. However, it is important to note that the composition of plasma membranes varies not only among cell types but also both spatially (e.g., leaflet asymmetry and lipid rafts) and temporally (e.g., metabolic state and growth phase) within a single cell. Therefore, it is likely that membrane proteins, in particular MS channels, may respond differently in various lipid environments as discussed in previous sections. Hence, it is essential to develop an accurate representation of the plasma membrane, e.g. an *E. coli* membrane for use in further simulation studies of ion channels. Complex membranes consisting of a more diverse lipid population have been developed to model *Chlamydia*, yeast, *E. coli* membranes and mammalian membranes [169, 171–173]. These systems provide the opportunity to perform more accurate simulations which can further our understanding of the role of lipid-protein interactions during the gating cycle of MS channels.

Continuum and hybrid (MD-continuum) approaches are usually adopted in mechanobiology for two main reasons: i) they are not usually computationally expensive, and ii) it is more straight forward to describe and interpret the parametric dependence of a quantity (e.g., free energy) on its possible variable(s) (e.g., membrane thickness). However, this is usually done at the expense of atomistic detail, which may be overlooked in modeling by using only continuum approaches. Nevertheless, there are several applications of continuum mechanics approaches to the study of MS channels [156, 157, 174, 175]. In hybrid techniques, atomistic calculations of energy and/or force (e.g., obtained from MD simulation) can be implemented to mesoscopic (e.g., bead-spring) and continuum models (e.g., finite element models [176, 177]) of channels to increase the accuracy of their prediction [174]. Most of these models used MscL as a prototypical channel because of the relative wealth of experimental data available on the structure and function of MscL compared to other MS channels [155, 157, 177, 178]. This combined computational approach promises to have a greater predictive value for future experimental work compared to currently used individual approaches.

4.8.2 Experimental Approaches

To detect ionic currents from MS channels in response to mechanical force, several experimental methods have been developed that apply various mechanical stimuli to cells. All these methods are either coupled with patch-clamp electrophysiology techniques or calcium (Ca^{2+}) imaging assays. Patch-clamp electrophysiology is the most versatile approach and "gold standard" technique for functional studies of ion channels, as it provides quantitative information on the relationship between membrane tension and channel open probability [179, 180]. Calcium imaging assays are widely utilized to optically probe intracellular calcium flux *in vivo* and *ex vivo* by using various Ca^{2+}-sensitive indicators.

In a usual patch-clamp experiment, the traditional approach is to generate force on a membrane patch through a syringe or via a computer-controlled high-speed pressure clamp. After the formation of a tight seal ("giga-seal") between the membrane and the micropipette [1, 181, 182], the membrane patch can be stretched by generating a pressure difference across the patch (Fig. 4.9a). This membrane manipulation technique was essential for the discovery of mechanosensitive (MS) channels [183–185]. The key equation to calculate mechanical force in patch-clamp electrophysiology experiments is the Young-Laplace equation.

$$\tau = pr/2 \tag{4.7}$$

where τ is membrane tension; p is pressure difference (negative or positive); r is radius of patch curvature [1]. In this setup, cells can be patched in different configurations such as cell-attached, whole-cell and excised patch modes. The rheometry of membrane patches in different configurations are not similar and care should be taken when comparing between configurations [129]. The choice of patch configuration usually depends on the number of channels (i.e., single channel or cohort current) [185, 186] and on the accessibility to either side of a patch (intra- or extracellular).

Alternatively, in whole-cell recordings, osmotic pressure gradients can be applied across the cell membrane (Fig. 4.9b) by changing the osmolarity of the extracellular medium. Under hypoosmotic conditions, the osmotic difference causes cell swelling due to net influx of water, which in turn leads to membrane (and/or cytoskeletal network) stretch and MS channel activation [44, 187–192]. Osmotic pressure Π can be estimated using the Van't Hoff equation as follows;

$$\Pi = cRT \tag{4.8}$$

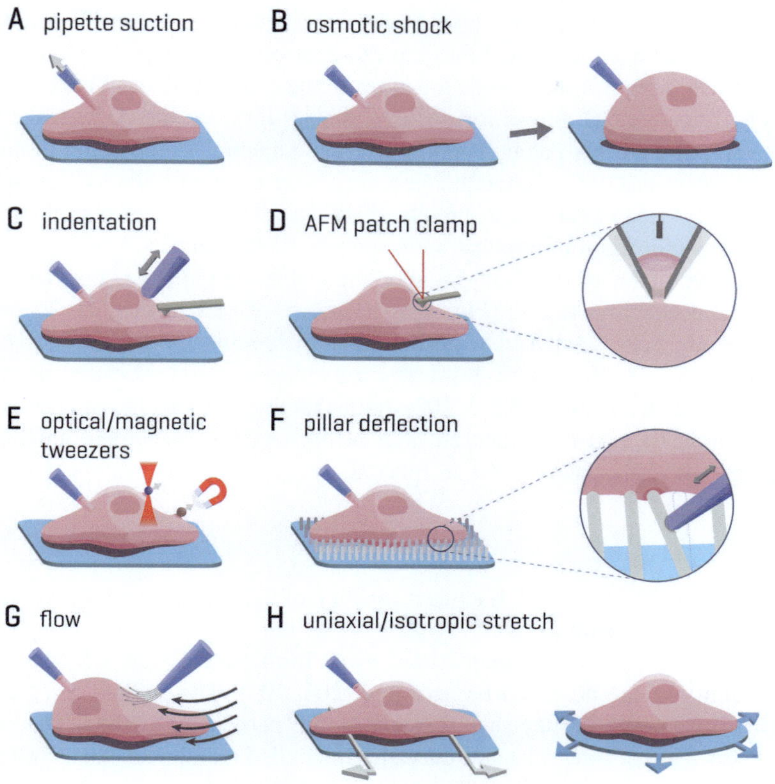

Fig. 4.9 Experimental techniques for the study of mechanosensitive ion channels. Schematics of (**a**) High speed pressure clamp (HSPC) experiment (*stretch*), (**b**) hypo-osmotic shock assay, (**c**) microindentation by a rounded glass or by an AFM tip, (**d**) AFM patch-clamp electrophysiology, (**e**) optical/magnetic tweezers, (**f**) pillar (*deflection*) arrays, (**g**) normal and shear flow, (**h**) uniaxial (*left*) and isotropic stretcher (*right*)

where c is the molar concentration of solutes; R is the gas constant; and T is absolute temperature. Interested readers should refer to Jiang and Sun [193] for detailed equations linking the open probability of MS channels to the magnitude of osmotic pressure on the cell.

Whole-cell patches can also be 'poked' by an indenter to provoke MS channel current (Fig. 4.9c). A microindenter can either be an AFM tip or a rounded glass tip that is usually controlled by a piezoelectric device. Ideally, the relationship between the applied force F and the indentation depth δ is:

$$\boldsymbol{F} = \frac{4}{3}\frac{\boldsymbol{E}}{1-\boldsymbol{\nu}^2}\boldsymbol{R}^{0.5}\boldsymbol{\delta}^{1.5} \tag{4.9}$$

where ν is the Poisson's ratio; E is the cell Young's modulus; and R the radius of the spherical indenter. In addition to cell indentation, AFM tips can also be used to apply pulling/stretching force on the cell (tether assay) [194–196]. Furthermore, the AFM tip has been converted into a piezo-controlled nanopipette which can be used to detect current from specific locations in the cell (Fig. 4.9d). Feedback from the AFM force control allows for a stable contact between the AFM tip and the cell membrane enabling patch clamping and nanoinjection without compromising the cell's integrity [197]. Although some initial successes in utilizing this method have been reported [197], the technique is still in its infancy and requires further improvements for rigorous application in studies of MS channels.

Alternatively, the local pulling force on the cell membrane can be generated by using optical tweezers (traps) (Fig. 4.9e). The use of optical methods to deform cells is already well-established in mechanobiology specifically in the context of cell mechanics [15, 198–201]. Optical tweezers can either be used to perturb a portion of a cell or to stretch a whole cell [15, 202]. In this method, a dielectric microspherical bead is stably trapped by a focused laser beam. Optical forces here are linearly related to *P,* the laser power [203], as follows:

$$\boldsymbol{F} = \boldsymbol{CP} \tag{4.10}$$

where *C* is a coefficient which is a linear function of light speed and refraction index of the suspending medium.

By pulling the membrane-bound bead away from the cell, the MS channels in the proximity of the bead can be activated [204, 205]. The applied force can be estimated by changes in the displacement of the particle from the center of the optical trap. Thus, a force-distance curve for the pulled membrane can be related to the MS channel open probability. This method has already been utilized for studying the role of actin filaments in conveying the mechanical force to MS channels in plants [204].

Magnetic tweezers offer a similar scenario whereby a magnetic particle attached to the cell (e.g., integrins with fibronectin coated beads) is pulled by changing the magnetic field (Fig. 4.9e) [206].

$$\boldsymbol{F} = \boldsymbol{d}\nabla\boldsymbol{B} \tag{4.11}$$

where ∇B is the magnetic field gradient and *d* is the magnetic moment. Relevant forces can be applied by adjusting the magnetic field and distance.

It is also possible to apply torsional stress by rotating the magnet [207] demonstrating the usefulness of applying magnetic fields in studies of biological cells [208, 209]. Despite its common usage in cell mechanics assays, this method has not yet been employed to examine any MS channel function. Nevertheless, magnetically sensitive actuator "Magneto" has already been designed to genetically control the nervous system. In this elegant study, TRPV4 channels were fused to the paramagnetic protein Ferritin. The channels could be activated by applying magnetic force while simultaneously monitoring the experiment using *in vitro* calcium imaging assays and electrophysiological recordings in brain slices [210]. Similarly, a high-frequency magnetic field was used to heat the iron core of GFP-tagged Ferritin, leading to a local temperature increase sufficient to open the TRPV1 channels. This presents the first example of "temporally regulated neural modulation" [211]. Also, Wu et al. (2016) [86] attached magnetic nanoparticles directly to different parts of Piezo1 channel, and in combination with pressure-clamp electrophysiology, it was shown that applying magnetic field can modulate Piezo1 function. Similar experiments were previously carried out with MscL channels [208].

Amongst new experimental approaches the so-called "deflection-mediated" assays, such as pillar arrays, have been designed to stretch on specific locations of cells (Fig. 4.9f). This method was initially developed for cell mechanics assays (i.e., cell-surface interactions) [212] which was later optimised for electrophysiological purposes by Poole et al. (2014) [89]. Pillars can be deflected by a piezoelectric indenter to control the displacement with nanometric precision. Local force F can subsequently be measured based on the pillar displacement d and elastomeric mechanical properties k.

$$\boldsymbol{F = kd} \tag{4.12}$$

Using this system, distinct populations of MS channel currents in dorsal root ganglia neurons were recorded by substrate deflection at both the level of the soma and the neurite [89]. More recently, the same group has used pillar arrays and high speed pressure clamp to measure TRPV4 and Piezo1 current in primary murine chondrocytes [36].

Given the wide-reaching implications of shear flow and stress in biology, there is a growing demand for a tool that can be used to interrogate MS channel activity under physiological shear stress. To generate stress from fluid flow *in vitro*, a perfusion tube with a small opening at the tip is placed near the cell and bath solution is ejected onto the cell membrane [6], alternatively, a larger fluid flow can be generated through a microfluidic chamber (Fig. 4.9g) [213]. Shear flow instruments are better suited for studying and recapitulating hair cell mechanotransduction [12, 105, 214]. In the first configuration, the stimulus is not significantly different from generating a positive hydrostatic pressure or poking as the flow mainly generates a perpendicular stress on the cell membrane. However, the latter is more suitable to induce physiologically relevant shear stress. The shear stress τ_s caused by a Newtonian flow with viscosity of μ and strain rate of $\dot{\gamma}$ is:

$$\boldsymbol{\tau_s = \mu\dot{\gamma}} \tag{4.13}$$

Many studies have indicated the sensitivity of Piezo1, TRPV4 and PKD channels to shear stress in red blood cells, vascular endothelial cells as well as renal epithelial and bladder urothelial cells [74, 214–218].

Another challenge in mechanobiology is to stretch different cell types (e.g., cardiomyocytes) in an isotropic manner, reproducing their physiological contact forces. The most commonly used stretch devices apply longitudinal uniaxial stretch to adherent cell types [219] (Fig. 4.9h). More physiologically appropriate devices, so called "iso-stretchers", have recently been developed to apply isotropic stretch to various primary cell types [220, 221]. These set ups are often coupled with calcium imaging, both of which were used to produce ionic currents from a population of somatosensory neurons [220, 221].

Although all of the aforementioned experimental approaches apply different mechanical stimuli, it should be noted that the type of mechanical stress that is eventually sensed by MS channels could be similar or even identical. This is because there are only three distinct types of mechanical stress in nature: (i) tension-compression, (ii) shear/torsion and (3) bending (Fig. 4.6). Moreover, as previously mentioned, MS channels are either gated by force-from-lipids or force-from-filament paradigm. Both attributes beg the question: why do we need all these various paradigms? The answer is that these tools are designed to generate tunable mechanical stress that mimics the physiological stress in terms of type, amplitude, duration, localization and frequency. Moreover, each technique has certain advantages in terms of spatial and optical resolution, applicable force range, as well as ease of experimental procedure and data analysis. Hence, the suitability of each technique varies depending on its specific application. With these limits in mind, it is expected that many more experimental tools will be developed in the future that are particularly useful for advancing our fundamental biophysical understanding of MS channels and their role in development, physiology, and disease. All the examples above are tools that can generate different types of mechanical stimuli, yet they fail to precisely report on the extent of the applied forces. It will be important to design devices that can boost the accuracy of the molecular force/stress measurements (stress quantification) to avoid the simplifications due to limited theoretical models. Several groups have focused on genetically encoded force sensors for measuring mechanical forces in cytoskeletal scaffold proteins [222–226]. There is also a long-standing need for a molecular probe that can accurately measure the membrane tension upon different stimuli, synchronous with recording current from MS channel(s).

4.9 MscL Structure and Function: A Prototypical MS Channel

Interactions with the lipid bilayer are particularly important for bacterial MS channel function because they are gated by bilayer tension without the involvement of extra- or intracellular structural elements [32]. This is the case despite the fact that the bacterial cell-wall in cells exposed to hypo-osmotic shock serves as a powerful attenuator of the exerted physical force [227]. As previously established [228, 229], the bacterial MscL and MscS play crucial roles in the protection of bacterial cells against hypo-osmotic shock. The functional characteristics of MscL have been extensively studied using liposomal reconstitution [1, 61, 63, 230–233]. This channel displays no selectivity for molecules smaller than 6.5 kDa as a consequence of its very large unitary conductance of ~3 nS (Fig. 4.3). MscL has a C-terminal domain, which aligns co-axially with the transmembrane pore and an N-terminal amphipathic helix aligned parallel to the membrane bilayer. The cytoplasmic C-terminal domain functions as a "cytoplasmic sieve" to prevent large molecules from passing during the gating and to avoid losing vital components inside the cell, as well as acting as a stabilizer of the oligomeric structure of the channel [234, 235]. A number of studies indicate that the N-terminal helix of MscL, linked to the TM1 transmembrane pore helix, resides at the lipid-solvent interface [141, 236, 237]. The N-terminus serves as a force-conveying element that upon bilayer expansion drags the TM1 pore helix resulting in channel gating [141].

4.9.1 Lessons from MscL Applied to MS Channels of Eukaryotes

While 3D structures of five MS channels have been resolved to date, the key structural element that links mechanical stress from the membrane to MS channel dynamics remains a mystery [141]. By utilizing MscL as a prototypical MS channel, it has been previously shown that short amphipathic N-terminus is a crucial structural element during tension-induced gating. Despite its structural simplicity, MscL has continued to provide novel structural insight into basic principles of mechanotransduction [116, 128]. From recently resolved structures lipid binding domains, similar to the MscL N-terminus, were identified in different families of MS ion channels, including the MscS like channels, members of the human K2P family of channels (TREK/TRAAK), TRPV4 and most recently PIEZO ion channels (Fig. 4.10) [238–240]. More specifically, the horizontal force coupling helices, which are juxtaposed to the pore-lining helices via a flexible linker, have recently been proposed to be the hallmark of mechanosensitivity in other MS channel families [128, 141]. Although speculative at this point, providing an architectural blueprint for MS channel mechanosensitivity (e.g., conserved amphipathic anchor domain), aids in identifying and classifying genes which encode MS channel proteins based on their primary structure.

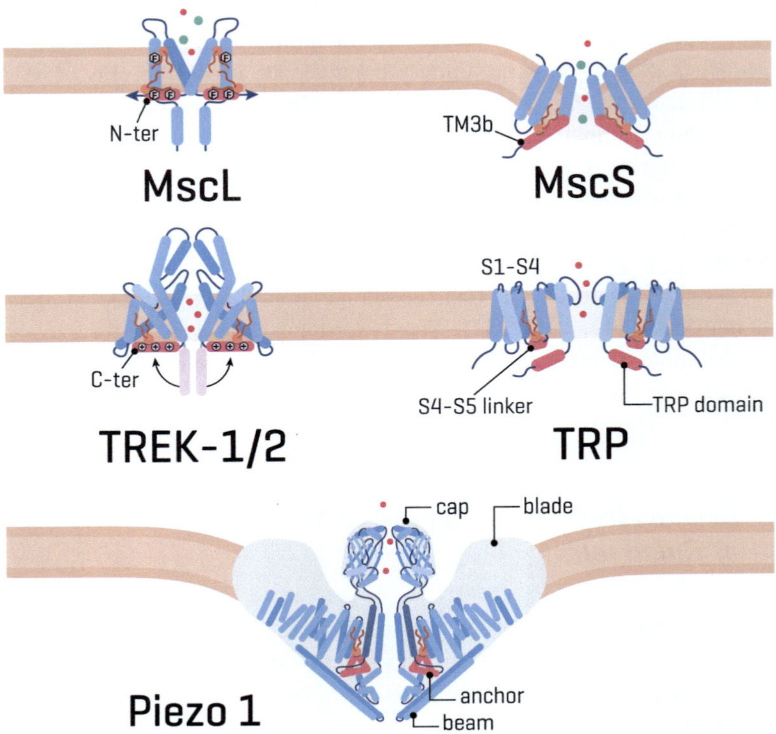

Fig. 4.10 Schematic of MS ion channels with their putative conserved mechanosensing domain shown in red. MscL, MscS, TREK-1/2 (K_{2P} members), TRP and Piezo1 channels are shown from *left to right* and *top to bottom*, respectively

4.10 Conclusions

In this chapter, the basic principles of mechanosensitivity and significance of mechanotransduction processes at the cell membrane interface are discussed. Among all the mechanotransducers found in living systems, MS channels are the fastest signaling molecules acting at the source of different signaling pathways in the physiology of cell mechanotransduction. MS channels can be gated by stretching the membrane bilayer (force-from-lipids principle) and/or by the force conveyed to the channels from the cytoskeleton/extracellular matrix (force-from-filament).

There is strong evidence from numerous studies that the mechanical properties of the lipid bilayer affect and modulate a large variety of membrane proteins, in particular MS channels. A simple case of a bilayer property affecting MS channels is bilayer thickness, which, for example, affects MscL to a much greater extent than MscS. Membrane lipid composition also alters the transbilayer pressure profile and hence, the line tension, surface tension, membrane spontaneous curvature, area compressibility modulus, bending stiffness and first and second moments of the pressure profile. Determining which of these parameters has a greater impact on a particular channel function depends on the type and structure of the MS channel.

Most of the present and potential theoretical and experimental paradigms for the study of MS channels have been covered in this chapter. Although great advances have been made in study of MS channels, there is still need for more techniques to accurately address central questions relevant to basic science as well as to human physiology and pathology.

Finally, by using bacterial MscL as a prototypical MS channel, determination of structurally common features underlying channel mechanosensitivity became possible. Importantly, a specific feature became apparent in the form of a horizontal force-conveying helix among all bilayer activated MS channels that has been preserved throughout evolution [128, 141]. Given the involvement of MS ion channels in numerous physiological processes and major human diseases, basic and translational research in this area of mechanobiology promises to pave the way for discovering novel therapies contributing to the betterment of human health.

Acknowledgements N.B. has been supported by a University International Postgraduate Award (UIPA) and P.R. has been supported by a University International Tuition Award (UITA) from the University of New South Wales, whereas Y.N. has been supported by a UIPA from the University of Newcastle. This project was supported by the Australian Research Council and Principal Research Fellowship to B.M. from the National Health and Medical Research Council of Australia. This work was also supported in part by funds from the Office of Health and Medical Research, NSW State Government.

References

1. Hamill OP, Martinac B (2001) Molecular basis of mechanotransduction in living cells. Physiol Rev 81:685–740
2. Hamilton ES, Schlegel AM, Haswell ES (2015) United in diversity: mechanosensitive ion channels in plants. Annu Rev Plant Biol 66:113–137. doi:10.1146/annurev-arplant-043014-114700
3. Haswell ES, Verslues PE (2015) The ongoing search for the molecular basis of plant osmosensing. J Gen Physiol 145:389–394. doi:10.1085/jgp.201411295
4. Monshausen GB, Haswell ES (2013) A force of nature: molecular mechanisms of mechanoperception in plants. J Exp Bot 64:4663–4680. doi:10.1093/jxb/ert204
5. Martinac B, Cox CD (2016) Mechanosensory transduction: focus on ion channels. In: Comprehensive biophysics. Elsevier
6. Ranade SS, Syeda R, Patapoutian A (2015) Mechanically activated ion channels. Neuron 87:1162–1179. doi:10.1016/j.neuron.2015.08.032
7. Nilius B, Honore E (2012) Sensing pressure with ion channels. Trends Neurosci 35:477–486. doi:10.1016/j.tins.2012.04.002
8. Arnadottir J, Chalfie M (2010) Eukaryotic mechanosensitive channels. Annu Rev Biophys 39:111–137. doi:10.1146/annurev.biophys.37.032807.125836
9. Chalfie M (2009) Neurosensory mechanotransduction. Nat Rev Mol Cell Biol 10:44–52., doi:nrm2595 [pii]10.1038/nrm2595
10. Kocer A (2015) Mechanisms of mechanosensing - mechanosensitive channels, function and re-engineering. Curr Opin Chem Biol 29:120–127. doi:10.1016/j.cbpa.2015.10.006

11. Nikolaev YA, Dosen PJ, Laver DR, Van Helden DF, Hamill OP (2016) Biophysical factors that promote mechanically-induced action potentials in neocortical and hippocampal pyramidal neurons. Biophys J 110(3):349a
12. Vollrath MA, Kwan KY, Corey DP (2007) The micromachinery of mechanotransduction in hair cells. Annu Rev Neurosci 30:339
13. Fettiplace R, Kim KX (2014) The physiology of mechanoelectrical transduction channels in hearing. Physiol Rev 94:951–986
14. Pruitt BL, Dunn AR, Weis WI, Nelson WJ (2014) Mechano-transduction: from molecules to tissues. PLoS Biol 12:e1001996
15. Moeendarbary E, Harris AR (2014) Cell mechanics: principles, practices, and prospects. Wiley Interdiscip Rev Syst Biol Med 6:371–388
16. Iadecola C, Davisson RL (2008) Hypertension and cerebrovascular dysfunction. Cell Metab 7:476–484
17. Intengan HD, Schiffrin EL (2001) Vascular remodeling in hypertension roles of apoptosis, inflammation, and fibrosis. Hypertension 38:581–587
18. Martinac B (2012) Mechanosensitive ion channels: an evolutionary and scientific tour de force in mechanobiology. Channels (Austin) 6:211–213. doi:10.4161/chan.22047
19. Wang W et al (2010) Fluid shear stress stimulates phosphorylation-dependent nuclear export of HDAC5 and mediates expression of KLF2 and eNOS. Blood 115:2971–2979
20. Na S et al (2008) Rapid signal transduction in living cells is a unique feature of mechanotransduction. Proc Natl Acad Sci 105:6626–6631
21. Morris CE, Horn R (1991) Failure to elicit neuronal macroscopic mechanosensitive currents anticipated by single-channel studies. Science 251:1246–1249
22. Honore E, Martins JR, Penton D, Patel A, Demolombe S (2015) The piezo mechanosensitive ion channels: may the force be with you! Rev Physiol Biochem Pharmacol 169:25–41. doi:10.1007/112_2015_26
23. Freund JB, Vermot J (2014) The wall-stress footprint of blood cells flowing in microvessels. Biophys J 106:752–762
24. Woo SH et al (2015) Piezo2 is the principal mechanotransduction channel for proprioception. Nat Neurosci 18:1756–1762. doi:10.1038/nn.4162
25. Nikolaev YA, Dosen PJ, Laver DR, van Helden DF, Hamill OP (2015) Single mechanically-gated cation channel currents can trigger action potentials in neocortical and hippocampal pyramidal neurons. Brain Res 1608:1–13. doi:10.1016/j.brainres.2015.02.051
26. Jensen MO, Mouritsen OG (2004) Lipids do influence protein function-the hydrophobic matching hypothesis revisited. Biochim Biophys Acta 1666:205–226. doi:10.1016/j.bbamem.2004.06.009
27. Marsh D (2007) Lateral pressure profile, spontaneous curvature frustration, and the incorporation and conformation of proteins in membranes. Biophys J 93:3884–3899. doi:10.1529/biophysj.107.107938
28. Cantor RS (1998) The lateral pressure profile in membranes: a physical mechanism of general anesthesia. Toxicol Lett 100-101:451–458
29. Anishkin A, Loukin SH, Teng J, Kung C (2014) Feeling the hidden mechanical forces in lipid bilayer is an original sense. Proc Natl Acad Sci U S A 111:7898. doi:10.1073/pnas.1313364111
30. Kung C, Martinac B, Sukharev S (2010) Mechanosensitive channels in microbes. Annu Rev Microbiol 64:313–329. doi:10.1146/annurev.micro.112408.134106
31. Brohawn SG, Su Z, MacKinnon R (2014) Mechanosensitivity is mediated directly by the lipid membrane in TRAAK and TREK1 K+ channels. Proc Natl Acad Sci U S A 111:3614–3619. doi:10.1073/pnas.1320768111
32. Cox CD, Bavi N, Martinac B (2017) Origin of the force: the force-from-lipids principle applied to piezo channels. Curr Top Membr 79:59–96
33. Fettiplace R (2016) Is TMC1 the hair cell Mechanotransducer Channel? Biophys J 111:3–9

34. Beurg M, Kim KX, Fettiplace R (2014) Conductance and block of hair-cell mechanotransducer channels in transmembrane channel-like protein mutants. J Gen Physiol 144:55–69. doi:10.1085/jgp.201411173
35. Poole K, Moroni M, Lewin GR (2015) Sensory mechanotransduction at membrane-matrix interfaces. Pflugers Arch 467:121–132. doi:10.1007/s00424-014-1563-6
36. Servin-Vences MR, Moroni M, Lewin GR, Poole K (2017) Direct measurement of TRPV4 and PIEZO1 activity reveals multiple mechanotransduction pathways in chondrocytes. elife 6:e21074
37. Mukherjee N et al (2014) The activation mode of the mechanosensitive ion channel, MscL, by lysophosphatidylcholine differs from tension-induced gating. FASEB J 28:4292–4302. doi:10.1096/fj.14-251579
38. Katta S, Krieg M, Goodman MB (2015) Feeling force: physical and physiological principles enabling sensory mechanotransduction. Annu Rev Cell Dev Biol 31:347–371. doi:10.1146/annurev-cellbio-100913-013426
39. Martinac B (2013) The ion channels to cytoskeleton connection as potential mechanism of mechanosensitivity. Biochim Biophys Acta 1838:682. doi:10.1016/j.bbamem.2013.07.015
40. Pivetti CD et al (2003) Two families of mechanosensitive channel proteins. Microbiol Mol Biol Rev 67, 66–85, table of contents
41. Berrier C, Besnard M, Ajouz B, Coulombe A, Ghazi A (1996) Multiple mechanosensitive ion channels from Escherichia coli, activated at different thresholds of applied pressure. J Membr Biol 151:175–187. doi:10.1007/s002329900068
42. Perozo E, Rees DC (2003) Structure and mechanism in prokaryotic mechanosensitive channels. Curr Opin Struct Biol 13:432–442
43. Schumann U et al (2010) YbdG in Escherichia Coli is a threshold-setting mechanosensitive channel with MscM activity. Proc Natl Acad Sci U S A 107:12664–12669. doi:10.1073/pnas.1001405107
44. Prager-Khoutorsky M, Khoutorsky A, Bourque CW (2014) Unique interweaved microtubule scaffold mediates osmosensory transduction via physical interaction with TRPV1. Neuron 83:866–878. doi:10.1016/j.neuron.2014.07.023
45. Teng J, Loukin SH, Anishkin A, Kung C (2015) L596-W733 bond between the start of the S4-S5 linker and the TRP box stabilizes the closed state of TRPV4 channel. Proc Natl Acad Sci U S A 112:3386–3391. doi:10.1073/pnas.1502366112
46. Huynh KW et al (2016) Structure of the full-length TRPV2 channel by cryo-EM. Nat Commun 7:11130. doi:10.1038/ncomms11130
47. Zhang W et al (2015) Ankyrin repeats convey force to gate the NOMPC Mechanotransduction Channel. Cell 162:1391–1403. doi:10.1016/j.cell.2015.08.024
48. Ronan D, Gillespie P (2005) Metazoan mechanotransduction mystery finally solved. Nat Neurosci 8:7–8
49. Gillespie PG, Muller U (2009) Mechanotransduction by hair cells: models, molecules, and mechanisms. Cell 139:33–44. doi:10.1016/j.cell.2009.09.010
50. Nilius B, Voets T (2013) The puzzle of TRPV4 channelopathies. EMBO Rep 14:152–163. doi:10.1038/embor.2012.219
51. Albuisson J et al (2013) Dehydrated hereditary stomatocytosis linked to gain-of-function mutations in mechanically activated PIEZO1 ion channels. Nat Commun 4:1884. doi:10.1038/ncomms2899
52. Peyronnet R, Nerbonne JM, Kohl P (2016) Cardiac Mechano-Gated Ion Channels and Arrhythmias. Circ Res 118:311–329. doi:10.1161/CIRCRESAHA.115.305043
53. Seo K et al (2014) Combined TRPC3 and TRPC6 blockade by selective small-molecule or genetic deletion inhibits pathological cardiac hypertrophy. Proc Natl Acad Sci U S A 111:1551–1556. doi:10.1073/pnas.1308963111
54. Zubcevic L et al (2016) Cryo-electron microscopy structure of the TRPV2 ion channel. Nat Struct Mol Biol 23:180–186. doi:10.1038/nsmb.3159

55. Liao M, Cao E, Julius D, Cheng Y (2013) Structure of the TRPV1 ion channel determined by electron cryo-microscopy. Nature 504:107–112. doi:10.1038/nature12822
56. Martinac B, Adler J, Kung C (1990) Mechanosensitive ion channels of E. coli activated by amphipaths. Nature 348:261–263. doi:10.1038/348261a0
57. Markin VS, Martinac B (1991) Mechanosensitive ion channels as reporters of bilayer expansion. A theoretical model Biophys J 60:1120–1127. doi:10.1016/S0006-3495(91)82147-6
58. Anishkin A, Kung C (2013) Stiffened lipid platforms at molecular force foci. Proc Natl Acad Sci U S A 110:4886–4892. doi:10.1073/pnas.1302018110
59. Teng J, Loukin S, Anishkin A, Kung C (2015) The force-from-lipid (FFL) principle of mechanosensitivity, at large and in elements. Pflugers Arch 467:27. doi:10.1007/s00424-014-1530-2
60. Cox CD, Nakayama Y, Nomura T, Martinac B (2015) The evolutionary 'tinkering' of MscS-like channels: generation of structural and functional diversity. Pflugers Arch 467:3. doi:10.1007/s00424-014-1522-2
61. Sukharev SI, Blount P, Martinac B, Blattner FR, Kung C (1994) A large-conductance mechanosensitive channel in E. coli encoded by mscL alone. Nature 368:265–268. doi:10.1038/368265a0
62. Sukharev S (2002) Purification of the small mechanosensitive channel of Escherichia coli (MscS): the subunit structure, conduction, and gating characteristics in liposomes. Biophys J 83:290–298. doi:10.1016/S0006-3495(02)75169-2
63. Hase CC, Le Dain AC, Martinac B (1995) Purification and functional reconstitution of the recombinant large mechanosensitive ion channel (MscL) of Escherichia Coli. J Biol Chem 270:18329–18334
64. Koprowski P et al (2015) Cytoplasmic domain of MscS interacts with cell division protein FtsZ: a possible Non-Channel function of the Mechanosensitive Channel in Escherichia Coli. PLoS One 10:e0127029. doi:10.1371/journal.pone.0127029
65. Delcour AH, Martinac B, Adler J, Kung C (1989) Modified reconstitution method used in patch-clamp studies of Escherichia Coli ion channels. Biophys J 56:631–636. doi:10.1016/S0006-3495(89)82710-9
66. Sukharev SI, Martinac B, Arshavsky VY, Kung C (1993) Two types of mechanosensitive channels in the Escherichia Coli cell envelope: solubilization and functional reconstitution. Biophys J 65:177–183. doi:10.1016/S0006-3495(93)81044-0
67. Brohawn SG, Campbell EB, MacKinnon R (2014) Physical mechanism for gating and mechanosensitivity of the human TRAAK K+ channel. Nature 516:126–130. doi:10.1038/nature14013
68. Berrier C et al (2013) The purified mechanosensitive channel TREK-1 is directly sensitive to membrane tension. J Biol Chem 288:27307–27314. doi:10.1074/jbc.M113.478321
69. Dong YY et al (2015) K2P channel gating mechanisms revealed by structures of TREK-2 and a complex with Prozac. Science 347:1256–1259. doi:10.1126/science.1261512
70. Syeda R et al (2016) Piezo1 channels are inherently mechanosensitive. Cell Rep 17:1739–1746
71. Cox CD et al (2016) Removal of the mechanoprotective influence of the cytoskeleton reveals PIEZO1 is gated by bilayer tension. Nat Commun 7:10366. doi:10.1038/ncomms10366
72. Ge J et al (2015) Architecture of the mammalian mechanosensitive Piezo1 channel. Nature 527:64–69. doi:10.1038/nature15247
73. Ranade SS et al (2014) Piezo1, a mechanically activated ion channel, is required for vascular development in mice. Proc Natl Acad Sci 111:10347–10352
74. Li J et al (2014) Piezo1 integration of vascular architecture with physiological force. Nature 515:279–282. doi:10.1038/nature13701
75. Retailleau K et al (2015) Piezo1 in smooth muscle cells is involved in hypertension-dependent arterial remodeling. Cell Rep 13:1161–1171. doi:10.1016/j.celrep.2015.09.072
76. Cahalan SM et al (2015) Piezo1 links mechanical forces to red blood cell volume. Elife 4. doi:10.7554/eLife.07370

77. Eisenhoffer GT et al (2012) Crowding induces live cell extrusion to maintain homeostatic cell numbers in epithelia. Nature 484:546–549. doi:10.1038/nature10999
78. Gudipaty SA et al (2017) Mechanical stretch triggers rapid epithelial cell division through Piezo1. Nature 543:118–121. doi:10.1038/nature21407
79. Martins JR et al (2016) Piezo1-dependent regulation of urinary osmolarity. Pflügers Archiv-European Journal of Physiology 468:1197–1206
80. Wang S et al (2016) Endothelial cation channel PIEZO1 controls blood pressure by mediating flow-induced ATP release. J Clin Invest 126:4527–4536. doi:10.1172/JCI87343
81. Lukacs V et al (2015) Impaired PIEZO1 function in patients with a novel autosomal recessive congenital lymphatic dysplasia. Nat Commun 6:8329. doi:10.1038/ncomms9329
82. Pathak MM et al (2014) Stretch-activated ion channel Piezo1 directs lineage choice in human neural stem cells. Proc Natl Acad Sci U S A 111:16148–16153. doi:10.1073/pnas.1409802111
83. Koser DE et al (2016) Mechanosensing is critical for axon growth in the developing brain. Nat Neurosci 19:1592–1598. doi:10.1038/nn.4394
84. Hulce JJ, Cognetta AB, Niphakis MJ, Tully SE, Cravatt BF (2013) Proteome-wide mapping of cholesterol-interacting proteins in mammalian cells. Nat Methods 10:259–264. doi:10.1038/nmeth.2368
85. Qi Y et al (2015) Membrane stiffening by STOML3 facilitates mechanosensation in sensory neurons. Nat Commun 6:8512. doi:10.1038/ncomms9512
86. Wu J, Goyal R, Grandl J (2016) Localized force application reveals mechanically sensitive domains of Piezo1. Nat Commun 7:12939
87. Armstrong CM, Bezanilla F (1977) Inactivation of the sodium channel. II. Gating current experiments. J Gen Physiol 70:567–590
88. Honore E, Patel AJ, Chemin J, Suchyna T, Sachs F (2006) Desensitization of mechano-gated K2P channels. Proc Natl Acad Sci U S A 103:6859–6864. doi:10.1073/pnas.0600463103
89. Poole K, Herget R, Lapatsina L, Ngo HD, Lewin GR (2014) Tuning piezo ion channels to detect molecular-scale movements relevant for fine touch. Nat Commun 5:3520. doi:10.1038/ncomms4520
90. Peyronnet R et al (2012) Mechanoprotection by polycystins against apoptosis is mediated through the opening of stretch-activated K(2P) channels. Cell Rep 1:241–250. doi:10.1016/j.celrep.2012.01.006
91. Nonomura K et al (2017) Piezo2 senses airway stretch and mediates lung inflation-induced apnoea. Nature 541:176–181. doi:10.1038/nature20793
92. Zheng J (2013) Molecular mechanism of TRP channels. Compr Physiol 3:221–242. doi:10.1002/cphy.c120001
93. O'Conor CJ, Leddy HA, Benefield HC, Liedtke WB, Guilak F (2014) TRPV4-mediated mechanotransduction regulates the metabolic response of chondrocytes to dynamic loading. Proc Natl Acad Sci U S A 111:1316–1321. doi:10.1073/pnas.1319569111
94. Lieben L, Carmeliet G (2012) The involvement of TRP channels in bone homeostasis. Front Endocrinol (Lausanne) 3:99. doi:10.3389/fendo.2012.00099
95. Loukin S, Zhou X, Su Z, Saimi Y, Kung C (2010) Wild-type and brachyolmia-causing mutant TRPV4 channels respond directly to stretch force. J Biol Chem 285:27176–27181. doi:10.1074/jbc.M110.143370
96. Liedtke W et al (2000) Vanilloid receptor-related osmotically activated channel (VR-OAC), a candidate vertebrate osmoreceptor. Cell 103:525–535., doi:S0092–8674(00)00143–4 [pii]
97. Sabass B, Stone HA (2016) Role of the membrane for mechanosensing by tethered channels. Phys Rev Lett 16(25):258101. arXiv preprint arXiv:1603.05751
98. Pan B et al (2013) TMC1 and TMC2 are components of the mechanotransduction channel in hair cells of the mammalian inner ear. Neuron 79:504–515. doi:10.1016/j.neuron.2013.06.019
99. Kim KX, Fettiplace R (2013) Developmental changes in the cochlear hair cell mechanotransducer channel and their regulation by transmembrane channel-like proteins. J Gen Physiol 141:141–148. doi:10.1085/jgp.201210913

100. Gao X et al (2013) Novel compound heterozygous TMC1 mutations associated with autosomal recessive hearing loss in a Chinese family. PLoS One 8:e63026. doi:10.1371/journal.pone.0063026
101. Kawashima Y, Kurima K, Pan B, Griffith AJ, Holt JR (2015) Transmembrane channel-like (TMC) genes are required for auditory and vestibular mechanosensation. Pflugers Arch 467:85–94. doi:10.1007/s00424-014-1582-3
102. Chatzigeorgiou M, Bang S, Hwang SW, Schafer WR (2013) Tmc-1 encodes a sodium-sensitive channel required for salt chemosensation in C. elegans. Nature 494:95–99. doi:10.1038/nature11845
103. Sotomayor M, Weihofen WA, Gaudet R, Corey DP (2012) Structure of a force-conveying cadherin bond essential for inner-ear mechanotransduction. Nature 492:128–132
104. Delling M et al (2016) Primary cilia are not calcium-responsive mechanosensors. Nature 531:656–660. doi:10.1038/nature17426
105. Hudspeth A (2014) Integrating the active process of hair cells with cochlear function. Nat Rev Neurosci 15:600–614
106. Zhao B, Müller U (2015) The elusive mechanotransduction machinery of hair cells. Curr Opin Neurobiol 34:172–179
107. Zhang W, Yan Z, Jan LY, Jan YN (2013) Sound response mediated by the TRP channels NOMPC, NANCHUNG, and INACTIVE in chordotonal organs of drosophila larvae. Proc Natl Acad Sci U S A 110:13612–13617. doi:10.1073/pnas.1312477110
108. Yan Z et al (2013) Drosophila NOMPC is a mechanotransduction channel subunit for gentle-touch sensation. Nature 493:221–225. doi:10.1038/nature11685
109. Gao Y, Cao E, Julius D, Cheng Y (2016) TRPV1 structures in nanodiscs reveal mechanisms of ligand and lipid action. Nature 534:347
110. O'Hagan R, Chalfie M, Goodman MB (2005) The MEC-4 DEG/ENaC channel of Caenorhabditis elegans touch receptor neurons transduces mechanical signals. Nature Neurosci 8:43–50
111. Kellenberger S, Schild L (2002) Epithelial sodium channel/degenerin family of ion channels: a variety of functions for a shared structure. Physiol Rev 82:735–767. doi:10.1152/physrev.00007.2002
112. Stockand JD (2015) In: Zheng J, Trudeau MC (eds) Handbook of Ion channels Ch. 26. Taylor & Francis Group, Boca Raton
113. Ross S, Fuller CM, Bubien JK, Benos DJ (2007) Amiloride-sensitive Na channels contribute to regulatory volume increases in human glioma cells. Am J Physiol Cell Physiol 293:C1181–C1185
114. Eastwood AL et al (2015) Tissue mechanics govern the rapidly adapting and symmetrical response to touch. Proc Natl Acad Sci U S A 112:E6955–E6963. doi:10.1073/pnas.1514138112
115. Hill K, Schaefer M (2007) TRPA1 is differentially modulated by the amphipathic molecules trinitrophenol and chlorpromazine. J Biol Chem 282:7145–7153
116. Perozo E, Kloda A, Cortes DM, Martinac B (2002) Physical principles underlying the transduction of bilayer deformation forces during mechanosensitive channel gating. Nat Struct Biol 9:696–703. doi:10.1038/nsb827
117. Atkins P, De Paula J (2011) Physical chemistry for the life sciences. Oxford University Press, Oxford
118. Maksaev G, Milac A, Anishkin A, Guy HR, Sukharev S (2011) Analyses of gating thermodynamics and effects of deletions in the mechanosensitive channel TREK-1: comparisons with structural models. Channels (Austin) 5:34–42
119. Sukharev SI, Sigurdson WJ, Kung C, Sachs F (1999) Energetic and spatial parameters for gating of the bacterial large conductance mechanosensitive channel. MscL J Gen Physiol 113:525–540
120. Sukharev S, Corey DP (2004) Mechanosensitive channels: multiplicity of families and gating paradigms. Sci STKE 2004:re4. doi:10.1126/stke.2192004re4
121. Phillips R, Ursell T, Wiggins P, Sens P (2009) Emerging roles for lipids in shaping membrane-protein function. Nature 459:379–385

122. Battle AR et al (2015) Lipid–protein interactions: Lessons learned from stress. Biochimica et Biophysica Acta (BBA)-Biomembranes 1848:1744–1756
123. Boal D, Boal DH (2012) Mechanics of the cell. Cambridge University Press, Cambridge
124. Gruenheid S, Finlay BB (2003) Microbial pathogenesis and cytoskeletal function. Nature 422:775–781
125. Bertaud J, Hester J, Jimenez DD, Buehler MJ (2009) Energy landscape, structure and rate effects on strength properties of alpha-helical proteins. J Phys Condens Matter 22:035102
126. Rodriguez ML, McGarry PJ, Sniadecki NJ (2013) Review on cell mechanics: experimental and modeling approaches. Appl Mech Rev 65:060801
127. Pliotas C et al (2015) The role of lipids in mechanosensation. Nat Struct Mol Biol 22:991–998. doi:10.1038/nsmb.3120
128. Bavi N, Cox CD, Perozo E, Martinac B (2016) Toward a structural blueprint for bilayer-mediated channel mechanosensitivity. Channels 11(2):91–93
129. Bavi N et al (2014) Biophysical implications of lipid bilayer rheometry for mechanosensitive channels. Proc Natl Acad Sci U S A 111:13864–13869. doi:10.1073/pnas.1409011111
130. Laganowsky A et al (2014) Membrane proteins bind lipids selectively to modulate their structure and function. Nature 510:172–175. doi:10.1038/nature13419
131. Zhong D, Blount P (2013) Phosphatidylinositol is crucial for the mechanosensitivity of Mycobacterium tuberculosis MscL. Biochemistry 52:5415–5420. doi:10.1021/bi400790j
132. Cantor RS (1999) Lipid composition and the lateral pressure profile in membranes. Biophys J 76:A58–A58
133. Cantor RS (1999) Lipid composition and the lateral pressure profile in bilayers. Biophys J 76:2625–2639. doi:10.1016/S0006-3495(99)77415-1
134. Cantor RS (2000) Membrane composition and the lateral pressure profile. Biophys J 78:329a
135. Evans E, Rawicz W, Smith B (2013) Concluding remarks back to the future: mechanics and thermodynamics of lipid biomembranes. Faraday Discuss 161:591–611
136. Nomura T et al (2012) Differential effects of lipids and lyso-lipids on the mechanosensitivity of the mechanosensitive channels MscL and MscS. Proc Natl Acad Sci U S A 109:8770–8775. doi:10.1073/pnas.1200051109
137. Mukhin SI, Baoukina S (2005) Analytical derivation of thermodynamic characteristics of lipid bilayer from a flexible string model. Phys Rev E 71:061918
138. Evans E, Rawicz W, Smith BA (2013) Back to the future: mechanics and thermodynamics of lipid biomembranes. Faraday Discuss 161:591–611
139. Gullingsrud J, Schulten K (2004) Lipid bilayer pressure profiles and mechanosensitive channel gating. Biophys J 86:3496–3509. doi:10.1529/biophysj.103.034322
140. Cantor RS (1997) The lateral pressure profile in membranes: a physical mechanism of general anesthesia. Biochemistry 36:2339–2344. doi:10.1021/bi9627323
141. Bavi N et al (2016) The role of MscL amphipathic N terminus indicates a blueprint for bilayer-mediated gating of mechanosensitive channels. Nat Commun 7:11984. doi:10.1038/ncomms11984
142. Lee AG (2004) How lipids affect the activities of integral membrane proteins. Biochim Biophys Acta 1666:62–87. doi:10.1016/j.bbamem.2004.05.012
143. Lundbaek JA, Collingwood SA, Ingolfsson HI, Kapoor R, Andersen OS (2010) Lipid bilayer regulation of membrane protein function: gramicidin channels as molecular force probes. J R Soc Interface 7:373–395. doi:10.1098/rsif.2009.0443
144. Gullingsrud J, Schulten K (2003) Gating of MscL studied by steered molecular dynamics. Biophys J 85:2087–2099
145. Gullingsrud J, Kosztin D, Schulten K (2001) Structural determinants of MscL gating studied by molecular dynamics simulations. Biophys J 80:2074–2081. doi:10.1016/S0006-3495(01)76181-4
146. Templer R, Castle S, Curran A, Klug D (1999) Sensing isothermal changes in the lateral pressure in model membranes using di-pyrenyl phosphatidylcholine. Faraday Discuss 111:41–53
147. Gawrisch K, Holte LL (1996) NMR investigations of non-lamellar phase promoters in the lamellar phase state. Chem Phys Lipids 81:105–116

148. Cantor RS (1997) Membrane lateral pressures: a physical mechanism of general anesthesia. Biophys J 72:Mpo98
149. Cantor RS (1996) Theory of lipid monolayers comprised of mixtures of flexible and stiff amphiphiles in athermal solvents: fluid phase coexistence. J Chem Phys 104:8082–8095. doi:10.1063/1.471524
150. Harries D, Ben-Shaul A (1997) Conformational chain statistics in a model lipid bilayer: comparison between mean field and Monte Carlo calculations. J Chem Phys 106: 1609–1619
151. Lindahl E, Edholm O (2000) Spatial and energetic-entropic decomposition of surface tension in lipid bilayers from molecular dynamics simulations. J Chem Phys 113:3882–3893
152. Baoukina S, Marrink SJ, Tieleman DP (2010) Lateral pressure profiles in lipid monolayers. Faraday Discuss 144:393–409
153. Yoo J, Cui Q (2009) Curvature generation and pressure profile modulation in membrane by lysolipids: insights from coarse-grained simulations. Biophys J 97:2267–2276
154. Elmore DE, Dougherty DA (2003) Investigating lipid composition effects on the mechanosensitive channel of large conductance (MscL) using molecular dynamics simulations. Biophys J 85:1512–1524
155. Markin VS, Sachs F (2004) Thermodynamics of mechanosensitivity. Phys Biol 1:110–124. doi:10.1088/1478-3967/1/2/007
156. Wiggins P, Phillips R (2005) Membrane-protein interactions in mechanosensitive channels. Biophys J 88:880–902. doi:10.1529/biophysj.104.047431
157. Wiggins P, Phillips R (2004) Analytic models for mechanotransduction: gating a mechanosensitive channel. Proc Natl Acad Sci 101:4071–4076
158. Zhan H, Lazaridis T (2013) Inclusion of lateral pressure/curvature stress effects in implicit membrane models. Biophys J 104:643–654
159. Im W, Feig M, Brooks CL (2003) An implicit membrane generalized born theory for the study of structure, stability, and interactions of membrane proteins. Biophys J 85:2900–2918
160. Bavi O, Vossoughi M, Naghdabadi R, Jamali Y (2014) The effect of local bending on gating of MscL using a representative volume element and finite element simulation. Channels 8:0–1
161. Sonne J, Hansen FY, Peters GH (2005) Methodological problems in pressure profile calculations for lipid bilayers. J Chem Phys 122:124903
162. Berendsen HJ, van der Spoel D, van Drunen R (1995) GROMACS: a message-passing parallel molecular dynamics implementation. Comput Phys Commun 91:43–56
163. Lindahl E, Hess B, Van Der Spoel D (2001) GROMACS 3.0: a package for molecular simulation and trajectory analysis. J Mol Model 7:306–317
164. Phillips JC et al (2005) Scalable molecular dynamics with NAMD. J Comput Chem 26:1781–1802
165. Sansom MS, Scott KA, Bond PJ (2008) Coarse-grained simulation: a high-throughput computational approach to membrane proteins. Biochem Soc Trans 36:27–32
166. Sansom M, Hedger G (2016) Lipid interaction sites on channels, transporters and receptors: recent insights from molecular dynamics simulations. BBA-Biomembranes 1858(10):2390–2400
167. Zhuang X, Dávila-Contreras EM, Beaven AH, Im W, Klauda JB (2016) An extensive simulation study of lipid bilayer properties with different head groups, acyl chain lengths, and chain saturations. Biochimica et Biophysica Acta (BBA)-Biomembranes 1858:3093–3104
168. Pavlova A, Hwang H, Lundquist K, Balusek C, Gumbart JC (2016) Living on the edge: simulations of bacterial outer-membrane proteins. Biochimica et Biophysica Acta (BBA)-Biomembranes 1858:1753–1759
169. Pandit KR, Klauda JB (2012) Membrane models of E. coli containing cyclic moieties in the aliphatic lipid chain. Biochimica et Biophysica Acta (BBA)-Biomembranes 1818:1205–1210
170. Masetti M et al (2016) Multiscale simulations of a two-pore potassium channel. J Chem Theory Comput 12:5681–5687

171. Jo S, Lim JB, Klauda JB, Im W (2009) CHARMM-GUI membrane builder for mixed bilayers and its application to yeast membranes. Biophys J 97:50–58
172. Lim JB, Klauda JB (2011) Lipid chain branching at the iso-and anteiso-positions in complex chlamydia membranes: a molecular dynamics study. Biochimica et Biophysica Acta (BBA)-Biomembranes 1808:323–331
173. Ingólfsson HI et al (2014) Lipid organization of the plasma membrane. J Am Chem Soc 136:14554–14559
174. Bavi N et al (2016) Nanomechanical properties of MscL alpha helices: a steered molecular dynamics study. Channels 11(3):209–223
175. Anishkin A, Sukharev S (2017) Channel disassembled: pick, tweak, and soak parts to soften. Channels 11(3):173–175
176. Bavi O, Vossoughi M, Naghdabadi R, Jamali Y (2016) The combined effect of hydrophobic mismatch and bilayer local bending on the regulation of mechanosensitive ion channels. PLoS One 11:e0150578. doi:10.1371/journal.pone.0150578
177. Zhu L et al (2016) Gating mechanism of mechanosensitive channel of large conductance: a coupled continuum mechanical-continuum solvation approach. Biomass Model Mechanobiol 15(6):1557–1576
178. Bavi O et al (2016) Influence of Global and Local Membrane Curvature on Mechanosensitive Ion Channels: A Finite Element Approach. Membranes (Basel) 6:14. doi:10.3390/membranes6010014
179. Sakmann B (2013) Single-channel recording. Springer Science & Business Media, New York
180. Nilius B (2003) Pflugers Archiv and the advent of modern electrophysiology. From the first action potential to patch clamp. Pflugers Arch 447:267–271. doi:10.1007/s00424-003-1156-2
181. Nakayama Y, Slavchov RI, Bavi N, Martinac BT (2016) Energy of liposome patch adhesion to the pipette glass determined by confocal fluorescence microscopy. J Phys Chem Lett 7:4530
182. Suchyna TM, Markin VS, Sachs F (2009) Biophysics and structure of the patch and the gigaseal. Biophys J 97:738–747. doi:10.1016/j.bpj.2009.05.018
183. Guharay F, Sachs F (1984) Stretch-activated single ion channel currents in tissue-cultured embryonic chick skeletal muscle. J Physiol 352:685–701
184. Brehm P, Kullberg R, Moody-Corbett F (1984) Properties of non-junctional acetylcholine receptor channels on innervated muscle of Xenopus Laevis. J Physiol 350:631–648
185. Hamill OP, Marty A, Neher E, Sakmann B, Sigworth FJ (1981) Improved patch-clamp techniques for high-resolution current recording from cells and cell-free membrane patches. Pflugers Arch 391:85–100
186. Hamill O (2006) Twenty odd years of stretch-sensitive channels. Pflugers Arch 453:333–351
187. Zhang Y, Gao F, Popov VL, Wen JW, Hamill OP (2000) Mechanically gated channel activity in cytoskeleton-deficient plasma membrane blebs and vesicles from Xenopus oocytes. J Physiol 523(1):117–130
188. Zhang Y, Hamill OP (2000) Calcium-, voltage-and osmotic stress-sensitive currents in Xenopus oocytes and their relationship to single mechanically gated channels. J Physiol 523:83–99
189. Hamill OP, McBride DW (1995) Mechanoreceptive membrane channels. Am Sci 83:30–37
190. Blount P, Schroeder MJ, Kung C (1997) Mutations in a bacterial mechanosensitive channel change the cellular response to osmotic stress. J Biol Chem 272:32150–32157
191. Nilius B (2007) Transient receptor potential (TRP) cation channels: rewarding unique proteins. Bull Mem Acad R Med Belg 162:244–253
192. Buday T, Kovacikova L, Ruzinak R, Plevkova J (2017) TRPV4 antagonist GSK2193874 does not modulate cough response to osmotic stimuli. Respir Physiol Neurobiol 236:1–4
193. Jiang H, Sun SX (2013) Cellular pressure and volume regulation and implications for cell mechanics. Biophys J 105:609–619
194. Sun M et al (2005) Multiple membrane tethers probed by atomic force microscopy. Biophys J 89:4320–4329
195. Chaudhuri O, Parekh SH, Lam WA, Fletcher DA (2009) Combined atomic force microscopy and side-view optical imaging for mechanical studies of cells. Nat Methods 6:383–387

196. Mescola A et al (2012) Probing cytoskeleton organisation of neuroblastoma cells with single-cell force spectroscopy. J Mol Recognit 25:270–277
197. Ossola D et al (2015) Force-controlled patch clamp of beating cardiac cells. Nano Lett 15:1743–1750
198. Hochmuth F, Shao J-Y, Dai J, Sheetz MP (1996) Deformation and flow of membrane into tethers extracted from neuronal growth cones. Biophys J 70:358–369
199. Dao M, Lim CT, Suresh S (2003) Mechanics of the human red blood cell deformed by optical tweezers. J Mech Phys Solids 51:2259–2280
200. Zhang H, Liu K-K (2008) Optical tweezers for single cells. J R Soc Interface 5:671–690
201. Hénon S, Lenormand G, Richert A, Gallet F (1999) A new determination of the shear modulus of the human erythrocyte membrane using optical tweezers. Biophys J 76:1145–1151
202. Iskratsch T, Wolfenson H, Sheetz MP (2014) Appreciating force and shape [mdash] the rise of mechanotransduction in cell biology. Nat Rev Mol Cell Biol 15:825–833
203. Svoboda K, Block SM (1994) Biological applications of optical forces. Annu Rev Biophys Biomol Struct 23:247–285
204. Tatsumi H et al (2014) Mechanosensitive channels are activated by stress in the actin stress fibres, and could be involved in gravity sensing in plants. Plant Biol 16:18–22
205. Hayakawa K, Tatsumi H, Sokabe M (2008) Actin stress fibers transmit and focus force to activate mechanosensitive channels. J Cell Sci 121:496–503. doi:10.1242/jcs.022053
206. Neuman KC, Nagy A (2008) Single-molecule force spectroscopy: optical tweezers, magnetic tweezers and atomic force microscopy. Nat Methods 5:491
207. Laurent VM et al (2002) Assessment of mechanical properties of adherent living cells by bead micromanipulation: comparison of magnetic twisting cytometry vs optical tweezers. J Biomech Eng 124:408–421
208. Nakayama Y et al (2015) Magnetic nanoparticles for "smart liposomes". Eur Biophys J 44:647–654
209. Meister M (2016) Physical limits to magnetogenetics. elife 5:e17210
210. Wheeler MA et al (2016) Genetically targeted magnetic control of the nervous system. Nat Neurosci 19:756–761
211. Stanley SA et al (2016) Bidirectional electromagnetic control of the hypothalamus regulates feeding and metabolism. Nature 531:647–650
212. Schoen I, Hu W, Klotzsch E, Vogel V (2010) Probing cellular traction forces by micropillar arrays: contribution of substrate warping to pillar deflection. Nano Lett 10:1823–1830
213. Lee LM, Liu AP (2015) A microfluidic pipette array for mechanophenotyping of cancer cells and mechanical gating of mechanosensitive channels. Lab Chip 15:264–273
214. Ranade SS et al (2014) Piezo1, a mechanically activated ion channel, is required for vascular development in mice. Proc Natl Acad Sci U S A 111:10347–10352. doi:10.1073/pnas.1409233111
215. Wu J, Lewis AH, Grandl J (2017) Touch, tension, and transduction–the function and regulation of Piezo Ion channels. Trends Biochem Sci 42:57–71
216. Miyamoto T et al (2014) Functional role for Piezo1 in stretch-evoked Ca2+ influx and ATP release in urothelial cell cultures. J Biol Chem 289:16565–16575
217. Wu L, Gao X, Brown RC, Heller S, O'Neil RG (2007) Dual role of the TRPV4 channel as a sensor of flow and osmolality in renal epithelial cells. Am J Physiol Ren Physiol 293:F1699–F1713
218. Nauli SM et al (2003) Polycystins 1 and 2 mediate mechanosensation in the primary cilium of kidney cells. Nat Genet 33:129–137
219. Peyronnet R, Tran D, Girault T, Frachisse J-M (2013) Mechanosensitive channels: feeling tension in a world under pressure. Front Plant Sci 5:558–558
220. Bhattacharya MR et al (2008) Radial stretch reveals distinct populations of mechanosensitive mammalian somatosensory neurons. Proc Natl Acad Sci 105:20015–20020
221. Schürmann S et al (2016) The IsoStretcher: an isotropic cell stretch device to study mechanical biosensor pathways in living cells. Biosens Bioelectron 81:363–372

222. Meng F, Sachs F (2011) Measuring strain of structural proteins in vivo in real time. In: Kohl P, Sachs F, Franz MR (eds) Cardiac mechano-electric coupling and arrhythmia: from pipette to patient. Oxford University Press, Oxford, pp 431–434
223. Meng F, Suchyna TM, Lazakovitch E, Gronostajski RM, Sachs F (2011) Real time FRET based detection of mechanical stress in cytoskeletal and extracellular matrix proteins. Cell Mol Bioeng 4:148–159. doi:10.1007/s12195-010-0140-0
224. Wang Y, Meng F, Sachs F (2011) Genetically encoded force sensors for measuring mechanical forces in proteins. Commun Integr Biol 4:385–390. doi:10.4161/cib.4.4.15505
225. Grashoff C et al (2010) Measuring mechanical tension across vinculin reveals regulation of focal adhesion dynamics. Nature 466:263–266
226. Yamashita S, Tsuboi T, Ishinabe N, Kitaguchi T, Michiue T (2016) Wide and high resolution tension measurement using FRET in embryo. Sci Rep 6:28535
227. Silhavy TJ, Kahne D, Walker S (2010) The bacterial cell envelope. Cold Spring Harb Perspect Biol 2:a000414
228. Levina N et al (1999) Protection of Escherichia Coli cells against extreme turgor by activation of MscS and MscL mechanosensitive channels: identification of genes required for MscS activity. EMBO J 18:1730–1737. doi:10.1093/emboj/18.7.1730
229. Bialecka-Fornal M, Lee HJ, Phillips R (2015) The rate of osmotic downshock determines the survival probability of bacterial mechanosensitive channel mutants. J Bacteriol 197:231–237. doi:10.1128/JB.02175-14
230. Nomura T, Cox CD, Bavi N, Sokabe M, Martinac B (2015) Unidirectional incorporation of a bacterial mechanosensitive channel into liposomal membranes. FASEB J 29:4334–4345. doi:10.1096/fj.15-275198
231. Martinac B et al (2013) Bacterial mechanosensitive channels: models for studying mechanosensory transduction. Antioxid Redox Signal 20:952. doi:10.1089/ars.2013.5471
232. Blount P, Sukharev SI, Moe P, Kung C (1997) Mechanosensitive channels of E. coli: a genetic and molecular dissection. Biol Bull 192:126–127
233. Vasquez V, Sotomayor M, Cordero-Morales J, Schulten K, Perozo E (2008) A structural mechanism for MscS gating in lipid bilayers. Science 321:1210–1214. doi:10.1126/science.1159674
234. Anishkin A et al (2003) On the conformation of the COOH-terminal domain of the large mechanosensitive channel MscL. J Gen Physiol 121:227–244
235. Ando C, Liu N, Yoshimura K (2015) A cytoplasmic helix is required for pentamer formation of the Escherichia coli MscL mechanosensitive channel. J Biochem 158:109–114. doi:10.1093/jb/mvv019
236. Iscla I, Wray R, Blount P (2008) On the structure of the N-terminal domain of the MscL channel: helical bundle or membrane interface. Biophys J 95:2283–2291. doi:10.1529/biophysj.107.127423
237. Steinbacher S, Bass R, Strop P, Rees DC (2007) Structures of the prokaryotic mechanosensitive channels MscL and MscS. In: Hamill OP (ed) Mechanosensitive ion channels, part A, San Diego, Elsevier Academic Press, Inc, pp 1–24
238. Honore E (2007) The neuronal background K2P channels: focus on TREK1. Nat Rev Neurosci 8:251–261., doi:nrn2117 [pii]10.1038/nrn2117
239. Patel AJ, Honore E (2001) Properties and modulation of mammalian 2P domain K+ channels. Trends Neurosci 24:339–346
240. Patel AJ et al (1998) A mammalian two pore domain mechano-gated S-like K+ channel. EMBO J 17:4283–4290. doi:10.1093/emboj/17.15.4283

Chapter 5
Lipid Domains and Membrane (Re)Shaping: From Biophysics to Biology

Catherine Léonard, David Alsteens, Andra C. Dumitru, Marie-Paule Mingeot-Leclercq, and Donatienne Tyteca

Abstract The surface of living cells provides an interface that not only separates the outer and inner environments but also contributes to several functions, including regulation of solute influx and efflux, signal transduction, lipid metabolism and trafficking. To fulfill these roles, the cell surface must be tough and plastic at the same time. This could explain why cell membranes exhibit such a large number of different lipid species and why some lipids form membrane domains. Besides the transient nanometric lipid rafts, morphogical evidence for stable submicrometric domains, well-accepted for artificial and highly specialized biological membranes, has been recently reported for a variety of living cells. Such complexity in lipid distribution could play a role in cell physiology, including in cell shaping and reshaping upon deformation and vesiculation. However, this remains to be clearly demonstrated. In this chapter, we highlight the main actors involved in cell (re)shaping, including the cytoskeleton, membrane-bending proteins and membrane biophysical properties. Based on integration of theoretical work and data obtained on model membranes, highly specialized cells and living cells (from prokaryotes to yeast and mammalian cells), we then discuss recent evidences supporting the existence of submicrometric lipid domains and documented mechanisms involved in their control. We also provide key recent advances supporting the role of lipid

C. Léonard
Louvain Drug Research Institute, Université catholique de Louvain (UCL), Brussels, Belgium

de Duve Institute, Université catholique de Louvain (UCL), Brussels, Belgium

D. Alsteens • A.C. Dumitru
Institute of Life Sciences, Université catholique de Louvain (UCL), Louvain-La-Neuve, Belgium

M.-P. Mingeot-Leclercq
Louvain Drug Research Institute, Université catholique de Louvain (UCL), Brussels, Belgium

D. Tyteca (✉)
de Duve Institute, Université catholique de Louvain (UCL), Brussels, Belgium
e-mail: donatienne.tyteca@uclouvain.be

R.M. Epand, J.-M. Ruysschaert (eds.), *The Biophysics of Cell Membranes*, Springer Series in Biophysics 19, DOI 10.1007/978-981-10-6244-5_5

domains in cell (re)shaping. We believe that the surface of living cells is made of a variety of lipid domains that are differentially controlled and remodelled upon cell (re)shaping.

Keywords Biological membranes • Membrane lateral structure • Lipid domains • Model membranes • Living cells • Imaging • Lipid probes • Membrane shaping • Cell deformation • Atomic force microscopy • Micropipette • Microfluidics • Cytoskeleton • Curvature • Fluidity • Asymmetry • Membrane dipole • Calcium

Abbreviations

AFM	atomic force microscopy
BODIPY	4,4-difluoro-5,7-dimethyl-4-bora-3a,4a-diaza-*s*-indacene
Ca^{2+}	calcium ion
Chol	cholesterol
CTxB	cholera toxin B subunit
ER	endoplasmic reticulum
ERM	ezrin, radixin, moesin
FCS	fluorescence correlation spectroscopy
FRAP	fluorescence recovery after photobleaching
FRET	fluorescence resonance energy transfer
GPI	glycosylphosphatidylinositol
GPMV	giant plasma membrane vesicle
GSL	glycosphingolipid
GUV	giant unilamellar vesicle
Ld	liquid-disordered
Lo	liquid-ordered
mβCD	methyl-β-cyclodextrin
MV	microvesicle
PC	phosphatidylcholine
PE	phosphatidylethanolamine
PI	phosphatidylinositol
PIP_2	$PI(4,5)P_2$, phosphatidylinositol-4,5-bisphosphate
PIPs	phosphoinositides
PM	plasma membrane
PS	phosphatidylserine
RBC	red blood cell
SDS	sodium dodecyl sulfate
SIM	structured illumination microscopy
SIMS	secondary ion mass spectrometry
SM	sphingomyelin
SMase	sphingomyelinase
STED	stimulated emission depletion microscopy

TCR T cell receptor
T*m* melting temperature

5.1 Introduction

The surface of living cells is a complex assembly of a variety of molecules that provides an interface separating the outer and the inner environments. It is also responsible for a number of important functions, including regulation of solute influx and efflux, signal transduction, lipid metabolism and trafficking, and represents the target of infectious agents such as bacteria and their associated toxins, viruses and parasites, *a.o.* To fulfill these functions, the cell surface must be tough and plastic at the same time. This could explain why cell membranes exhibit a so large number of different lipid species that are heterogeneously distributed, both transversally and laterally.

Glycerophospholipids, sphingolipids and sterols are the main lipids found in biological membranes. Glycerophospholipids include phosphatidylcholine (PC), phosphatidylethanolamine (PE), phosphatidylserine (PS), phosphatidic acid (PA) and phosphatidylinositol (PI) and its phosphorylated derivatives (PIP, PIP_2 and PIP_3; collectively PIPs). Sphingolipids are derived from ceramide, which is decorated with a phosphocholine headgroup in the case of sphingomyelin (SM) or with saccharides in the case of glycosphingolipids (GSLs). Sterols are constituted by an inflexible core formed by four fused rings, with cholesterol predominating in mammals. Several features indicate remarkable membrane lipid diversity. Thus, even within the same lipid class, lipids can differ regarding headgroup structures and length and degree of unsaturation of the hydrophobic chains, creating thousands of combinations. Whether each lipid species has a defined biological function or whether cell function is modulated by structural organization of membrane lipids remain to be elucidated. In favour of the second hypothesis, lipid diversity could guarantee a much more stable, robust membrane that can withstand changes in the surrounding pH, temperature and osmolarity [1]. In this context, diversity could be intrinsically related to the different functions assumed by cell membranes including ligand binding, endocytosis, intracellular transport, cell migration or squeezing *e.g.*, Whatever the hypothesis, the lipid diversity allows for different non-covalent forces, *i.e.* van der Waals, electrostatic, solvation (hydration, hydrophobic), steric, entropic, *e.g.*, which are critical for membrane structure, organization and functions through modulation of biophysical membrane properties including lipid packing, membrane curvature and asymmetry [2–4].

Current views on structural and dynamical aspects of biological membranes have been strongly influenced by the homogenous fluid mosaic model of Singer and Nicolson in 1972 [5]. Today, this basic model remains relevant [6], although it is widely accepted that it cannot explain the role of mosaic, aggregate and domain structures in membranes as well as the lateral mobility restriction of many membrane proteins [7]. In the 90's, Simons and coll. proposed the lipid raft hypothesis [8], where GSLs form detergent-resistant membranes (DRMs) enriched in cholesterol and glycosylphosphatidylinositol (GPI)-anchored proteins in cold non-ionic detergents. In 2006, lipid rafts were redefined as: "small (20–100 nm), heterogeneous, highly dynamic, sterol- and sphingolipid-enriched domains that compartmentalize cellular processes. Small rafts can sometimes be stabilized to form larger platforms through protein-protein and protein-lipid interactions" [9]. However, the raft hypothesis is still up for debate [10, 11]. This could arise from its original definition experiments using the controversial detergent extraction method in combination with cholesterol and/or sphingolipid depletion. It should be nevertheless noticed that recent progress in microscopy, such as combined fluorescence correlation spectroscopy (FCS) with stimulated emission depletion microscopy (STED) [12] or super-resolution microscopy [13], provides strong evidence for the existence of transient, nanoscale, cholesterol- and sphingolipid-enriched membrane clusters, giving new insight in the raft hypothesis. Controversial opinions regarding the raft hypothesis could also arise from its restricted definition as compared to the high diversity of lipid composition among cellular membranes and the wide amount of factors regulating lipid clustering. Yet, the original definition of rafts is often revisited. In 2010, lipid rafts were redefined as "fluctuating nanoscale assemblies of sphingolipids, cholesterol and proteins that can be stabilized to coalesce, forming platforms that function in membrane signaling and trafficking" [14], taking into account the dynamical aspect of membranes.

In addition to rafts, other nanoscale domains, *i.e.* < 100 nm in diameter, have been described at the plasma membrane (PM) of eukaryotes: caveolae [15] and tetraspanin-rich domains [16], *a.o.* Moreover, morphological evidence for stable (min *vs* sec) submicrometric (> 200 nm *vs* < 100 nm) lipid domains was reported in artificial [17–19] and highly specialized biological membranes [18, 20]. In the past decades, owing to the development of new probes and imaging methods, several groups have presented evidence for submicrometric domains in a variety of cells from prokaryotes to yeast and mammalian cells [21–27].

Such complexity in lipid distribution could play a role in cell physiology, including in cell shaping and reshaping processes. However, whereas the traction by the cytoskeleton, the action of membrane-bending proteins and membrane transversal asymmetry have been shown to contribute to cell reshaping [28–30], the importance of membrane lateral heterogeneity remains to be clearly determined. In this chapter, we highlight the physiopathological importance of membrane (re)shaping (Sect. 5.2), describe some methods to measure it (Sect. 5.3) and summarize the main actors involved in its regulation (Sect. 5.4). We then provide evidence for lipid domains in living cells (Sect. 5.5) and a summary of their regulation mechanisms (Sect. 5.6). We then integrate the importance of lipid lateral heterogeneity in membrane (re)shaping (Sect. 5.7).

5.2 Membrane Shaping & Reshaping – Role in Physiopathology

Membranes are at the centre of cell shaping and reshaping processes. For examples, red blood cell (RBC) exhibits a biconcave shape needed for its optimal deformation and function (see Sect. 5.4), while bacteria can be cocci, rods and spirochetes

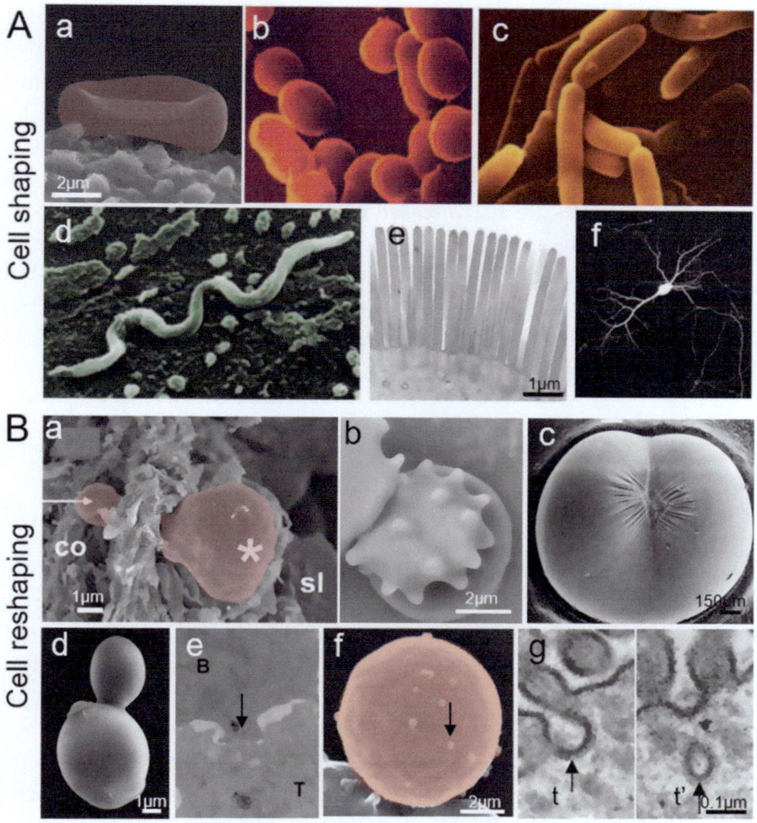

Fig. 5.1 Physiological importance of cell (re)shaping. **A**. Shape of (a) a biconcave RBC; (b–d) cocci, rods and spirochetes; (e) enterocyte brush border of mouse intestinal explant; (f) growing neuron branches into dendrite (Adapted from: (a) our unpublished data; (b–d) [31]; (e) [32]; (f) [33]). **B**. Reshaping upon (a) RBC crossing from the splenic cord to sinus; (b) platelet activation; (c) cleavage furrow initiation in *Xenopus*; (d) yeast *S. cerevisiae* budding; (e) formation of the immunological synapse (*arrow*); (f) RBC vesiculation upon senescence (*arrow*); (g) vesicle endocytosis (*arrow*) in mouse intestinal explants during fat absorption (Adapted from: (a) [34]; (c) [35]; (d) [36]; (e) [37]; (f) our unpublished data; (g) [32])

(Fig. 5.1Aa–d). Epithelial cells and neurons are other examples of cells showing 'special' shapes needed for their functions (Fig. 5.1Ae,f). In their environment, cells face a variety of stimuli and stresses, either chemical/biochemical (*e.g.* hormones, ligands, toxins, ions) or physico-mechanical (*e.g.* temperature, pH, pressure, shear stress, stretching). Examples include squeezing of RBCs in the narrow pores of spleen sinusoids, pressure exerted by tumors on surrounding cells, shear stress by the blood stream on endothelial cells, stretching of muscle cells during contraction, gathering of blood platelets to stop bleeding, cell division and formation of the immunological synapse (Fig. 5.1Ba-e). As opposed to global cell deformability, local budding and vesicle formation can also occur from the PM, allowing for endocytosis (Fig. 5.1Bg) or microvesicle (MV) formation. While considered for a long time as inert cellular fragments, MVs are nowadays recognized to play crucial roles in both physiological and pathological processes, such as intercellular communication [38], coagulation [39], inflammation [40], tumorigenesis [41] and migration [42], *a.o.* MVs are also released from RBCs upon normal senescence (Fig. 5.1Bf), a process accelerated in RBC membrane fragility diseases, leading to loss of biconcavity and deformability. It is thus important to decipher molecular details of membrane structure and mechanisms involved in cell (re)shaping. In Sect. 5.7, we focus on the potential implication of lipid domains in cell shaping, squeezing, vesiculation and division.

5.3 Membrane Shaping & Reshaping – Measurement

Cell mechanical properties involved in deformation can be studied by several biophysical methods. These can be classified into two categories, based on measurements on individual or multiple cells. Biophysical techniques devoted to individual cells include micropipette aspiration, atomic force microscopy (AFM), optical tweezers or microfluidics, *a.o.* Measurement on multiple cells can be performed by cell separation (by filtration through polycarbonate membrane or a mixture of microbeads) or microfluidics. Based on their ability to image lipid domains in relation to cell deformation, we decided to focus in this Section on the micropipette aspiration, AFM and microfluidics. For optical tweezers, we recommend [47]; for cell separation by filtration through polycarbonate membrane or a mixture of microbeads, see [34, 48].

Micropipette aspiration was initially developed to measure the RBC elastic properties [49]. Briefly, a micropipette is manipulated towards a cell, usually in suspension, and a small suction pressure is applied, partially aspirating the cell inside the micropipette. Upon increasing the suction pressure, the cell deforms, flows into the micropipette and increases the length of projection of the aspirated portion (Fig. 5.2Aa). This deformation is then analyzed to determine the cell elastic property, *i.e.* the Young's modulus. The technique can be used to apply forces over a range of 10 pN to 1 nN [50]. It has been applied to various cells including red cell membranes [51] and cancer cells [52, 53]. In the context of lipid domains, it can be used to test whether these exhibit a gradient along the deformation projection (Fig. 5.2Aa,b).

AFM was invented 30 years ago to image non-conductive samples at high-resolution. Key developments now allow AFM to investigate biological sample, *i.e.* imaging in buffer solution and maintaining the native state of the biological system [54]. Briefly the principle of AFM is to scan a tip over the sample while using an optical detection system to measure with high-sensitivity the force applied on the

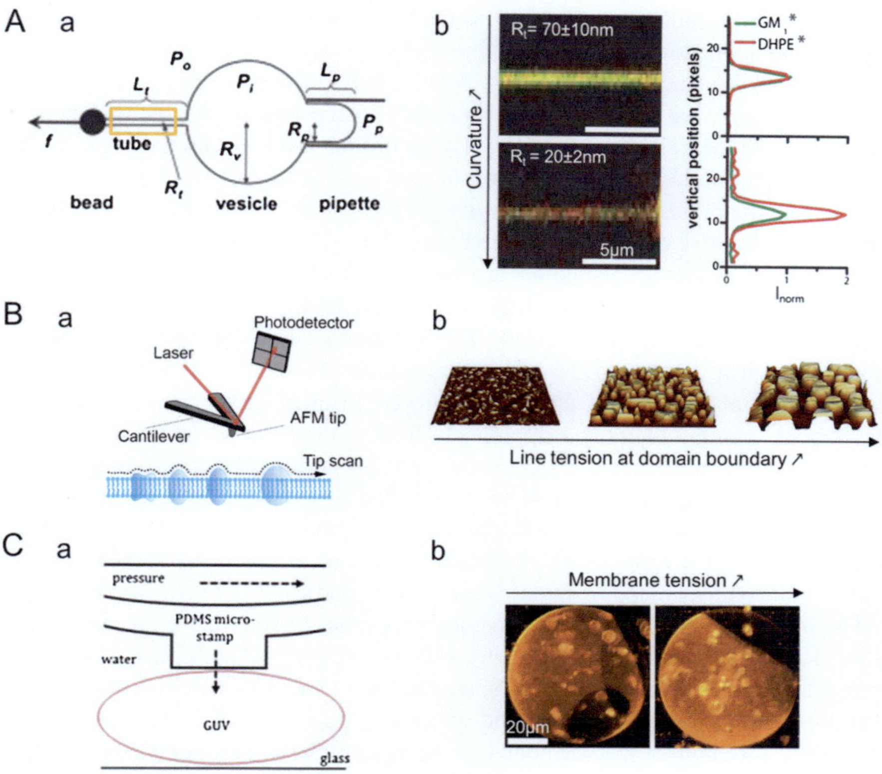

Fig. 5.2 Principles of some methods used to evaluate lipid domain organization upon deformation. **A**. Micropipette aspiration. (a) Experimental apparatus for GUV aspiration and tube pulling. *Pi*, *Po* & *Pp* pressures inside vesicle, outside vesicle & in the micropipette, *Rt*, *Rv* & *Rp* radius of tube, vesicle & pipette, *Lt* & *Lp* length of tube and vesicle projection in pipette, *f* pulling force, I_{norm} normalized intensity [43]. (b) Confocal equatorial section showing the partitioning of fluorescent lipids in membrane nanotubes of different radii (*yellow* box in a) pulled out from a GUV [44]. **B**. AFM. (a) The AFM consists in an AFM tip positioned at the end of a flexible cantilever and brought in contact with the sample. During scanning, the applied force is maintained constant thanks to an optical system based on a laser focused at the end of the cantilever and reflected into a photodetector. The tip contours the surface and its movement results in a height image. (b) Three-dimensional AFM images of phase-separated membranes of differential composition and domain height mismatch [45]. **C**. Microfluidics. (a) Side view of a microfluidic device composed of a pressure layer with integrated microstamps situated above a fluidic layer containing a GUV. (b) Induced tension by the microfluidic device causes lipid lateral sorting in GUVs: 0 min, one Ld and two Lo phases (*left*); after 45 min, lipid sorting into one Lo patch to reduce the tension (*right*) [46]

sample [55]. This detection system is based on a laser focused at the end of the cantilever and deflected into a photodiode. This electric signal is then converted into a force using calibrated parameters. By maintaining the applied force constant, the height of the tip is adjusted and its movement results in the height image that resembles the sample topography with the resolution depending on the radius of the tip, the applied force, the physical properties of the sample, and how precisely the feedback system acts (Fig. 5.2B). AFM rapidly evolved from an imaging tool to a multifunctional tool that, simultaneously to the topography, is also capable to probe biophysical properties [56]. By recording force-distance curves, *i.e.* by monitoring the variation of the force while approaching the AFM tip and retracting it away from the biological sample, various properties can be quantified either during the approach curve or during the retraction curve. The approach curve allows the extraction of properties including mechanical deformation of the sample, elastic modulus and energy dissipation. The dissipation is the area of the hysteresis between the approach and the retract curve. The retraction curve can quantify adhesion forces established between the tip and the sample. Modern AFM instruments can acquire several hundreds of thousands of force-distance curves while imaging the biological sample, allowing mapping physical properties and interactions to the sample topography. Force-distance based AFM (or multiparametric imaging) opens the door to image complex biological systems and simultaneously quantify and map their properties. Nowadays, multiparametric imaging allows investigating native biosystems with a resolution approaching 1 nm on purified membrane proteins and simultaneously mapping their mechanical properties [57]. Force-distance based AFM was also used to map the mechanical properties of heterogeneous membranes [58], RBCs [59], human keratinocytes [60] or bacteria [61]. Besides mechanical properties, force-distance based AFM also enables to map specific receptors. Using functionalized AFM tips with specific chemical groups or ligands, the adhesion and mechanical strength of specific bonds can be measured. Furthermore recording these forces while imaging the biological systems allows detecting and localizing specific interaction of biological samples ranging from antibodies to living cells [62–64]. Biospecific AFM mapping has proven useful to map receptor sites on animal cells [62, 65, 66].

Microfluidic technologies can also be used to investigate cell mechanical properties upon deformation (Fig. 5.2Ca). These can be classified according to the mechanical stimuli used to deform the cell (for review [67]): constriction channel, shear stress, voltage shock, optical stretcher, electric field or micropipette aspiration. Microfluidics have been used to measure the deformability of RBCs [68], leukocytes [69], human cancer cell lines [70] and patient oral squamous cells [71]. Furthermore, their development in pathological contexts gives promising perspectives for labelled-free clinical diagnostic. For example, breast cancer cells were distinguished from non-malignant cells [72], malignant cells were identified in human pleural fluid sample [73] and RBCs with deficiencies of cytoskeletal protein network were detected [74]. While cost-effective, microfluidics is biocompatible, requires small sample volume and gives fast responses. In addition, its "tune-ability" makes it adaptable to any cell type, permits to recreate specific environmental deformation conditions and to record electrical or biochemical properties besides mechanical parameters. Furthermore, as microfluidics can be coupled to fluorescence microscopy, it has been used to evaluate reorganization of lipid domains under stretching of model membranes (Fig. 5.2Cb) [46, 75].

5.4 Membrane Shaping & Reshaping – Regulation

In this Section we provide an integrated view on documented mechanisms that govern cell (re)shaping, first focusing on the simplest, best characterized and highly deformable RBCs (Sect. 5.4.1). We will then deepen three key determinants for cell (re)shaping, *i.e.* the cytoskeleton (Sect. 5.4.2), membrane shaping proteins (Sect. 5.4.3) and intrinsic membrane properties (Sect. 5.4.4).

5.4.1 Main Determinants of Cell (Re)Shaping – A Focus on RBCs

The unique ability of RBC to deform is allowed by (i) its particular constitutive biconcave geometry, (ii) a finely controlled cytoplasmic viscosity, and (iii) communication with its environment.

RBC exhibits a particular geometry characterized by a high membrane surface-to-volume ratio. Indeed, by comparison to a sphere of the same volume, RBC presents a membrane surface excess of ~40%, explaining its biconcave shape. Three factors allow to maintain and adapt this shape. First, the RBC membrane is supported by a particularly dense and stable spectrin network. Second, this cytoskeleton is strongly linked to the RBC PM, thus preventing membrane area loss by vesiculation [28] (see Sect. 5.4.2). Third, the RBC cellular volume is tightly regulated by several ionic transports. The Na^+/K^+-ATPase and the Ca^{2+}-ATPase (also called PMCA) set up the major cation gradients across the RBC PM [76, 77]. In addition, the RBC is endowed with a large variety of ion channels that are nowadays proposed to play a dynamic role (reviewed in [78]). Thus, upon shear stress in the circulation, a reversible increase in Ca^{2+} permeability occurs. Piezo1, a mechanosensitive non-selective cation channel, has been recently identified as the link between mechanical forces, Ca^{2+} influx and RBC volume homeostasis. This study clearly indicates a role for mechanotransduction in cell volume regulation *via* Ca^{2+} influx through Piezo1 and subsequent RBC dehydration *via* downstream activation of the Gardos channel [79], a Ca^{2+}-dependent K^+ efflux channel [77].

RBC cytoplasmic viscosity, determined by hemoglobin concentration, is finely regulated and comprised between 30–35 g/dL [80]. This characteristic allows RBCs to rapidly adapt their shape upon shear stress. Aged and sickle RBCs exhibit a hemoglobin concentration > 37 g/dL, resulting into cell rigidity and reduced deformability [80].

RBCs are now known to communicate with their local environment, serving as both recipients and producers of extracellular stimuli. Among the various RBC molecules that contribute to such signaling is ATP, a regulatory molecule for both intracellular and extracellular functions. Intracellular ATP represents an energy source needed for Na^+/K^+- and Ca^{2+}-ATPases, ATP-dependent glucose transporters and flippases *a.o.* and for modulation of the compliance of the membrane with the cytoskeleton [81–84] (see Sect. 5.4.2). The release of ATP from RBCs to the extracellular space, which occurs in response to small changes in osmotic pressure, O_2 concentration and pH, is proposed to be triggered by the retraction of the spectrin-actin network [85]. Released ATP then induces the release of the potent vasodilator nitric oxide from endothelial cells into the surrounding smooth muscle cells. Among the most important signaling molecules able to regulate RBC properties is O_2. It has been recently demonstrated that the RBC oxygenation state regulates membrane mechanical stability, glucose metabolism and ATP release *via* the reversible association of deoxyhemoglobin with the anion transport protein Band 3 that acts as an O_2-triggered molecular switch to regulate RBC properties [86].

Whether RBC membrane intrinsic properties, in particular lateral membrane heterogeneity, contribute along with the above factors to deform the RBC remains poorly understood. Arguments in favor of this hypothesis are discussed at Sects. 5.7.2 and 5.7.3.

5.4.2 Cytoskeleton

As highlighted above, the RBC cytoskeleton strengthens the lipid bilayer and endows the membrane with durability and flexibility to survive in the circulation [28]. It is made of a pseudohexagonal meshwork of spectrin, actin, protein 4.1R, ankyrin and actin-associated proteins, attached to the membrane via multiprotein complexes, centered on ankyrin and protein 4.1R. For readers interested in the anatomy of the red cell membrane skeleton, an excellent recent review is recommended [28].

It is well established that the RBC membrane is not static and ATP allows maintaining the RBC biconcave shape and dynamic characteristics. For instance, intracellular ATP increases the compliance of the membrane, as revealed by AFM upon small compression of subcellular components [81] and through fluctuation analysis [82, 83]. Mechanical measurements using optical tweezers has however led to the opposite conclusion [84]. Thus, while it is clear that phosphorylation directly modulates the mechanical stability of the RBC membrane, the response is complex due to the fact that many cytoskeleton components are phosphoproteins. Among those, phosphorylation of the 4.1R by PKC triggers the relaxation of the RBC membrane through loosening the link between membrane and spectrin [87] whereas phosphorylation of β-spectrin through the membrane-bound casein kinase I [88] could enhance the maintenance of stable network interaction by increasing the spectrin-ankyrin affinity [81]. Besides phosphoproteins, PIPs are alternative candidates. While PIP_2 enhances the binding of 4.1R to glycophorin C, it inhibits the binding to Band 3 *in vitro* [89].

In nucleated mammalian cells, the cytoskeletal network scaffolding the PM at the macroscopic level is a heterogeneous system consisting of actin fibers, intermediate filaments and microtubules. This elaborate protein organization mediates and controls membrane shaping and organization through the continuous dynamic interplay between the PM and the cortical network underlying it. Adhesion between the cytoskeleton and the lipid bilayer maintains membrane tension, while membrane shape and cytoskeletal assembly/disassembly processes are also strongly intertwined. The cytoskeleton is known to regulate several essential cell processes as follows: cortical actin supports the macroscopic curvature of the membrane during mitosis, membrane ruffling is involved in phagocytosis, while actin dynamics (treadmilling, branching and bundling) provides the mechanical force needed for endocytosis, migration (formation of filopodia and lamellopodia) and morphogenesis. Additionally, molecular motors such as kinesins, dynein and myosin support some organelle morphologies and promote the reorganization of the membrane [30].

Cortical actin is tightly bound to the PM *via* actin binding proteins, such as α-actinin domain, the ERM (Ezrin/Radixin/Moesin) domain [90, 91], the calponin homology (CH) domain [92] and the Wiskott-Aldrich syndrome (WASP) homology domain-2 (WH2) [93]. More commonly, linker proteins act as crosslinkers between the actin architecture and membrane proteins by means of protein-protein interactions motifs such as PDZ or ankyrin. PIPs are major regulatory factors of actin-membrane interactions [94, 95], playing several roles in actin dynamics regulation by controlling the localization and activity of actin-binding proteins, *a.o.* Interactions between proteins and PIPs are mainly mediated by pleckstrin homology (PH) domains [96–99]. In addition, many actin binding and actin modulating proteins get activated through the interaction with PIP_2-containing membranes (*e.g.* α-actinin, vinculin, talin or ezrin). Moreover, PIP_2 stimulates actin polymerization activators (*e.g.* WASP, WAVE), while it inhibits proteins that break and depolymerize actin (*e.g.* gelsolin, cofilin, villin, profilin) [100]. Some studies also proposed the connection between actin cytoskeleton and PS to induce nanoclustering of GPI-anchored proteins *via* transbilayer lipid interactions [101].

5.4.3 Membrane Shaping Proteins

Remodeling of cell shape is accomplished by recruiting specialized proteins, which contain motifs able to generate, sense or stabilize membrane curvature. The synergistic actions of membrane shaping proteins along with changes in the lipid bilayer and the cytoskeleton enable numerous cellular processes like division, migration and intracellular trafficking.

Three key mechanisms underlying membrane shaping are currently known. The first mechanism acts at the nanoscopic level and is a result of protein crowding and partitioning of transmembrane domains. Molecular crowding by protein-protein interactions has been recently pinpointed as a mechanism for altering the effective bending modulus and the curvature of the membrane [102]. Transmembrane proteins with a conical or inverted-conical shape can also mold their associated membranes around their shapes, as in voltage-dependent K^+ channels [103] or the nicotinic acetylcholine receptor [104]. Additionally, transmembrane receptor clustering, such as transferrin or low-density lipoprotein, results in endocytic clathrin-coated pit formation [105].

The second mechanism involves the direct insertion of small hydrophobic protein motifs between the lipid headgroups. The hydrophobic surface insertion into the membrane hemilayer enlarges the surface of the inner leaflet, thus causing membrane curvature. Numerous proteins playing key roles in membrane shaping are known to present amphiphatic helices (endophilin, amphipysin, epsin, Bin2 *a.o* [106]), but the most well-known classes leading to this type of electrostatic interactions are the endosomal sorting complexes required for transport (ESCRT) and Bin/Amphiphysin/Rvs (BAR) domain containing proteins. The electrostatic interactions between ESCRT proteins are involved in membrane budding during virus infection, membrane scission in the multivesicular body pathway and cytokinesis [107, 108]. The BAR domain protein superfamily includes dimeric banana shaped structures, which bind electrostatically to the membrane through their concave face. Binding is thought to be mediated by the interaction between positively-enriched areas of the BAR module (membrane contact site) and negatively-charged lipids like PIPs [109, 110]. BAR proteins can also target negative PIPs through pleckstrin homology (PH) or PhoX (PX) domains. The fact that BAR domains interact preferentially with curved membranes makes them a sensor of high positive curvature [111, 112]. The F-BAR (FCH-BAR) domains recognize shallow positive curvature, while I-BAR (Inverse-BAR) domains interact with shallow negatively curved membranes. BAR domain proteins have been implicated in many cellular functions involving sensing or induction of membrane curvature, such as endocytosis and membrane trafficking, podosome and filopodia formation, or mitochondria and autophagosome shape (reviewed in [113]). BAR domain proteins can be seen as a signaling 'hub' connecting membrane geometry and/or lipid composition to actin cytoskeleton regulation and to different signaling pathways [114, 115].

The last mechanism of membrane deformation is the scaffold mechanism by peripheral proteins at the nanoscopic level and their oligomeric assemblies at the microscopic level. Clathrin, COPI and COPII are coat proteins recruited from the cytosol during vesicle budding. They have the capacity to bend membranes by relying on adaptor proteins. After the spherical coated vesicle pinches off, these proteins are released back into the cytosol and can be recycled [116, 117]. Oligomerization of caveolin is linked to the formation of caveolae [15], while reticulons and flotilins stabilize the ER curvature [118].

5.4.4 Intrinsic Membrane Properties

Several intrinsic membrane properties contribute to cell (re)shaping. We here provide information on their regulation and physiological implication, with specific focus on the molecular level. Membrane properties and regulation at the larger scale of lipid domains and resulting from collective lipid behaviour will be discussed in more details in Sect. 5.6.

The cellular membrane exhibits transbilayer asymmetry, first hypothesized in the 70's by Bretscher [119]. This asymmetry contributes to PM complexity and diversity by the differential repartition between the two leaflets of lipid (i) order and packing (Sect. 5.4.4.1), (ii) charge and dipole (Sect. 5.4.4.2) and (iii) molecular shape (Sect. 5.4.4.3), thereby leading to optimal physiological output. The inner monolayer contains most of PS and PE whereas PC and SM are mostly located within the outer leaflet. Whereas lipid PM asymmetry has been largely reported including in human RBCs [120] and platelets [121], it is cell type-dependent [122, 123]. The asymmetric distribution of phospholipids is accompanied by asymmetry of fatty acid chains. For example, in human RBCs, the double bond index is 1.54 for the inner face *vs* 0.78 for the outer face [124]. In contrast to phospholipids, transbilayer distribution of cholesterol is highly debated [125]. Recently, cholesterol has been shown to inhibit phospholipid scrambling [126], an unsuspected function that could be critical for cell deformation. Membrane proteins (with their preferred orientation) and communication with the exterior and interior aqueous compartments (which contain different concentrations of ions, small molecules, and/or proteins) also contribute to the bilayer asymmetry. Rapid exchanges between leaflets are presumed to be prohibited by the large enthalpic barrier associated with translocating hydrophilic materials, such as a charged lipid headgroup, through the hydrophobic membrane core. The mechanism underlying transbilayer asymmetry involves specific flippase (inward moving), floppase (outward moving) and scramblase (bidirectional) enzymes that assist in the movement of lipids between the two leaflets of cellular membranes [127]. The transbilayer coupling may be also an intrinsic property of the lipid themselves [128] *via* interdigitation through long acyl chain (C22-C24) [101, 129].

5.4.4.1 Lipid Order and Packing

Lipid packing depends on the ratio between small and large polar heads and the ratio between unsaturated and saturated acyl chains. The usual cis-unsaturated oleyl chain (C18:1) occupies a larger volume than the palmitoyl chain (C16:0) because the double bond induces a “kink” in the middle of the chain which lowers the packing density of the acyl chains, thereby increasing membrane fluidity [130]. Owing to its acyl chain composition, SM forms a taller, narrower cylinder than PC, increasing its packing density in the membrane. Consequently, at physiological temperature, a SM bilayer exists in a solid gel phase with tightly packed, immobile acyl chains [131, 132]. By interfering with acyl chain packing, sterols inhibit the transition of the membrane to the solid gel state. At the same time, sterols rigidify fluid membranes by reducing the flexibility of neighbouring unsaturated acyl chains, thereby increasing membrane thickness and impermeability to solutes (the so-called condensing effect of sterols) [133].

Based on membrane packing criteria, membrane can be viewed as a patchwork with areas characterized by differences in membrane fluidity. The areas of low fluidity are named the solid phase (Lβ) (or solid-ordered (So) phase). In these areas the lipid acyl chains are tightly packed and there is a low rate of lateral diffusion. In contrast, the more fluid areas are named the liquid crystalline (Lα)

phase [134] (more commonly called the liquid-disordered (Ld) state) which exhibits both low packing and high lateral diffusion. In addition, at the proper concentration, cholesterol may facilitate lateral segregation of lipids into cholesterol-depleted and -enriched regions, such as liquid-ordered (Lo) lipid domains, which expose high packing and high lateral diffusion. Lipid phase behavior is temperature-dependent and the Lβ phase transition into Ld phase occurs when temperature increases. The temperature at which this transition occurs is known as the gel-to-liquid transition temperature (T*m*) and depends on lipid acyl chains. Lipids with long saturated fatty acyl chains (*e.g.* most sphingolipids) have high T*m*, whereas lipids with fatty acids having cis double bonds (*e.g.* most phospholipids) have low T*m*.

Membrane fluidity is critical to warrant proper protein sorting and membrane trafficking required during adaptive responses. For example, organelles of the secretory pathway differ in lipid composition, resulting into gradual increase of molecular packing density and membrane rigidity from the ER toward the PM [131, 135]. Thus, modulation of lipid composition and fluidity seems critical for adaptive responses even though cytosolic proteins and integral membrane sensors also contribute to regulate fluidity [2, 136, 137]. As a consequence of membrane transbilayer asymmetry, membrane physical properties are also asymmetrical, the outer monolayer being more packed and rigid than the inner one [124]. How differential order of lipid domains in one leaflet can affect the order of the opposite leaflet is highlighted in Sect. 5.6.2.

5.4.4.2 Lipid Dipole Potential

Most of the phospholipids and sphingolipids are zwitterionic and exhibit a significant permanent electric dipole moment [138]. Cell membrane transversal asymmetry thus creates a permanent dipole potential, leading to a significant difference in electric potential across the membrane that can vary from 100 to 400 mV and is positive in the membrane interior [139]. The dipole potential arises from the water dipole of the hydrated lipid bilayer [138, 140], the fatty acid carbonyl groups [138, 141] and the lipid headgroup [142]. Cholesterol increases the membrane dipole potential by impacting the orientation, strength and packing density of the molecular dipoles at the membrane surface [143, 144] and *via* its own dipole moment, which depends on its membrane orientation and membrane packing [145, 146]. How membrane and domain dipole potential can affect lipid domain size and topography and how electrostatic interactions modulate and reorganize lipid domains are developed in Sects. 5.6.2 and 5.6.3.

5.4.4.3 Lipid Molecular Shape

Moving to lipid molecular shape, *i.e.* ratio of head-to-tail area, some lipids like PC show comparable lateral areas in the head and tail regions with an overall cylindrical molecular geometry, forming a planar bilayer. In contrast, PE have a small headgroup relative to the cross-sectional area of the hydrocarbon tails (conical shape) whereas lysophospholipids are characterized by tail regions of bigger lateral cross-section than the headgroups (inverted conical shape). The cylindrical, conical and inverted-conical lipids have zero, negative and positive spontaneous curvatures, respectively.

Generation of membrane shape by lipids is generally attributed to intrinsic lipid molecular shapes (lipid morphism [147]) and lipid membrane transversal asymmetry (bilayer couple hypothesis [148]) and this specific molecular lipid sorting is usually associated to a substantial and persistent energy input mediated by proteins [31]. The asymmetry between the inner and outer leaflets results in spontaneous bending of originally flat membrane. This can be ascribed by an elastic parameter, named the spontaneous curvature which corresponds to the curvature that an unconstrained monolayer would adopt. It can be positive (if the membrane prefers to bulge toward the exterior compartment) or negative (the opposite). When a system is forced to adopt a curvature different from the spontaneous curvature, the curvature elastic stress is considered. Examples include cellular processes that require membrane bending like endocytosis, budding or cell deformation. Since these processes are highly sensitive to changes in lipid composition and to the presence of specific lipids [149, 150], subtle modifications in lipid composition may have major implications for lipid and protein sorting under a curvature-based membrane-sorting model [30, 151, 152]. How membrane curvature can affect lipid domain sorting and topography is discussed in Sect. 5.6.2, and how lipid domains could be involved in the generation of membrane shape is discussed in Sect. 5.7.1.

5.5 Lipid Domains – Evidence

The concept of lipid rafts is used to describe unstable nanoscale assemblies (20–100 nm) enriched in sphingolipids, cholesterol and GPI-anchored proteins [8, 14]. Besides rafts, there are various types of membrane domains that are characterized by their enrichment in specific proteins, such as caveolae and tetraspanin-enriched domains [15, 16]. Rafts can sometimes be stabilized to form larger platforms through protein:protein and protein:lipid interactions [9]. Morphological evidence for stable (min *vs* sec for rafts) submicrometric domains (> 200 nm in diameter *vs* < 100 nm) has been reported in artificial [17–19] (Sect. 5.5.1) and highly specialized biological membranes [18, 20] (Sect. 5.5.2).

However, there is an intensified debate on the real existence of stable submicrometric lipid domains in cells. This can result from three main features. First, whereas some groups have provided evidences for stable submicrometric lipid domains in physiological conditions (Sects. 5.5.3 and 5.5.4 for examples), there are cases in which they have not been detected. For example, whereas submicrometric domains enriched in sphingolipids have been demonstrated by secondary ion mass spectrometry (SIMS) at the fibroblast PM, cholesterol is uniformly distributed throughout [153, 154]. Likewise, using protein micropatterning combined with single-molecule tracking, Schutz and coll. have shown that GPI-anchored proteins do not reside in ordered domains at the PM of living cells [155]. Differences between studies can be explained by several reasons. For instance, analysis of lipid lateral heterogeneity suffers from technical issues such as lipid unresponsiveness to chemical fixation, fast translational movement, small molecular size and high packing density. As a consequence, it is a big challenge to design small specific fluorescent tools that can be used to analyze lipid organization by microscopic methods with resolution approaching the nanometer-scale under poor lipid fixation. Moreover, imaging artefacts could arise from non-resolved membrane projections and domain abundance strongly varies with temperature, another possible cause of non-reproducibility. Despite these limitations, which are also discussed elsewhere [156, 157], novel specific probes have recently been developed and validated (reviewed in [156, 158–160]). Membrane composition and biophysical properties also strongly influence lipid lateral distribution. Finally, living cells are far from equilibrium and are instead constantly reorganized by energy-driven processes, including motor-driven constriction of the cytoskeleton, membrane trafficking, lipid metabolism and exchanges of ions and molecules with the environment. A second reason alimenting the debate is that submicrometric lipid domains have sometimes been reported under non-physiological conditions: (i) in RBCs after alteration of membrane ceramide or cholesterol contents upon treatment with a toxin from *Pseudomonas aeruginosa* [161] or methyl-β-cyclodextrin (mβCD) [162], respectively; and (ii) in CHO cells upon cholesterol depletion [163]. Third, lipid domains could be not stably present but transiently generated by the hydrolysis of specific lipids. One can cite the ceramide-rich domains with diameters of ~200 nm up to several micrometers that can be formed upon SM degradation by acid SMase in response to stress [164, 165].

Therefore, one major challenge will be to evaluate whether submicrometric lipid domains can be generalized or if they are restricted to cells exhibiting particular membrane lipid composition, biophysical properties and membrane:cytoskeleton anchorage. The rest of this book chapter is dedicated to this crucial question.

5.5.1 Membrane Models

Different types of model membranes have been developed to study phase separation, including planar supported bilayers [170] and giant unilamellar vesicles (GUVs) [171] made from lipid mixtures as well as giant PM vesicles (GPMVs) isolated from cellular PMs after chemical treatment [172]. All these models are useful to perform systematic analysis of the impact of lipid composition on phase separation, like in model membranes mimicking the composition of the PM outer leaflet (Fig. 5.3a,b). They show liquid-liquid phase separations, with domains of variable sizes depending on lipid composition and temperature (Fig. 5.3c,d). Planar supported bilayers are useful for methods requiring rigid planar surfaces like AFM and ToF-SIMS.

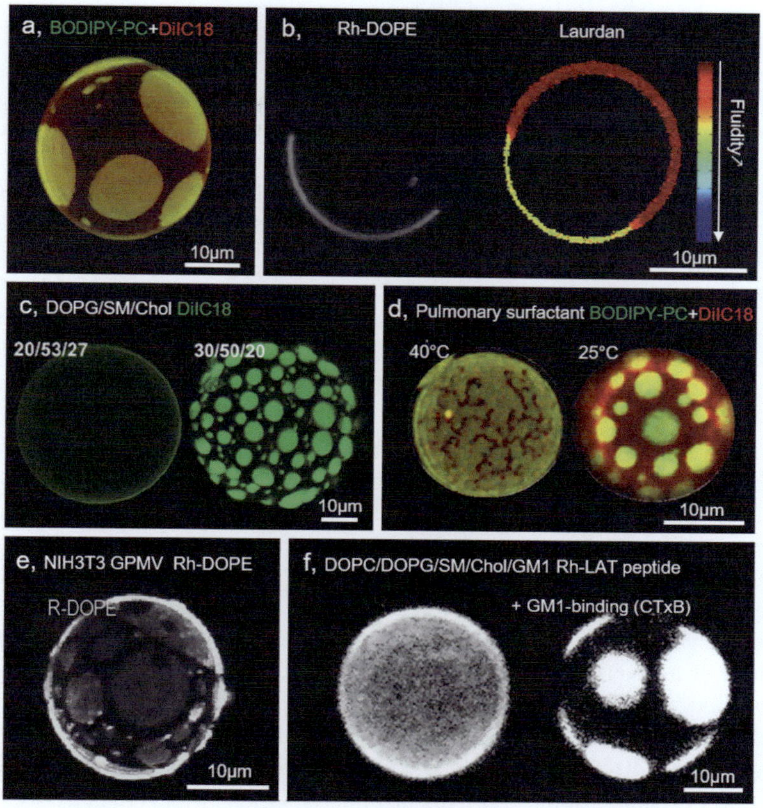

Fig. 5.3 Visualization of phase coexistence in model membranes. (**a**) GUV (DOPC/DPPC/Chol) labelled with BODIPY-PC and DiIC18 showing Lo and Ld phases. (**b**) GUV (DOPC/stearyl-SM/Chol) labelled with Rhodamine-DOPE (Ld phase) or Laurdan (fluidity). (**c**) GUVs of DOPG/egg SM/Chol in different ratios labeled with the Ld phase marker DiIC18. (**d**) GUV produced from native pulmonary surfactant labelled with BODIPY-PC and DiIC18 and examined at 40 °C and 25 °C. (**e**) GPMV from NIH 3 T3 cells labeled with Rhodamine-DOPE, showing fluid/fluid phase coexistence. (**f**) GUV (DOPC/DOPG/SM/Chol/GM1) labelled with Rhodamine-LAT peptides examined before and after addition of CTxB (*dark areas*, Lo phase) (Adapted from: (**a**) [18]; (**b**) [166]; (**c**) [167]; (**d**) [168]; (**e**); [17]; (**f**) [169])

With a size of 15–30 μm, GUVs are more suitable to approach the PM morphology. However, even if proteins can be incorporated [173], GUV composition remains far from the complex PM composition. GPMVs also exhibit phase separation (Fig. 5.3e). However, whereas they contain both PM lipids and proteins, several factors known to modulate phase separation are still missing, including cytoskeleton anchorage, active cellular processes and cross-binding proteins (Fig. 5.3f). These differences, which must be considered when studying phase separation, are reflected in the differential T*m* in living cells and isolated GPMVs [174].

5.5.2 *Highly-Specialized Biological Membranes*

The question is whether and how lipid organization in model membranes can be extrapolated to PMs. Indeed, in contrast to model membranes, PMs (i) are complex in lipid composition and intrinsic membrane properties, (ii) exhibit a high diversity of membrane proteins and a more or less anchored cytoskeleton, and (iii) are out of thermodynamic equilibrium. It is thus interesting to first describe lipid lateral distribution in membranes in which local equilibrium conditions are prone to occur, *i.e.* the pulmonary surfactant and the skin stratum corneum membranes, due to a relatively slow molecular turnover [18, 20]. Pulmonary surfactant membranes contain an important quantity of DPPC, cholesterol and unsaturated lipids and a low fraction of membrane proteins. These membranes show the coexistence of two liquid domains at physiological temperature that was linked with their spreading capacity at the air-water interface [18, 168]. Whereas extraction of the surfactant proteins does not alter phase coexistence, partial cholesterol depletion leads to elongated irregular domains, typical of gel/fluid phase coexistence. Domain organization is also strongly affected by temperature [18]. In the skin stratum corneum membrane, the lipid composition is also unique with mainly unusually long chain ceramides and free fatty acids as well as cholesterol [175, 176]. Using GUVs composed of lipid mixtures extracted from human skin stratum corneum, Plasencia et al. have shown a pH- and temperature-dependent membrane lateral organization. At pH 5, membranes exhibit a Ld phase at temperature > 70 °C, a Ld/gel phase coexistence between 40 and 70 °C and a gel/gel-like phase coexistence at temperature < 40 °C (relevant since skin physiological temperature is ~30 °C). At pH 7, the coexistence of these two distinct micrometric gel-like domains disappears and has been linked to the permeability properties of the skin stratum corneum [20].

Thus, these two specialized membranes highlight three important features regarding membrane lateral distribution. First, they represent alternative models besides model membranes in which equilibrium thermodynamic lipid phases have been evidenced. Second, lipid domains could be favored by the exceptional lipid composition of these membranes. It is thus crucial to consider this parameter when discussing the existence of lipid domains. Third, domains seem to be physiologically relevant.

5.5.3 *Prokaryotes & Yeast*

In contrast to mammalian cells, membrane domains are understudied in bacteria. This could be explained by three main reasons. First, the presence of lipid domains in cell membranes was for a long time thought to be a step during the course of evolution of cellular complexity. Second, the formation of lipid rafts requires sterols, which are missing from the membrane of most bacteria. Third, taking into account the small size of bacteria and the resolution limits of conventional confocal microscopy, exploring lipid domains in bacteria is particularly difficult. Fluorescence resonance energy transfer (FRET) and fluorescence anisotropy have nevertheless provided a significant amount of information.

Cardiolipin-enriched domains have been evidenced in bacteria with the fluorescent dye 10-N-nonylacridine orange (NAO). This probe, which was initially developed to visualize cardiolipin-rich mitochondria in eukaryotic cells, was also used to localize cardiolipin at the polar and septal poles of *E. coli* and *B. subtilis* [177, 178]. More recently, it has been shown that bacteria have the capacity to organize protein transport, secretion and signal transduction cascade in functional membrane microdomains (FMMs) [179]. Whereas FMMs have been suggested to be equivalent to eukaryotic lipid rafts [180], they exhibit differential composition. Indeed, bacterial membranes are enriched in phospholipids, lipopolysaccharides and various lipoproteins [181, 182] but most of them do not have sphingolipids [183] and only a few contain sterols [184, 185]. It should be stressed that some bacteria synthesize hopanoids, which have a chemical structure similar to that of cholesterol [186] and which could form nanometric domains by self-aggregation [180]. A nanoSIMS technique was employed to probe the existence of hopanoid lipid domains in cyanobacterium *Nostoc punctiforme* [187]. Bacterial flotillin FloT and FloA proteins along with squalene biosynthesis were found to play key roles in the formation of lipid domains. Heterogeneous distribution of flotillin-like proteins in *B. subtilis* was directly visualized by fluorescence microscopy upon labelling with the translational fusion FloT-GFP [188, 189]. The question of lipid composition of the flotillin-enriched structures still remains.

In contrast to bacteria, yeast represents a powerful system to explore lipid domain organization based on genetic approaches. In addition, like plants, yeast exhibits membranes which appear highly heterogeneous and can be imaged with conventional methods [190]. As indirect evidence for lipid domains in yeast, a Lo/Ld phase coexistence has been shown on model membranes either prepared from yeast total lipid extracts or with defined composition including ergosterol and inositolphosphoceramide [191]. Then, sterol-enriched submicrometric compartments containing the eisosome protein Sur7 and proton symporters Can1, Fur4, Tat2 and HUP1 have been evidenced thanks to filipin labelling [192]. More recently, major redistribution of PIP_2 into membrane clusters has been evidenced upon osmotic stress in both fission and budding yeast cells [193, 194]. Such PIP_2 clusters are spatially organized by eisosomes, protein-based structures of the yeast PM. After perturbation of sphingolipids, sterol, PS or PIP_2 levels, patchwork protein distribution is modified, suggesting a relation between proteins and lipids at the yeast PM domains. Besides PM, the yeast vacuole membrane proteins also segregate in two large stable membrane domains exhibiting differential ordering properties in response to nutrient deprivation, changes in pH of the medium and other stresses [195].

5.5.4 Animal Cells

In the past decades, submicrometric lipid domains have been documented at the outer and/or inner PM leaflet of various cell types, using several tools and methods. A substantial, albeit non-exhaustive, list of examples is presented in [156]. We will here select some cells based on their need to reshape during essential physiopathological processes, *i.e.* RBC (squeezing in narrow pores), platelet (spreading during coagulation), neutrophil (chemotaxis), neuron and glial cell (shape adaptation), epithelial cell (polarization) and cancer cell (squeezing to invade tissues). Images are provided in Figs. 5.4 and 5.5.

Human RBCs are the simplest and best characterized eukaryotic cell system both at lipid and protein levels [196, 197]. Hence, they are biconcave and are submitted to strong deformability during their 120-days lifetime. Moreover, for practical purposes, RBCs are a model of choice to explore PM lipid lateral heterogeneity because they (i) are easily available and robust, (ii) exhibit high homogeneity in size and shape due to rapid clearance of damaged RBCs by the spleen, (iii) present a flat surface without membrane projections or protrusions, avoiding confusion between domains and lipid enrichment in membrane ruffles, and (iv) do not make endocytosis, avoiding any confusion between domains and endosomes. We first revealed submicrometric domains by vital confocal imaging of spread RBCs upon trace insertion in the outer PM leaflet of fluorescent lipid analogs (*e.g.* BODIPY-SM) [25, 198]. Similar domains have then been observed upon direct labeling of endogenous SM and cholesterol using specific fluorescent toxin derivatives, Lysenin and Theta [21, 24] (Fig. 5.4). Double labeling of RBCs with the SM-specific Lysenin, then with BODIPY-SM, reveals perfect colocalization, suggesting the relevance of BODIPY-SM to study its native counterpart [21]. In contrast,

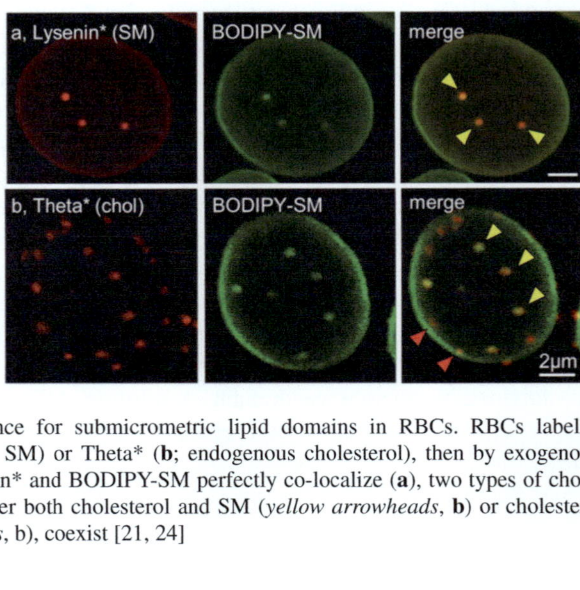

Fig. 5.4 Evidence for submicrometric lipid domains in RBCs. RBCs labelled by Lysenin* (**a**; endogenous SM) or Theta* (**b**; endogenous cholesterol), then by exogenous BODIPY-SM. Whereas Lysenin* and BODIPY-SM perfectly co-localize (**a**), two types of cholesterol domains, enriched in either both cholesterol and SM (*yellow arrowheads*, **b**) or cholesterol (chol) mainly (*red arrowheads*, b), coexist [21, 24]

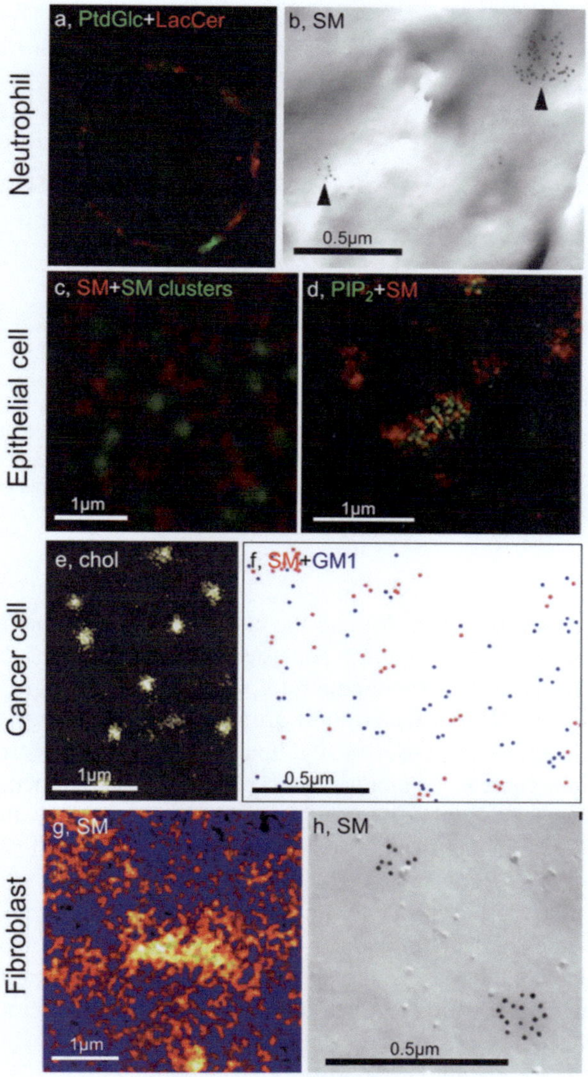

Fig. 5.5 Evidence for submicrometric lipid domains in nucleated mammalian cells. (**a**) Human neutrophil stained for phosphatidylglucoside (PtdGlc) and lactosylceramide (LacCer) and examined by STED microscopy. (**b**) Human neutrophil analyzed for SM at the inner face by SDS-digested freeze-fracture replica labelling with a Lysenin fragment. (**c**) LLC-PK1 cell labeled for SM (Equinatoxin) and SM clusters (Lysenin) and analyzed at its apical surface by structured illumination microscopy (SIM). (**d**) LLC-PK1 cell expressing Dronpa-PH (PIP_2), stained with Lysenin (SM) and analyzed by PALM/dSTORM. (**e**) HeLa cell labeled for cholesterol (Theta toxin fragment) and analyzed by PALM. (**f**) Jurkat cell labeled with a Lysenin fragment (SM) and biotinylated CTxB (GM1). (**g**) Fibroblast labeled with ^{15}N–sphingolipid precursors and examined by SIMS combined with TIRF. (**h**) Human skin fibroblast labeled and imaged as in (**b**) (Adapted from (**a**) [199]; (**b**) [122]; (**c**) [200]; (**d**) [201]; (**e**) [202]; (**f**) [203]; (**g**) [154]; (**h**) [122])

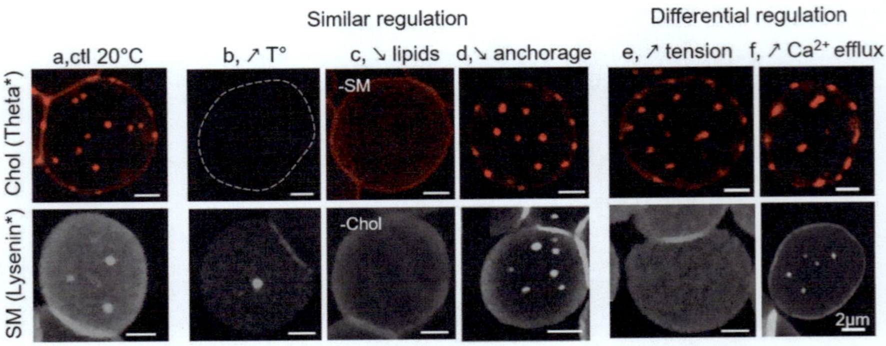

Fig. 5.6 Regulation of submicrometric lipid domains in RBCs. Illustration for cholesterol (chol)- and SM-enriched domains (respectively labelled by Theta* & Lysenin*). (**a**) Control (ctl) RBCs at 20 °C. (**b–d**) Similar regulation: (**b**) increased temperature (42 °C), (**c**) specific lipid depletion (−SM or -Chol), (**d**) acute uncoupling of membrane:cytoskeleton anchorage at 4.1R complexes (PKC activation). (**e**,**f**) Differential regulation: (**e**) increased tension (increased spreading on coverslip), (**f**) increased Ca^{2+} efflux (Ca^{2+}-free medium containing EGTA) (Adapted from [21, 24, 204])

double labeling with BODIPY-SM and the cholesterol-specific probe Theta leads to partial dissociation, indicating the coexistence of two types of domains at the RBC surface (see below). Submicrometric lipid domains have been confirmed on RBCs suspended in a 3D–gel, thus without artificial stretching, suggesting a genuine feature of RBCs *in vivo*. Mechanistically, lipid domains of RBCs are governed by temperature, lipid content, membrane:cytoskeleton anchorage, membrane tension and Ca^{2+} exchanges [21, 24, 204] (see Fig. 5.6.). In agreement with our confocal imaging data, Scheuring and coll. revealed by AFM the structural and mechanical heterogeneity of the RBC membrane [81]. These studies contrast with the random distribution of SM clusters observed by Kobayashi and coll. using SDS-digested freeze-fracture replica labelling [122, 123]. Whether this discrepancy reflects differences in methodology (imaging approach, fixation, labelling efficiency) or in parameters that are known to regulate domains (temperature, membrane tension, bending) remains to be determined.

Human platelets are central to hemostasis. Using 4-dimensional live-cell imaging and electron microscopy, Agbani et al. have recently shown that platelets adherent to collagen are transformed into PS-exposing balloon-like structures with expansive macro/microvesiculate contact surfaces, by a process called procoagulant spreading [205]. Whereas platelet activation is known to critically depend on PS surface exposure [206], the importance of lipid lateral distribution is less understood. On one hand, using the artificial lipid probe DiIC18, Gousset et al. have shown submicrometric domains in platelets upon activation, suggesting regulated raft coalescence into larger domains under appropriate conditions [207, 208]. On the other hand, random distribution of SM clusters has been revealed by SDS-digested freeze-fracture replica labelling in both non-stimulated and stimulated platelets [122, 123].

Upon recruitment to sites of inflammation *via* chemotaxis, neutrophils rapidly change their morphology, from roughly spherical resting to migratory cells with distinct leading and trailing edges. Maxfield and coll. have proposed that membrane lipid organization is critical for human neutrophil, through the formation of submicrometric domains that help in amplifying the chemoattractant gradient and maintaining cell polarization [209]. However, this suggestion was mainly based on disruption of lipid organization using mβCD and morphological evidence for lipid domains was not provided. Recently, STED has revealed that the two main neutrophil glycolipids, phosphatidylglucoside and lactosylceramide, form distinct domains in their outer PM (Fig. 5.5a). Moreover, lactosylceramide domains associate with the Src family kinase Lyn and could thereby participate in chemotaxis [199, 210]. Evidence for domains in the inner leaflet of neutrophil PM has also been provided, with diameter > 200 nm and enrichment in SM [122] (Fig. 5.5b).

Neurons can also adopt a variety of shapes to adapt to the region and function in the nervous system. By characterizing the elastic properties of specific membrane domains in living hippocampal neurons by AFM, it was demonstrated that GPI-anchored proteins reside within domains of ~70 nm size that are stiffer than the surrounding membrane. Upon inhibition of actin filament formation, the size of the GPI-enriched domains increases without change in stiffness [66]. During the development of the central nervous system, the reciprocal communication between neurons and oligodendrocytes is essential for the generation of myelin. Oligodendrocytes exhibit a differential relative abundance of specific lipids during differentiation [211] and a high global lipid content [212]. Several reports have shown that some of these lipids cluster into domains. First, galactosylceramide and sulfatides form submicrometric domains [213], mutually interacting at the apposed membranes of wrapped myelin [214], regulating PM organization and myelin protein lateral diffusion [215]. Second, GM1-enriched domains are essential for oligodendrocyte precursor survival by providing signaling platforms for growth factor-mediated integrin activation [216].

Lipid domains could also play a role in epithelial cell polarization. By FRAP of several membrane proteins, Meder and coll. revelead the coexistence of at least two different lipid phases in the apical PM of epithelial cells, but not in fibroblasts [217]. In differentiated MDCK cells, SM-enriched domains have been evidenced at the basolateral membrane [200]. In contrast, the SM-specific probe Lysenin selectively stains the apical PM of Eph4 cells, a cell line derived from mouse mammary gland epithelial cells [218], and SIM evidences SM clusters in the apical PM of LLC-PK1 cells [200] (Fig. 5.5c). Such differences in lipid lateral distribution should be discussed in light of epithelial cell biochemical and morphological characteristics, as proposed in [219, 220].

Lipid domains are also relevant to cancer cells. First, imaging by AFM of membranes purified by ultracentrifugation from human breast cancer cells (MDA-MB-231) has revealed the presence of submicrometric domains [221]. Second, super-resolution fluorescence microscopy of HeLa cells labeled with fluorescent Lysenin and Theta has demonstrated two types of lipid domains of ~250 nm in diameter and differentially enriched in cholesterol and SM [202] (Fig. 5.5e).

Third, electron microscopy of Jurkat T-cells (an immortalized line of human T lymphocyte cells) double labeled with Lysenin and CTxB indicates the coexistence of SM- and GM1-enriched domains [203] (Fig. 5.5f). Upon labeling of the same cells with Laurdan, Dinic and coll. have shown the association of TCR with small ordered actin-dependent PM domains in resting T cells that can aggregate upon TCR engagement [222]. Acquisition of a motile phenotype in T lymphocytes results in the redistribution of ganglioside GM3- and GM1-enriched raft domains to the leading edge and to the uropod, respectively, in a cholesterol- and actin-dependent process. It was suggested that segregation of membrane proteins between distinct lipid domains allows mediating redistribution of specialized molecules needed for T cell migration [223].

Lipid domains have also been observed in other cells such as fibroblasts and myoblasts. At the PM of fixed mouse embryo fibroblasts labeled with Lysenin, SM clusters that appear to be membrane lipid trafficking-dependent have been observed [224]. In contrast to cholesterol which is uniformly distributed throughout, evidence for submicrometric sphingolipid-enriched domains has been provided at the NIH3T3 mouse embryo fibroblast PM using SIMS (Fig. 5.5g). These domains are only reduced in abundance upon cholesterol depletion but fully eliminated upon cytoskeleton disruption, suggesting they are not lipid rafts [153, 154]. SM domains [122] (Fig. 5.5h) and restriction of PIP_2 diffusion have also been shown in the inner leaflet of fibroblasts [225]. At the lateral PM of living C2C12 myoblasts, which exhibit a high level of cholesterol [226] and a strong membrane:cytoskeleton anchorage, we revealed heterogeneous distribution of cholesterol upon decoration by Theta [24].

Thus, stable lipid domains can be evidenced in a large diversity of living cells but the concept is still difficult to generalize. As recently proposed by Kobayashi and coll. for asymmetric lipid distribution across the PM, we suggest that the lateral distribution of lipids is highly regulated and cell-dependent. It is thus crucial to integrate PM lipid composition and membrane properties, cytoskeleton:membrane coupling as well as membrane trafficking and lipid turnover while discussing lipid domains (see Sect. 5.6).

Another important challenge is to evaluate lipid domain diversity, both at the inner and the outer leaflets, and to establish whether there is a correspondence between lipid domains in the two leaflets. Several studies based on multiple labeling using validated probes, combined or not with specific lipid depletion, report for the coexistence of distinct lipid domains in the PM. First, sterol- and sphingolipid-enriched domains only partially overlap in several PMs. For instance, by double labeling experiments in RBCs, we showed the coexistence of two types of domains, one enriched in SM and cholesterol *vs* another mainly enriched in cholesterol [24]. SIMS in mouse fibroblasts revealed that partial cholesterol depletion does not eliminate the sphingolipid domains but reduces their abundance [154]. The structure and abundance of sphingolipid domains at the yeast PM seem independent of ergosterol [227]. Second, one (class of) lipid can even be distributed in several different pools. Thus, the dissociation of SM- and GM1-rich domains in the outer PM leaflet of Jurkat T-cells has been shown by electron microscopy [203] (Fig. 5.5f). GM1 and GM3 clusters at the fibroblast PM largely dissociate and are redistributed upon actin cytoskeleton disruption, indicating that their distribution not only depends on phase separation but also on cytoskeleton [228]. By confocal vital imaging of the RBC PM, we evidenced two types of cholesterol-enriched domains [24]. Using a toxin that binds to cholesterol-rich membranes, Das et al. have shown that the human fibroblast PM contains three types of cholesterol pools, *i.e.* a pool accessible to the toxin, a SM-sequestered pool that binds to the toxin only when SM is abrogated and a residual pool that does not bind the toxin even after SM abrogation [229]. Whether these pools represent real domains remains to be determined. Likewise, in *S. cerevisiae*, two ergosterol pools, one enriched in sphingolipids and the other not, are involved in two different aspects of yeast mating, pheromone signaling and PM fusion, respectively [230].

Regarding lipid domain transbilayer distribution, a superposition of SM clusters in the outer PM leaflet and PIP_2 in the inner leaflet has been shown by super-resolution microscopy of LLC-PK1 cells [201] (Fig. 5.5d). By delivering fluorescent PIP_2 and F-actin specific probes using synthetic vesicles and real time live cell imaging, Chierico and coll. have shown that PIP_2 domain formation during the early stage of cell adhesion correlates with rafts [231]. Mayor and co-workers provided experimental and simulation data showing that nanoclustering of GPI-anchored proteins at the outer PM leaflet by dynamic cortical actin is made by the interdigitation and transbilayer coupling of long saturated acyl chains [101].

5.6 Lipid Domains – Control

Biological membranes possess two characteristic, yet opposing, features. They present a fluid-like nature allowing for free movement of their constituents, while providing area for a variety of biological functions suggesting non-uniform distribution and formation of lipid/protein domains. A large variety of mechanisms, including energetic considerations (Sect. 5.6.1), intrinsic membrane properties (Sect. 5.6.2) and extrinsic factors (Sect. 5.6.3), could contribute to control domains. We believe that these mechanisms can differentially impact distinct lipid domains, as illustrated for RBCs at Fig. 5.6, resulting in a wide diversity of domains in cells (see Sects. 5.5.3 and 5.5.4).

5.6.1 Energetic Considerations

Phase transition temperature (Sect. 5.6.1.1) and line tension at phase boundary (Sect. 5.6.1.2) are potential energetic sources for cell control of domain size.

5.6.1.1 Phase Transition Temperature

Lipid immiscibility and Lo-Ld liquid phase separation, well-characterized on ternary mixture of polar lipids and cholesterol [232], are proposed to drive lipid domain biogenesis, according to the lipid raft hypothesis. However, recent experiments on GPMVs expose liquid-phase separation at lower temperature (~15–25 °C) than 37 °C [17, 174]. Since in the one-phase region micrometer-scale composition fluctuations occur and become increasingly large and long-lived as temperature is decreased to the transition, Veatch and coll. proposed that lipid rafts are the manifestation of transient compositional fluctuations and suggested that cells may exploit the low energy cost associated with (re)organizing components in membranes with critical composition [174]. By using the relatively long-range fluctuation-driven forces between membrane inclusions (called Casimir forces), Machta et al. proposed that cells may also take advantage of being close to the critical point to (re)organize the lateral segregation of membrane proteins [233].

In living cells, although no evidence of miscibility transition over a temperature range of 14–37 °C was observed, GPMVs derived from these cells do instead exhibit such a transition, pointing out that phase transition is not driven in living cells by temperature in a range of 14–37 °C. Groves and coll. therefore suggested that living cells maintain either the $T > Tm$ or $T < Tm$ through the wide temperature range and highlighted the robustness of the cellular membrane to temperature change [234], an opposite view from the models of Veatch and Machta.

The discrepancy between Veatch/Machta and Groves and coll. models could arise from the fact that GPMVs, as compared to the mother-cell PM, have lost cytoskeleton anchorage and transmembrane asymmetry [17] and are no longer connected with cross-linking components or lipid recycling, all known to modulate liquid phase separation (see Sects. 5.6.2 and 5.6.3). Indeed, Groves and coll. have shown differential tension between living cell PMs and derived GPMVs, suggesting that cell membranes may be maintained in a different region of the phase diagram avoiding a temperature-driven phase transition [234]. It is possible that interactions with cytoskeletal/membrane proteins or active cellular process dominate or even obliterate lipid miscibility effects.

Bagatolli and coworkers discussed the biological existence and significance of equilibrium thermodynamic phase and equilibrium critical points in biological membranes, which normally are in non-equilibrium conditions [10, 11]. They suggest that critical point phenomena are unlikely to be a major factor regulating biological phenomena since, for example, small mistuning near critical point could lead to drastic change in membrane structure and cell function. Another point of view is that biological system could exhibit a non-equilibrium critical behavior or a self-organized critical behavior [235]. The self-organized critical system exposes a critical state which is robust to perturbations and needs no tuning as it evolves itself towards the critical state thought self-organization [236]. This idea is appealing for biological system regarding its robustness as compared to system near critical points.

5.6.1.2 Line Tension at Phase Boundary

The line tension, *i.e.* the energy at domain boundary, results from the different phase properties of the lipid domains and its surrounding, leading to thickness mismatch at domain boundary and unfavorable exposure of lipid hydrocarbon regions to water. Theoretical work exposed the central role of line tension for lipid domain lateral sorting [237]. Different observations on model membranes also support this view. First, lipid domains are circular and rapidly return to a circular shape after external perturbation [238] to minimize the boundary length, supporting the importance of line tension at the phase interface. Second, degrading cholesterol (a key regulator of differential order and thickness between lipid domains and surrounding membrane) with cholesterol oxidase in one monolayer induces irregular domain boundary followed by domain disappearance [238]. Third, domain size increases with the extent of acyl chain unsaturation [237], another regulator of differential order and thickness between lipid domains and surrounding membrane. Fourth, domain size and the mismatch in bilayer thickness at phase boundary are directly correlated [237]. Fifth, with increasing temperature, GPMVs exhibit in their two-phase region a linear decrease of the line tension which approaches zero at the T*m* [174]. Experiments on living RBCs have confirmed some of these observations: (i) lipid domains are all circular [21, 24]; (ii) cholesterol or SM depletion induce irregular domain boundary (our unpublished data); and (iii) domain abundance and size are differently modulated by line-active agents [239].

The energy cost of the line tension depends on the size of the height mismatch and the length of the boundary. Mechanisms minimizing hydrophobic exposure at domain boundary by reducing both factors were suggested to take place in the membrane. The first one involves elastic lipid deformation to decrease the step-like change in thickness at domain boundary [240] while the second one favors domain coalescence to reduce domain boundary perimeter [241]. But, if the line tension at domain boundary was the only relevant energy consideration, any system with coexisting liquid domains would achieve equilibrium at one round domain. In fact, this is not the case since a stable distribution of lipid domain size can be observed in both model membranes and living cells. This suggests that other energy factors compete with the line tension, such as the entropic penalty for domain merge [241]. In addition, in living cells, the lipid composition is far more complex than in model membranes. Taking this in account, the step-like nature of lipid domain boundary found in a ternary mixture could be compensated in living cells by accumulation at domain boundary of (i) lipids of intermediate length, decreasing the abruptness of the boundary and the strength of the line tension [242], and (ii) proteins playing the role of line-active agents by accumulating at the interface [243].

5.6.2 *Intrinsic Membrane Properties*

Intrinsic membrane properties are also critical for controling lipid domains. These include membrane lipid:lipid interactions (Sect. 5.6.2.1), curvature (Sect. 5.6.2.2), transversal asymmetry (Sect. 5.6.2.3), dipole potential (Sect. 5.6.2.4) and protein:lipid interactions (Sect. 5.6.2.5).

5.6.2.1 Membrane Lipid:Lipid Interactions

The favorable cholesterol and SM interaction observed in biomimetic model membranes and leading to the coexistence of cholesterol/SM-enriched phase ("raft-like") with cholesterol/SM-poor phase ("non raft-like") is proposed in the seminal lipid raft definition to drive lipid domain biogenesis. Accordingly, cholesterol-enriched domains partially colocalize with SM-enriched domains at the living RBC PM and both lipid domains exhibit reciprocal dependence, as revealed by specific cholesterol or SM membrane depletion [21, 24]. However, several other studies indicate the opposite. First, depletion of homogenously distributed cholesterol in mouse fibroblast PM does not influence the morphology of sphingolipid-enriched domains [154]. Second, in yeast, sphingolipids do not accumulate in ergosterol-enriched domains [27] and sphingolipid-enriched domain structure and abundance do not depend on ergosterol metabolism [227]. Accordingly, stability of the proton-ATPase Pma1 at the yeast PM specifically requires sphingolipids but not sterols [27]. All these observations suggest that sterol-sphingolipid interactions are not sufficient to explain the formation of lipid domains in cellular membranes. Besides SM/cholesterol, GSLs and ceramides, which present very particular physico-chemical properties, have been proposed to contribute to generate and/or maintain lipid domains [199, 210, 244, 245] (see also Sects. 5.5.2 and 5.5.4).

5.6.2.2 Membrane Curvature

Using simulation of complex asymmetric PM model containing seven lipid species including GM3 and PIP_2, Koldso and coll. have shown that the concave regions of the bilayer surface are enriched in GM3 [246]. Likewise the increase in GM1 concentration in POPC bilayers induces tighter lipid packing, driven mainly by inter-GM1 carbohydrate-carbohydrate interactions, leading to a greater preference for the positive curvature of GM1-containing membranes and larger cluster sizes of ordered-lipid clusters [247]. These two studies suggest a relation between membrane curvature and lipid lateral sorting.

One step further, the observation of specific lipid sorting in vesicle and tubule budding from organelles involved in endocytosis [248, 249] has suggested that membrane curvature could provide a mechanism for the spatial sorting of lipids. This hypothesis has been tested by experiments pulling membrane tubes out of GUVs, confirming a curvature-driven lipid sorting [44, 250]. Mechanistically, it has been proposed that individual lipids are not effectively curvature-sorted according to their individual shape by membrane curvature differences of magnitudes found in intracellular membranes, but that cooperativity of lipid domains is needed to enable efficient curvature sorting in function of domain intrinsic curvature and bending stiffness [43, 251]. Besides domain bending stiffness and intrinsic curvature, a competition between domain bending rigidity and line tension at phase boundary is also proposed as driving force for lipid domain association to membrane curvature [252].

Thus, membrane curvature seems to provide a mechanism for lipid spatial sorting. It should be stressed that domain size and topography also imply on domain dimensionality that can switch from a flat to "dimpled" shape. This switch depends on the competition between (i) the 3D surface tension/mass ratio that favors small surface and then flat domain, and (ii) the 2D phase boundary line tension/mass ratio that prefers any domain morphology that reduces the boundary length [243, 253]. Ursell and coworkers used theoretical and experimental work to show that, when this competition results in a transition from a flat to dimpled domain shape, it leads to two dimpled domains that are able of repulsive elastic interaction, slowing domain merge and thus regulating domain size and topography [253].

5.6.2.3 Membrane Transversal Asymmetry

As explained in Sect. 5.4.4, PM exhibits transversal asymmetry. Lipid mixtures that are typically found in the outer leaflet tend to phase-separate in Lo and Ld liquid phases when reconstituted in model membranes [254]. In contrast, lipid mixtures that represent the inner leaflet do not undergo macroscopic phase separation and are in Ld state [255]. In cells, whereas lipid domains have been more documented on the outer PM leaflet, they have nevertheless been identified at the inner leaflet of various cell types [201, 256, 257], asking for a potential interleaflet coupling resulting in domain formation.

Theoretical works addressed physical mechanisms leading to fluid domain coupling across membranes. May and coll. focused on electrostatic coupling, cholesterol flip-flop and dynamic chain interdigitation as underlying mechanisms of interleaflet coupling, and argued that the latter likely provides the main contribution [258]. Other potential mechanisms are the van der Waals interactions and composition curvature coupling [259]. May also discussed the importance of a fine balance between interleaflet line tension at the bilayer midplane and intraleaflet line tension at domain interface within each leaflet as crucial energetic considerations for interleaflet coupling [258]. As a first line of evidence for interleaflet coupling in lipid bilayers, Sackmann and coworkers imaged the deposition of a DMPC (dimyristoy-PC) monolayer doped with green NBD-DMPE (dimyristoyl-PE) (Ld state) on a supported DMPE monolayer doped with Texas-Red DMPE (So state). They evidenced the formation in the DMPC monolayer of crystalline domains which appear to be in perfect register with the So domains of the DMPE monolayer [260].

Since asymmetric model membranes have long been difficult to obtain, experiments studying if and how one leaflet affects the structure and thermodynamic phase behavior of the apposed leaflet have generated controversial results. Recent preparation of asymmetric GUVs yielded significant insight and suggested that a Lo domain in one leaflet can induce a Lo domain in the apposed leaflet [255, 261].

However, phase-state across leaflets of asymmetric bilayers appears to be highly sensitive to lipid composition in one leaflet [261]. A recent elegant study, using fast Laurdan general polarization imaging on active planar supported bilayers and showing the formation of lipid domains upon lipase action, provides an example for the biological relevance of interleaflet coupling at non-equilibrium conditions [262].

Three lines of evidence on living cells support the reciprocal interaction between inner and outer leaflet domains: (i) the superposition of outer SM and inner PIP_2 clusters [201]; (ii) the colocalization of inner and outer leaflet proteins during signaling events [263]; and (iii) the colocalization of inner leaflet-associated proteins with outer leaflet rafts [264, 265].

5.6.2.4 Membrane Dipole Potential

Based on theoretical and experimental investigations on lipid monolayers, it has been shown that the size of domains results from balancing the line tension (which favors the formation of a large single circular domain) against the electrostatic cost of assembling the dipolar moments of the lipids (which prevents monolayers from reaching complete phase separation) [266]. Calculations were then extended to lipid bilayers. Hence, the work took in account ionic strength, showing that, at high ionic strength, the effects of dipole are short-ranged and the system is dominated by line tension, leading to domain size increase [267]. In biological membranes, the transmembrane voltage has been shown to significantly increase the phase transition temperature in squid axon membranes [268] and growing pollen tubes [269] and abundance of SM-enriched domains is decreased in living yeast following membrane depolarization [270]. Thus, it seems that depolarized membrane is more homogeneous than polarized membrane, but the mechanism underlying the induction of lipid domains by the transmembrane electric field is not clear yet [271].

5.6.2.5 Membrane Protein:Lipid Interactions

Since lipids diffuse fast in the membranes, the local synthesis of a given species is not sufficient to form lipid domains. Therefore, lipid diffusion must be confined by proteins to allow for domain formation and stabilization [263, 272, 273]. Thus, lipid domains can be captured and stabilized by lipid:protein interactions thanks to lipid-anchored proteins, such as GPI-anchored proteins [274], or transmembrane proteins. If the initial site for lipid:protein interaction is the boundary between the Lo domain and the adjacent Ld membrane, then proteins could function as surfactants. For example, confocal microscopy and AFM have revealed the preferential *in vitro* localization of lipid-anchored N-Ras to Lo–Ld domain boundaries [275] and the reorganization of phase-separated membranes into irregular domains by the reduction of line tension at phase boundary due to the binding of a membrane-active peptide derived from the apoptotic protein Bax at the domain interface [276].

Integral membrane proteins can also organize lipids, as the intramembrane protein needs to be solvated by the flexible disordered chains of phospholipids. Mouritsen's hydrophobic matching hypothesis proposes that integral membrane proteins perturb surrounding lipids so that bilayer thickness matches the length of the transmembrane domain [277]. Consistently, recent work indicates that proteins might be the most important determinants of membrane thickness, at least in the exocytic pathway [278]. Larger more stable lipid domains can be formed by protein:protein interactions. As examples, one can cite the T cell receptor (TCR) and IgE receptor signaling platforms [279, 280].

5.6.3 Extrinsic Factors

Besides membrane proteins (Sect. 5.6.2.5) are those that shape the membrane such as cytoskeleton (Sect. 5.6.3.1) and cross-binding proteins (Sect. 5.6.3.2). Moreover, lipid domains can be influenced by electrostatic interactions with cations (Sect. 5.6.3.3) and membrane/lipid turnover (Sect. 5.6.3.4).

5.6.3.1 Cytoskeleton

Proximity and direct interaction between the membrane and cytoskeleton *via* actin-binding proteins or complexes makes the cytoskeleton one of the most important extrinsic factor to influence PM lateral distribution. Kusumi and coll. suggested that the PM is compartmentalized into large areas containing smaller regions, resulting from an actin-based membrane cytoskeleton fence structure with anchored transmembrane proteins acting as pickets [256, 281].

However, membrane scaffolds can have strongly differential effects on lipid organization. Thus, Frisz and coll. demonstrated that actin depolymerization induces a randomization of ^{15}N–sphingolipids in fibroblasts, indicating that sphingolipid-enriched domains strongly depend on the actin-based cytoskeleton [154]. Thanks to a genetically encoded fluorescent PS biosensor (GFP–LactC2) and a fluorescent PS analog together with single-particle tracking and fluorescence correlation spectroscopy, Grinstein and coll. revealed that a sizable fraction of PS with limited mobility exists in the PM and that cortical actin contributes to this confinement [282]. More recently, Mayor and co-workers provided experimental and simulation data showing that nanoclustering of GPI-anchored proteins at the outer PM leaflet by dynamic cortical actin is made by the interdigitation and transbilayer coupling of long saturated acyl chains and that cholesterol can stabilize Lo domains over a length scale that is larger than the size of the immobilized cluster [101]. In RBCs, observations are more contrasting since acute membrane:cytoskeleton uncoupling at 4.1R and ankyrin complexes differentially modulate the abundance of lipid submicrometric domains [24]. Effect of cytoskeleton on lipid phase separation in model membranes also led to contrasting results: (i) polymerization of dendritic actin network on the membrane of GUVs induces phase separation [283]; (ii) actin fibers bound on supported lipid bilayer prevent lipid phase separation that occurs at low temperature [284]; and (iii) the prokaryotic tubulin homolog FtsZ attached to GUVs suppresses large-scale phase separation below the phase transition temperature but preserves phase separation above this temperature [285].

Besides temperature (see Sect. 5.6.1), two explanations can be provided for such differential effects. First, the properties of the anchoring type and pattern at the PM considerably vary between cell types and even within a same cell. Three types of membrane scaffold structures have been described so far and are well-summarized in [286]. First, the picket-and-fence model proposed by Kusumi and coll. [287–289] is based on actin fence formation and binding to transmembrane proteins and lipids *via* adaptor proteins. This model describes the PM organization into three domains of decreasing size and showing cooperative actions: (i) the membrane compartment (40–300 nm in diameter), corresponding to the PM partitioning mediated by the interactions with the actin-based membrane cytoskeleton (fence) and the transmembrane proteins anchored to the membrane cytoskeleton fence (pickets); (ii) the raft domains (2–20 nm) confined by the anchored transmembrane proteins; and (iii) the dynamic protein complex domains (3–10 nm), including dimers/oligomers and greater complexes of membrane-associated and integral membrane proteins. Among the specialized soluble proteins that can bind membrane bilayers *via* lipid-binding domains, allowing for interaction between inner leaflet lipids and cortical actin and contributing to compartimentalize the PM, one can cite the ERM proteins [290]. The second model is based on the active actin fiber polymerizing binding to the membrane constituents that drives clustering through aster formation [291, 292]. It is proposed that the living cell membrane is well-organized and that localization, clustering, transport and/or transformation of membrane molecules are allowed through the local engagement of the cortical actin machinery and need energy [291]. This model especially accounts for the transient clustering of molecules such as GPI-anchored proteins. Coupling of these proteins with the actin cytoskeleton involves long chain lipids which couple across the bilayer in the presence of cholesterol [101]. Third, some cells such as RBCs and neurons exhibit regular spectrin/actin/ankyrin-based membrane scaffolds that provide mechanical robustness. For information on the RBC cytoskeleton, see Sect. 5.4.2 and [28].

Second, we have to keep in mind that the cytoskeleton does not provide a satisfactory explanation for all membrane-associated phenomena and there is no universal model of the PM lateral organization [293]. At least, cytoskeleton should be integrated in a more global view including membrane curvature, as recently proposed by [284]. Based on computer simulations, super-resolution optical STED microscopy and FCS, it has been demonstrated that the actin fibers bound to the membrane help to organize the distribution of lipids and proteins at physiological temperatures (*i.e.* > Tm), while preventing lipid phase separation happening at low temperature. In the presence of curvature coupling, these two effects are enhanced [284]. The idea behind is an extension of the picket-fence model, by including a coupling of the local membrane curvature to the membrane composition, in a way that the actin fibers cause the membrane to curve reinforcing the influence of the picket-fence [284].

5.6.3.2 Cross-Binding Proteins

Several observations indicate that peripheral protein binding may represent an additional regulator of lateral heterogeneity. First, cross-linking components like upon CTxB (GM1 cross-linking) and Annexin V (PS binding) modulate phase transition temperatures in membrane models [294]. Second, GSL clustering induced by CTxB or Shiga toxin induce phase segregation in GUVs and GPMVs [295, 296]. Third, besides their recognized roles in generating membrane protrusions or invaginations through the sculpting of PI-rich membranes, elegant studies have shown a role for BAR domain proteins in generating stable PIP_2 domains by limiting their lateral diffusion, before inducing membrane curvature [297, 298]. These domains could play a role in various physiological processes including endocytosis, membrane protein trapping or storage of lipids in eisosomes [297].

5.6.3.3 Electrostatic Interactions of Charged Headgroups with Cations

The inner PM leaflet is the most negatively charged membrane of all cell bilayers, attributed to its high PI and PS contents. Localized negative membrane charge achieved by cations represents an alternative mechanism for domain formation and/or stabilization. It is indeed well established that the lateral organization of PIP_2 can be modulated by Ca^{2+}, as shown in GUVs, lipid monolayers and bilayers [299–301]. PIP_2 heterogeneous distribution has been confirmed in the PM and depends on the interaction between PIP_2 and polybasic protein domains (such as MARCKS) that can be modulated by Ca^{2+} and calmodulin [302, 303]. Contrasting with the PIP_2 domain formation by cations, PS (which can also modulate membrane charge locally) domains seem to preferentially rely on the association with protein complexes immobilized by the cytoskeleton [282] than on anionic domains, suggesting that the formation of Ca^{2+} induced domains depend on the high charge density of the lipid [299].

Such localized membrane charge can facilitate PM protein clustering to confined regions. For example, Ca^{2+} (but not Mg^{2+}) has been shown to promote the formation of syntaxin 1 (a SNARE protein) mesoscale domains through PIP_2 in PC12 cell sheets, indicating that this cation acts as a bridge that specifically and reversibly connects multiple syntaxin 1/PIP_2 complexes and suggesting a role for Ca^{2+} in PM reorganization during Ca^{2+}-regulated secretion [304]. Alternatively, localized membrane charge can induce conformational change of PM proteins [305], as shown during the activation of T cell receptor (TCR) upon antigen engagement. TCR interacts with acidic phospholipids through ionic interactions in quiescent T cells, resulting into deep membrane insertion of the tyrosine side chains. This renders TCR inaccessible to phosphorylation by the Src-kinase Lck. After antigen engagement of TCR, local Ca^{2+} concentration increases, leading to disruption of the ionic protein:lipid interaction, dissociation of tyrosines from the membrane and accessibility to Lck [305, 306].

In cells, a way to create localized membrane domains that differ in charge is through modification of local Ca^{2+}concentration by localized transient Ca^{2+} influx from membrane channels. Among these channels one can cite transient receptor potential ion channels that respond to mechanical stress induced by tension and trigger Ca^{2+} influx that interact with negatively-charged membrane lipids [305]. For example, TRMP7 has been shown to drive the formation of Ca^{2+} domains during invadosome formation in neuroblastoma cells [307] and at the leading edge of migrating cells [308]. It should be stressed that localized electrostatic interaction of charged lipid headgroups with cations could be linked to other mechanisms involved in membrane lipid lateral heterogeneity, such as actin dynamics [309]. It remains to be determined if such localized membrane charge in the inner PM can have consequences on the organization of the outer PM leaflet (see Sect. 5.5.4).

5.6.3.4 Membrane Recycling and Enzymatic Activity

A key difference between biological and model membranes is that the former are not at thermodynamic equilibrium but subjected to active processes such as membrane recycling and lipid turnover. Lipid recycling can be due to permanent exchange of lipids with the surrounding medium (membrane reservoir) where lipids are locally inserted at a constant rate everywhere along the membrane and removed at a rate proportional to their local concentration [310]. Lipid recycling can also occur *via* vesicular lipid transport events that can either specifically target lipid domains or random areas of the membrane [311]. For a review on this topic, please refer to [312]. In all living cells except RBCs, there is an active lipid recycling to and from the membrane that is proposed to limit lipid domain size [313]. Lipid domains in mature RBCs are larger, more stable and more round than in other living cells [21, 24], which could support the implication of active cellular processes in lipid domain destabilization in nucleated mammalian cells. In addition to lipid recycling, the molecular interactions that control phase behavior can also be dramatically affected by the activity of membrane lipases or kinases that generate phase-changing products [132]. A recent study on model membranes evidenced the formation of lipid domains upon addition of sphingomyelinase D [262].

5.7 Lipid Domains – Role in Membrane Shaping & Reshaping

The view of membrane organization into submicrometric domains could confer the size and stability required for PMs to deform. We here highlight how domains could contribute to cell shaping (Sect. 5.7.1), squeezing (Sect. 5.7.2), vesiculation (Sect. 5.7.3) and division (Sect. 5.7.4).

5.7.1 Cell Shaping

At short length scale, relationship between curvature and lipid molecular structure and lipid transbilayer sorting are well known (see Sect. 5.4.4). At long length scales, different mechanisms may participate but whether lipid domains play a role in this process is still unresolved. We here propose two mechanisms.

The first mechanism is based on the importance of cardiolipin-enriched domains in curvature maintenance in rod-shape bacteria. It was recently shown that cardiolipin localizes to the polar and septal regions of the inner membrane of *Escherichia coli* [177], *Bacillus subtilis* [178] and *Pseudomonas putida* [314]. Bacterial poles and septa are regions that have the largest curvature [315, 316]. As bacterial cardiolipin have a small ratio of head-to-tail surface areas [317], it is thus tempting to invoke the negatively curved regions of the inner leaflet of bacterial membrane poles relative to the cylindrical midcell to explain cardiolipin localization. However, the relative affinity of a single nanometer-sized cardiolipin molecule for the very slightly curved poles is likely insufficient for stable polar localization in a micrometer-sized bacterium. An alternative explanation is that cardiolipin localization is purely driven by lipid phase segregation. However, the observed rapid repartitioning of cardiolipin to the division site [177, 178] would be strongly disfavored if cardiolipin is segregated in a single, large cluster at one or both poles. Instead, Wingreen and coll. proposed stable finite-sized cardiolipin clusters which can spontaneously and independently target the two cell poles as well as the nascent division site [318, 319]. Weibel and coll. found that a cardiolipin synthase mutant of the rod-shaped *Rhodobacter sphaeroides* produces ellipsoid-shaped cells in a reversible process, and that bacteria with impaired MreB expose the same shape changes [320]. Huang and coll. recently demonstrated in *E. coli* that feedback between cell geometry and MreB cytoskeleton localization at the regions of negative curvature maintains rod-like shape by directing growth away from the poles and actively straightening locally curved cell regions [321]. In addition, cardiolipin has been shown to sense and transmit changes in inner membrane curvature to the bacterial phage shock protein (Psp) system (a cell envelope stress response system). Therefore, cardiolipin domains could be viewed as a membrane curvature sensor (due to its curvature-sorting properties) and as an indicator for membrane or cytoskeletal proteins that feedback on membrane curvature to maintain bacterial shape (due to its potential raft-like ability to segregate proteins).

A second mechanism suggests the direct stabilization of membrane curvature by lipid domains, as lipid lateral segregation into regions of preferential curvature could relax stresses in the membrane. The recruitment of lipid domains in areas of increased curvature would then result from a competition between the gains in the membrane elastic energy and the segregation-induced loss of entropy. This mechanism was proposed to stabilize the specific curvature of the Golgi cisternae [322], but evidence for this phenomenon is not currently available. Using a microfluidics to induce the deformation of GUVs with microstamps, Robinson and coworkers evidenced Lo phase merge due to the tension induced by the deformation and proposed that lipid domain merge is needed to reduce the line tension following the increase in membrane tension [46, 75]. Accordingly, theoretical work has shown that lateral tension applied on a membrane increases the line tension [323]. Hence, at a certain applied tension, the formation of a neck at Lo-Ld domains boundary allows to reduce even more the line tension. Other experiments on GUVs with different shapes also indicate the specific association of lipid domains with membrane curvature areas driven by line tension [252]. In living RBCs, the specific recruitment of cholesterol-enriched domains is observed in high curvature areas in RBC rim upon stretching (Fig. 5.7b,b′; 204). Whether this recruitment is directly required for membrane curvature stabilization, or indirectly *via* segregation of cytoskeleton or proteins that stabilize curvature (as cardiolipin domains in bacteria), remains to be investigated.

5.7.2 Cell Squeezing

RBCs are biconcave cells of ~8 μm in diameter that are strongly deformable, as tested upon crossing through small blood capillaries and splenic sinusoids, which exhibit diameters smaller than 5 and 1 μm respectively [324]. As highlighted in Sect. 5.4, RBC biconcavity, cytoskeleton strength and its PM anchorage are crucial for RBC deformation. However, whether and how the cytoskeleton interacts with specific membrane lipids and whether these interactions could have a regulatory and/or a structural role in RBC deformation remain to be elucidated [28]. Our data show that submicrometric lipid domains [21, 22, 24, 198] cluster upon membrane bending (Fig. 5.7b,b′) and provide platforms for Ca^{2+} exchanges needed for RBC shape recovery after deformation (Fig. 5.7c,c′), suggesting their role in two steps of RBC squeezing and highlighting the interplay between lipid domains, membrane curvature and Ca^{2+} in this process [204].

5.7.3 Cell Vesiculation

Upon senescence *in vivo* RBCs undergo multiple changes. These include the decrease of activities of multiple enzymes, the gradual accumulation of oxidative damage, the loss of membrane by vesiculation, the redistribution of ions and alterations in cell volume, density and deformability. For comprehensive reviews on aging mechanisms in healthy human RBCs, the reader is referred to [325, 326]. We will here focus on the release of vesicles that are generally classified into two groups, nanovesicles (~25 nm size) and microvesicles (MVs; ~60–300 nm). In contrast to nanovesicles, MVs seem influenced by partial membrane:cytoskeleton uncoupling. Thus, upon senescence, cytoskeleton stiffness and density both increase, leading to larger compressive forces on the cell membrane, that have been hypothesized to be accommodated by increased membrane curvature and vesicle detachment from the membrane [327–329]. MVs have been proposed to contribute to RBC senescence by two opposite mechanisms. They may (i) prevent the elimination of the senescent but

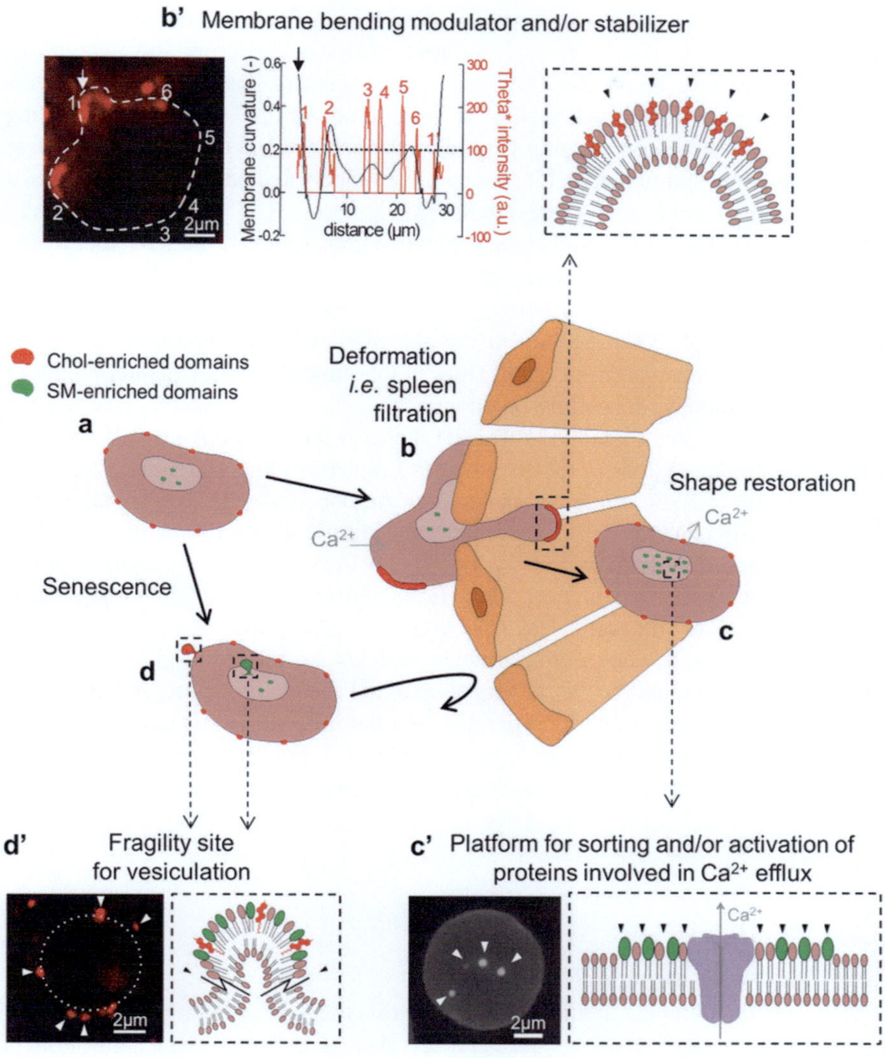

Fig. 5.7 Hypothetical model for the role of submicrometric lipid domains in RBC (re)shaping. Organization of cholesterol-enriched domains (red) and SM-enriched domains (green) in biconcave RBC in the circulation (**a**) and during reshaping upon either global deformation followed by shape restoration (**b**,**c**) or vesiculation during senescence (**d**). (**b′**) Theta*-labelled (endogenous cholesterol) RBC spread onto PDMS chamber and visualized by vital epifluorescence after deformation: *left*, imaging; *right*, relation between cholesterol domains (*red*) and membrane curvature (*black*). (**c′**) Lysenin*-labelled (endogenous SM) RBC upon Ca^{2+} efflux during shape restoration after deformation. (**d'**) Theta*-labelled RBC after extended storage at 4 °C to accelerate senescence (Adapted from [204])

yet functional RBCs, by elimination of Band 3 neoantigen, denatured hemoglobin and oxidized proteins [330]; or instead (ii) promote removal of senescent RBCs from the circulation, by elimination of CD47 [331], PS exposure to the outer PM leaflet and increased intracellular Ca^{2+} concentration [332].

Whether RBC lipid domains represent specific sites for local budding and vesiculation remains to be demonstrated. As direct evidences supporting this hypothesis, Prohaska et al. have shown that Ca^{2+}-induced vesicles are enriched in raft proteins [333]. We provide direct evidence for the vesiculation of lipid domains at the living RBC PM upon accelerated aging (Fig. 5.7d,d') [204]. Convincing arguments are also provided by simulation studies, membrane models and other cells. First, based on a two-component coarse-grained molecular dynamics RBC membrane model, Li and Lykotrafitis have revealed that the spontaneous curvature of the RBC membrane domains can cause the formation of nanovesicles and that lateral compression generates larger vesicles with heterogeneous composition, similar in size to the cytoskeleton corral [334]. Second, using a combination of mechanical modeling and GUV experiments, Phillips et al. showed that lipid domains can adopt a flat or dimpled morphology, depending on spontaneous curvature, boundary line tension of domains and domain size [253]. Third, Ld phases tend to spontaneously reside in curved membrane regions of GUVs whereas Lo phases are preferentially localized in flat regions [252]. Fourth, in living keratinocytes labeled by the Ld marker DiIC18 and the Lo GM1 marker CTxB, submicrometric lipid domain separation together with spontaneous vesiculation of the Ld domains occur. Such vesiculation is still increased by cholesterol depletion, which further enhances Lo/Ld domain separation and detachment of the cortical cytoskeleton from the membrane [335]. Fifth, in activated neutrophils, cholesterol-enriched vesicles are released [336], suggesting that lipid domains might be the starting point of the vesiculation process. Sixth, rafts are specifically selected and incorporated into the influenza virus envelope during the budding of enveloped viruses from the PM [337]. Finally, specific lipid sorting is observed in vesicle and tubule budding from organelles of the endocytic pathway [248, 249].

5.7.4 Cell Division

Prokaryotic membrane domains contribute to division and morphogenesis. Besides maintenance of membrane curvature and cell shape (see Sect. 5.7.1), cardiolipin is also involved in curvature changes occurring during bacterial division. Based on polar localization of chemotaxis receptors in *E. coli* [338] and septal localization of the division proteins MinCD and DivIVA in *B. subtilis* [339, 340], it was proposed that cardiolipin domains could play a role in sorting proteins. Renner and Weibel developed a microtechnology-based technique to confirm the relationship between bacteria curvature, cardiolipin domain localization at the poles and the positioning of amphiphilic cytoplasmic proteins. Thus, in giant *E. coli* spheroplasts confined in polymer microchambers, they demonstrated that cardiolipin domains localize to regions of large negative curvature. By expressing YFP fused to the N-terminal domain of the cytoplasmic division protein MinD, they showed the dependence of negative membrane curvature on MinD localization in spheroplasts and its colocalization with cardiolipin domains [341].

Membrane domains also participate to cell division in fungi. Elevated concentrations of sterols decorate developing membranes upon growth-induced elongation of the fungal cells *Candida albicans* and *Aspergillus nidulans* [342, 343] and septum or mating projection formation in the yeasts *Saccharomyces pombe* and *cerevisiae* [344, 345]. In fact, two sterol pools are required for two important aspects of mating in *S. cerevisiae*, pheromone signaling and PM fusion [230]. Regarding the implication of sphingolipid-enriched domains, it has been shown in *S. cerevisiae* that: (i) the inhibition of sphingolipid synthesis induces the formation of multinuclear cells due to a defect in cytokinesis [346]; and (ii) the organization in ordered domains at the mating projection depends on sphingolipids, as evidenced by microscopy with Laurdan [347]. Lipid gradients in the inner PM leaflet have also been revealed: (i) PIP_2 is densely distributed in the shmoo tip of *S. cerevisiae* [348]; (ii) the localized synthesis together with the restricted diffusion of PIP_2 in *C. albicans* result into a gradient from the tip of membrane protrusions to the neck [349]; and (iii) PE concentrates at polarized ends in budding yeast [350].

In mammalian cells, cholesterol-containing domains concentrate at the cleavage furrow and possess a signaling pathway that contributes to cytokinesis [351]. More recently, the transbilayer colocalization between the outer SM and the inner PIP_2 domains has been evidenced around the cleavage furrow and the midbody of HeLa cells by super-resolution fluorescence microscopy. This study highlights two key features of lipid domains. First, it shows the importance of SM domains in the regulation of cytokinesis, as revealed by PIP_2 domain dispersion, inhibition of the Rho GTPase RhoA recruitment to the cleavage furrow and regression of the cleavage furrow upon SMase treatment [201]. Second, it indicates that PIP_2 in the inner leaflet also form and remain in domains. Several reasons can explain the restricted localization of PIP_2 around SM clusters: (i) SM, $PIP_5K\beta$ and PIP_2 interact, restricting the diffusion of PIP_2; (ii) the Rho GTPase positively regulates the activity of $PIP_5K\beta$, enhancing the formation of PIP_2 domains at the inner leaflet; and (iii) the mobility of PIP_2 is restricted by protein fences [201]. Another possibility is that BAR domain proteins, besides their recognized role in membrane bending and curvature sensing, control the diffusion of PIP_2 through electrostatic interactions, thereby generating stable domains before inducing membrane deformation, as reported in [297]. Alternatively, PIP_2 domains could be stabilized by Ca^{2+} from the surrounding membrane. PIP_2 accumulation in the cleavage furrow of dividing cells has been confirmed by others [352, 353]. In contrast, PE, which is normally restricted to the inner leaflet, is exposed to the outer leaflet of the cleavage furrow during cytokinesis, contributing to regulation of contractile ring disassembly [352, 354].

5.8 Integration of Models and Observations

Since the mosaic fluid model of Singer and Nicolson in the 70s and the hypothesis of lipid rafts proposed by Simons and coll. in the 90s, numerous models trying to associate the two concepts were proposed. Some focused on the dynamic aggregation of small lipid rafts by proteins, others on the impact of immobile membrane proteins acting as picket-and-fence or of lipid recycling events. However, at the sight of the large list of intrinsic and extrinsic factors regulating lipid domains and the wide variety of lipid domain size, composition, shape, life-time and topological distribution observed in PMs, a new point of view rejecting these universal theories is now emerging. As emphasized in several key reviews in the membrane field, we need to accept that no simple and universal model can describe the complexity of the membrane [11, 293, 355, 356]. Membrane should instead be seen as a complex dynamic mosaic, where the composition, size, shape and topography of different domains depend on several intrinsic and extrinsic factors heterogeneously distributed along the membrane.

Such complex organization could be required for many cell reshaping processes like cell deformation, division and vesiculation, as highlighted in this Chapter. Three main, but still hypothetical, roles for lipid domains in cell reshaping can be proposed: (i) platform for membrane and skeletal protein sorting and/or activation, (ii) membrane bending modulator, and (iii) preferential fragility sites for membrane vesiculation (see Fig. 5.7). These roles emerge from the four main examples discussed in this Chapter. First, cardiolipin-enriched domains in rod-shape bacteria are sorted and reorganized by curvature, suggesting they could act both as curvature sensor and indicator platforms allowing for the cytoskeleton to maintain proper cell shape upon bacterial growth. Second, cholesterol-enriched domains in RBCs gather into areas of increased curvature upon deformation while SM-enriched domains are involved in Ca^{2+} exchanges needed for RBC shape recovery, suggesting the respective implication of these domains in membrane curvature and in Ca^{2+} exchanges providing platforms for protein recruitment/activation. Third, the specific vesiculation of lipid domains from the RBC PM could come from a change in the balance between lateral tension, domain bending and line tension, that could it-self result from modulations in cytoskeleton anchorage, protein organization or lipid asymmetry. Fourth, the recruitment of PIP_2/SM-enriched domains at the cleavage furrow is required for the progression of division through the recruitment of specific membrane proteins and cytoskeleton, suggesting again the role of lipid domains as platforms for protein sorting.

Despite these recent progresses, there are still many open questions and concerns regarding generalization, regulation and physiopathological importance of membrane lipid lateral distribution. We propose four directions for the future. First, the use of imaging methods with differential temporal and spatial resolutions, while taking into account the fixation issue, in combination with (multiple) labelling possibility using validated non-toxic relevant lipid probes and the dynamic perspective should help revealing lipid domains. Second, integrating theoretical predictions with experiments on model membranes and complex living cells will contribute to explore whether lipid domains can be generalized or not. Indeed, whether transient nanometric and stable submicrometric lipid domains evidenced on cells and those observed on model membranes are governed by the same mechanisms is currently unclear. This should be stressed by systematic study of lipid domain biophysical properties in living cell membranes, as it was done for model membranes during the past years. In addition, the development of active model membrane systems subject to transport, signal and enzymatic processes could help gain insight in how lipid domains are controlled by the non-equilibrium state of living cell membranes [11]. As a third issue, we need to explore whether and how lipid domains could modulate and/or stabilize membrane shape without involving proteins. Finally, membrane lipid domains can be of small size in resting state but become larger and more stable upon reshaping. Given that not all cells are subjected to extensive deformation, this represents a critical challenge that could be successfully approached using a cell model exhibiting intrinsic curvature and deformation ability.

Acknowledgements The authors acknowledge funding by UCL (FSR, ARC), the F.R.S-FNRS and the Salus Sanguinis foundation. We apologize to all colleagues whose work was not cited due to space constriction.

References

1. Thomas JA, Rana FR (2007) The influence of environmental conditions, lipid composition, and phase behavior on the origin of cell membranes. Orig Life Evol Biosph 37(3): 267–285
2. Bigay J, Antonny B (2012) Curvature, lipid packing, and electrostatics of membrane organelles: defining cellular territories in determining specificity. Dev Cell 23(5):886–895
3. Sodt AJ et al (2016) Nonadditive compositional curvature energetics of lipid bilayers. Phys Rev Lett 117(13):138104
4. Janmey PA, Kinnunen PK (2006) Biophysical properties of lipids and dynamic membranes. Trends Cell Biol 16(10):538–546
5. Singer SJ, Nicolson GL (1972) The fluid mosaic model of the structure of cell membranes. Science 175(4023):720–731
6. Nicolson GL (2014) The fluid-mosaic model of membrane structure: still relevant to understanding the structure, function and dynamics of biological membranes after more than 40 years. Biochim Biophys Acta 1838(6):1451–1466
7. Goni FM (2014) The basic structure and dynamics of cell membranes: an update of the singer-Nicolson model. Biochim Biophys Acta 1838(6):1467–1476
8. Simons K, Ikonen E (1997) Functional rafts in cell membranes. Nature 387(6633):569–572
9. Pike LJ (2006) Rafts defined: a report on the keystone symposium on lipid rafts and cell function. J Lipid Res 47(7):1597–1598
10. Bagatolli LA et al (2010) An outlook on organization of lipids in membranes: searching for a realistic connection with the organization of biological membranes. Prog Lipid Res 49(4):378–389
11. Bagatolli LA, Mouritsen OG (2013) Is the fluid mosaic (and the accompanying raft hypothesis) a suitable model to describe fundamental features of biological membranes? What may be missing? Front Plant Sci 4:457
12. Vicidomini G et al (2015) STED-FLCS: an advanced tool to reveal spatiotemporal heterogeneity of molecular membrane dynamics. Nano Lett 15(9):5912–5918

13. Stone MB, Shelby SA, Veatch SL (2017) Super-resolution microscopy: shedding light on the cellular plasma membrane. Chem Rev 117:7457
14. Lingwood D, Simons K (2010) Lipid rafts as a membrane-organizing principle. Science 327(5961):46–50
15. Parton RG, del Pozo MA (2013) Caveolae as plasma membrane sensors, protectors and organizers. Nat Rev Mol Cell Biol 14(2):98–112
16. Yanez-Mo M et al (2009) Tetraspanin-enriched microdomains: a functional unit in cell plasma membranes. Trends Cell Biol 19(9):434–446
17. Baumgart T et al (2007) Large-scale fluid/fluid phase separation of proteins and lipids in giant plasma membrane vesicles. Proc Natl Acad Sci USA 104(9):3165–3170
18. de la Serna JB et al (2004) Cholesterol rules: direct observation of the coexistence of two fluid phases in native pulmonary surfactant membranes at physiological temperatures. J Biol Chem 279(39):40715–40722
19. Kahya N et al (2003) Probing lipid mobility of raft-exhibiting model membranes by fluorescence correlation spectroscopy. J Biol Chem 278(30):28109–28115
20. Plasencia I, Norlen L, Bagatolli LA (2007) Direct visualization of lipid domains in human skin stratum corneum's lipid membranes: effect of pH and temperature. Biophys J 93(9):3142–3155
21. Carquin M et al (2014) Endogenous sphingomyelin segregates into submicrometric domains in the living erythrocyte membrane. J Lipid Res 55(7):1331–1342
22. D'Auria L et al (2013) Micrometric segregation of fluorescent membrane lipids: relevance for endogenous lipids and biogenesis in erythrocytes. J Lipid Res 54(4):1066–1076
23. Sanchez SA, Tricerri MA, Gratton E (2012) Laurdan generalized polarization fluctuations measures membrane packing micro-heterogeneity in vivo. Proc Natl Acad Sci USA 109(19):7314–7319
24. Carquin M et al (2015) Cholesterol segregates into submicrometric domains at the living erythrocyte membrane: evidence and regulation. Cell Mol Life Sci 72(23):4633–4651
25. Tyteca D et al (2010) Three unrelated sphingomyelin analogs spontaneously cluster into plasma membrane micrometric domains. Biochim Biophys Acta 1798(5):909–927
26. Bach JN, Bramkamp M (2013) Flotillins functionally organize the bacterial membrane. Mol Microbiol 88(6):1205–1217
27. Grossmann G et al (2007) Membrane potential governs lateral segregation of plasma membrane proteins and lipids in yeast. EMBO J 26(1):1–8
28. Lux SE (2016) Anatomy of the red cell membrane skeleton: unanswered questions. Blood 127(2):187–199
29. Waugh RE (1996) Elastic energy of curvature-driven bump formation on red blood cell membrane. Biophys J 70(2):1027–1035
30. McMahon HT, Gallop JL (2005) Membrane curvature and mechanisms of dynamic cell membrane remodelling. Nature 438(7068):590–596
31. Zimmerberg J, Kozlov MM (2006) How proteins produce cellular membrane curvature. Nat Rev Mol Cell Biol 7(1):9–19
32. Hansen GH et al (2007) Intestinal alkaline phosphatase: selective endocytosis from the enterocyte brush border during fat absorption. Am J Physiol Gastrointest Liver Physiol 293(6):G1325–G1332
33. Andreae LC, Burrone J (2015) Spontaneous neurotransmitter release shapes dendritic arbors via long-range activation of NMDA receptors. Cell Rep 10:873
34. Deplaine G et al (2011) The sensing of poorly deformable red blood cells by the human spleen can be mimicked in vitro. Blood 117(8):e88–e95
35. Danilchik MV, Brown EE, Riegert K (2006) Intrinsic chiral properties of the Xenopus egg cortex: an early indicator of left-right asymmetry? Development 133(22): 4517–4526
36. Osumi M (1998) The ultrastructure of yeast: cell wall structure and formation. Micron 29(2–3):207–233
37. Singleton K et al (2006) A large T cell invagination with CD2 enrichment resets receptor engagement in the immunological synapse. J Immunol 177(7):4402–4413

38. Turturici G et al (2014) Extracellular membrane vesicles as a mechanism of cell-to-cell communication: advantages and disadvantages. Am J Physiol Cell Physiol 306(7):C621–C633
39. Yuana Y, Sturk A, Nieuwland R (2013) Extracellular vesicles in physiological and pathological conditions. Blood Rev 27(1):31–39
40. Gupta A, Pulliam L (2014) Exosomes as mediators of neuroinflammation. J Neuroinflammation 11:68
41. Muralidharan-Chari V et al (2010) Microvesicles: mediators of extracellular communication during cancer progression. J Cell Sci 123(Pt 10):1603–1611
42. Shen B et al (2011) Biogenesis of the posterior pole is mediated by the exosome/microvesicle protein-sorting pathway. J Biol Chem 286(51):44162–44176
43. Tian A, Baumgart T (2009) Sorting of lipids and proteins in membrane curvature gradients. Biophys J 96(7):2676–2688
44. Sorre B et al (2009) Curvature-driven lipid sorting needs proximity to a demixing point and is aided by proteins. Proc Natl Acad Sci 106(14):5622–5626
45. Garcia-Saez AJ, Chiantia S, Schwille P (2007) Effect of line tension on the lateral Organization of Lipid Membranes. J Biol Chem 282(46):33537–33544
46. Robinson T et al (2012) Investigating the effects of membrane tension and shear stress on lipid domains in model membranes. 16th international conference on miniaturized systems for chemistry and life sciences
47. Henon S et al (1999) A new determination of the shear modulus of the human erythrocyte membrane using optical tweezers. Biophys J 76(2):1145–1151
48. Reid HL et al (1976) A simple method for measuring erythrocyte deformability. J Clin Pathol 29(9):855–858
49. Rand RP, Burton AC (1964) Mechanical properties of the red cell membrane. I membrane stiffness and intracellular pressure. Biophys J 4:115–135
50. Hochmuth RM (2000) Micropipette aspiration of living cells. J Biomech 33(1):15–22
51. Hosseini SM, Feng JJ (2012) How malaria parasites reduce the deformability of infected red blood cells. Biophys J 103(1):1–10
52. Lee LM, Liu AP (2015) A microfluidic pipette array for mechanophenotyping of cancer cells and mechanical gating of mechanosensitive channels. Lab Chip 15(1):264–273
53. Chivukula VK et al (2015) Alterations in cancer cell mechanical properties after fluid shear stress exposure: a micropipette aspiration study. Cell Health Cytoskelet 7:25–35
54. Müller DJ et al (2009) Force probing surfaces of living cells to molecular resolution. Nat Chem Biol 5(6):383–390
55. Gerber C, Lang HP (2006) How the doors to the nanoworld were opened. Nat Nanotechnol 1(1):3–5
56. Butt H-J, Cappella B, Kappl M (2005) Force measurements with the atomic force microscope: technique, interpretation and applications. Surf Sci Rep 59(1–6):1–152
57. Dufrêne YF et al (2013) Multiparametric imaging of biological systems by force-distance curve-based AFM. Nat Methods 10(9):847–854
58. Sullan RMA, Li JK, Zou S (2009) Direct correlation of structures and Nanomechanical properties of multicomponent lipid bilayers. Langmuir 25(13):7471–7477
59. Bremmell KE, Evans A, Prestidge CA (2006) Deformation and nano-rheology of red blood cells: an AFM investigation. Colloids Surf B: Biointerfaces 50(1):43–48
60. Heu C et al (2012) Glyphosate-induced stiffening of HaCaT keratinocytes, a peak force tapping study on living cells. J Struct Biol 178(1):1–7
61. Alsteens D et al (2013) Multiparametric atomic force microscopy imaging of single bacteriophages extruding from living bacteria. Nat Commun 4:2926
62. Grandbois M et al (2000) Affinity imaging of red blood cells using an atomic force microscope. J Histochem Cytochem 48(5):719–724
63. Baumgartner W et al (2000) Cadherin interaction probed by atomic force microscopy. Proc Natl Acad Sci USA 97(8):4005–4010
64. Thie M et al (1998) Interactions between trophoblast and uterine epithelium: monitoring of adhesive forces. Hum Reprod 13(11):3211–3219

65. Kim H et al (2006) Quantification of the number of EP3 receptors on a living CHO cell surface by the AFM. Ultramicroscopy 106(8–9):652–662
66. Roduit C et al (2008) Elastic membrane heterogeneity of living cells revealed by stiff nanoscale membrane domains. Biophys J 94(4):1521–1532
67. Zheng Y et al (2013) Recent advances in microfluidic techniques for single-cell biophysical characterization. Lab Chip 13(13):2464–2483
68. Lee WG et al (2007) On-chip erythrocyte deformability test under optical pressure. Lab Chip 7(4):516–519
69. Rosenbluth MJ, Lam WA, Fletcher DA (2008) Analyzing cell mechanics in hematologic diseases with microfluidic biophysical flow cytometry. Lab Chip 8(7):1062–1070
70. Guck J et al (2005) Optical deformability as an inherent cell marker for testing malignant transformation and metastatic competence. Biophys J 88(5):3689–3698
71. Remmerbach TW et al (2009) Oral cancer diagnosis by mechanical phenotyping. Cancer Res 69(5):1728–1732
72. Hou HW et al (2009) Deformability study of breast cancer cells using microfluidics. Biomed Microdevices 11(3):557–564
73. Gossett DR et al (2012) Hydrodynamic stretching of single cells for large population mechanical phenotyping. Proc Natl Acad Sci USA 109(20):7630–7635
74. Bao N et al (2011) Single-cell electrical lysis of erythrocytes detects deficiencies in the cytoskeletal protein network. Lab Chip 11(18):3053–3056
75. Robinson T, Kuhn P, Dittrich PS (2013) Reorganization of lipid domains in model membranes under deformation. 17th international conference on miniaturized systems for chemistry and life sciences 2013
76. Radosinska J, Vrbjar N (2016) The role of red blood cell deformability and Na,K-ATPase function in selected risk factors of cardiovascular diseases in humans: focus on hypertension, diabetes mellitus and hypercholesterolemia. Physiol Res 65(Suppl 1): S43–S54
77. Maher AD, Kuchel PW (2003) The Gardos channel: a review of the Ca2+−activated K+ channel in human erythrocytes. Int J Biochem Cell Biol 35(8):1182–1197
78. Thomas SL et al (2011) Ion channels in human red blood cell membrane: actors or relics? Blood Cells Mol Dis 46(4):261–265
79. Cahalan SM et al (2015) Piezo1 links mechanical forces to red blood cell volume. Elife 4:e07370
80. Evans E, Mohandas N, Leung A (1984) Static and dynamic rigidities of normal and sickle erythrocytes. Major influence of cell hemoglobin concentration. J Clin Invest 73(2):477–488
81. Picas L et al (2013) Structural and mechanical heterogeneity of the erythrocyte membrane reveals hallmarks of membrane stability. ACS Nano 7(2):1054–1063
82. Betz T et al (2009) ATP-dependent mechanics of red blood cells. Proc Natl Acad Sci USA 106(36):15320–15325
83. Park Y et al (2010) Metabolic remodeling of the human red blood cell membrane. Proc Natl Acad Sci USA 107(4):1289–1294
84. Yoon YZ et al (2008) The nonlinear mechanical response of the red blood cell. Phys Biol 5(3):036007
85. Wan J, Ristenpart WD, Stone HA (2008) Dynamics of shear-induced ATP release from red blood cells. Proc Natl Acad Sci USA 105(43):16432–16437
86. Chu H et al (2016) Reversible binding of hemoglobin to band 3 constitutes the molecular switch that mediates O2 regulation of erythrocyte properties. Blood 128(23):2708–2716
87. Manno S, Takakuwa Y, Mohandas N (2005) Modulation of erythrocyte membrane mechanical function by protein 4.1 phosphorylation. J Biol Chem 280(9):7581–7587
88. Manno S et al (1995) Modulation of erythrocyte membrane mechanical function by beta-spectrin phosphorylation and dephosphorylation. J Biol Chem 270(10):5659–5665
89. An X et al (2006) Phosphatidylinositol-4,5-biphosphate (PIP2) differentially regulates the interaction of human erythrocyte protein 4.1 (4.1R) with membrane proteins. Biochemistry 45(18):5725–5732

90. Saleh HS et al (2009) Properties of an Ezrin mutant defective in F-actin binding. J Mol Biol 385(4):1015–1031
91. Bretscher A et al (2000) ERM-Merlin and EBP50 protein families in plasma membrane organization and function. Annu Rev Cell Dev Biol 16(1):113–143
92. Gimona M et al (2002) Functional plasticity of CH domains. FEBS Lett 513(1):98–106
93. Paunola E, Mattila PK, Lappalainen P (2002) WH2 domain: a small, versatile adapter for actin monomers. FEBS Lett 513(1):92–97
94. Fukami K et al (1992) Requirement of phosphatidylinositol 4,5-bisphosphate for [alpha]-actinin function. Nature 359(6391):150–152
95. Fukami K et al (1996) Identification of a phosphatidylinositol 4,5-bisphosphate-binding site in chicken skeletal muscle α-Actinin. J Biol Chem 271(5):2646–2650
96. McKenna JMD, Ostap EM (2009) Kinetics of the interaction of myo1c with Phosphoinositides. J Biol Chem 284(42):28650–28659
97. Liu X et al (2016) Mammalian nonmuscle myosin II binds to anionic phospholipids with concomitant dissociation of the regulatory light chain. J Biol Chem 291(48):24828–24837
98. Feeser EA, Ostap EM (2010) Myo1e binds anionic phospholipids with high affinity. Biophys J 98(3):561a
99. Yu H et al (2012) PtdIns (3,4,5) P3 recruitment of Myo10 is essential for axon development. PLoS One 7(5):e36988
100. Köster DV, Mayor S (2016) Cortical actin and the plasma membrane: inextricably intertwined. Curr Opin Cell Biol 38:81–89
101. Raghupathy R et al (2015) Transbilayer lipid interactions mediate Nanoclustering of lipid-anchored proteins. Cell 161(3):581–594
102. Stachowiak JC et al (2012) Membrane bending by protein–protein crowding. Nat Cell Biol 14(9):944–949
103. MacKinnon R (2003) Potassium channels. FEBS Lett 555(1):62–65
104. Unwin N (2005) Refined structure of the nicotinic acetylcholine receptor at 4 Å resolution. J Mol Biol 346(4):967–989
105. Ehrlich M et al (2004) Endocytosis by Random Initiation and Stabilization of Clathrin-Coated Pits. Cell 118(5):591–605
106. Drin G, Antonny B (2010) Amphipathic helices and membrane curvature. FEBS Lett 584(9):1840–1847
107. Hurley JH, Hanson PI (2010) Membrane budding and scission by the ESCRT machinery: it's all in the neck. Nat Rev Mol Cell Biol 11(8):556–566
108. Raiborg C, Stenmark H (2009) The ESCRT machinery in endosomal sorting of ubiquitylated membrane proteins. Nature 458(7237):445–452
109. Mim C, Unger VM (2012) Membrane curvature and its generation by BAR proteins. Trends Biochem Sci 37(12):526–533
110. Rao Y, Haucke V (2011) Membrane shaping by the bin/amphiphysin/Rvs (BAR) domain protein superfamily. Cell Mol Life Sci 68(24):3983–3993
111. Hinshaw JE, Schmid SL (1995) Dynamin self-assembles into rings suggesting a mechanism for coated vesicle budding. Nature 374(6518):190–192
112. Peter BJ et al (2004) BAR domains as sensors of membrane curvature: the Amphiphysin BAR structure. Science 303(5657):495–499
113. Frost A, Unger VM, De Camilli P (2009) The BAR domain superfamily: membrane-molding macromolecules. Cell 137(2):191–196
114. Sit ST, Manser E (2011) Rho GTPases and their role in organizing the actin cytoskeleton. J Cell Sci 124(5):679–683
115. Aspenstrom P (2014) BAR domain proteins regulate Rho GTPase signaling. Small GTPases 5(2):7
116. McMahon HT, Boucrot E (2011) Molecular mechanism and physiological functions of clathrin-mediated endocytosis. Nat Rev Mol Cell Biol 12(8):517–533
117. Zanetti G et al (2012) COPII and the regulation of protein sorting in mammals. Nat Cell Biol 14(1):20–28

118. Shibata Y et al (2009) Mechanisms shaping the membranes of cellular organelles. Annu Rev Cell Dev Biol 25(1):329–354
119. Bretscher MS (1972) Phosphatidyl-ethanolamine: differential labelling in intact cells and cell ghosts of human erythrocytes by a membrane-impermeable reagent. J Mol Biol 71(3):523–528
120. Daleke DL (2008) Regulation of phospholipid asymmetry in the erythrocyte membrane. Curr Opin Hematol 15(3):191–195
121. Lhermusier T, Chap H, Payrastre B (2011) Platelet membrane phospholipid asymmetry: from the characterization of a scramblase activity to the identification of an essential protein mutated in Scott syndrome. J Thromb Haemost 9(10):1883–1891
122. Murate M et al (2015) Transbilayer distribution of lipids at nano scale. J Cell Sci 128(8):1627–1638
123. Murate M, Kobayashi T (2015) Revisiting transbilayer distribution of lipids in the plasma membrane. Chem Phys Lipids 194:58
124. Elani Y et al (2015) Measurements of the effect of membrane asymmetry on the mechanical properties of lipid bilayers. Chem Commun (Camb) 51(32):6976–6979
125. Liu SL et al (2016) Orthogonal lipid sensors identify transbilayer asymmetry of plasma membrane cholesterol. Nat Chem Biol 13:268
126. Arashiki N et al (2016) An unrecognized function of cholesterol: regulating the mechanism controlling membrane phospholipid asymmetry. Biochemistry 55(25):3504–3513
127. Pomorski TG, Menon AK (2016) Lipid somersaults: uncovering the mechanisms of protein-mediated lipid flipping. Prog Lipid Res 64:69–84
128. Chiantia S, London E (2012) Acyl chain length and saturation modulate interleaflet coupling in asymmetric bilayers: effects on dynamics and structural order. Biophys J 103(11):2311–2319
129. Lin Q, London E (2015) Ordered raft domains induced by outer leaflet sphingomyelin in cholesterol-rich asymmetric vesicles. Biophys J 108(9):2212–2222
130. Koynova R, Caffrey M (1998) Phases and phase transitions of the phosphatidylcholines. Biochim Biophys Acta 1376(1):91–145
131. van Meer G, Voelker DR, Feigenson GW (2008) Membrane lipids: where they are and how they behave. Nat Rev Mol Cell Biol 9(2):112–124
132. Slotte JP (2013) Biological functions of sphingomyelins. Prog Lipid Res 52(4):424–437
133. Brown DA, London E (1998) Functions of lipid rafts in biological membranes. Annu Rev Cell Dev Biol 14:111–136
134. McConnell HM, Vrljic M (2003) Liquid-liquid immiscibility in membranes. Annu Rev Biophys Biomol Struct 32:469–492
135. Holthuis JC, Menon AK (2014) Lipid landscapes and pipelines in membrane homeostasis. Nature 510(7503):48–57
136. Puth K et al (2015) Homeostatic control of biological membranes by dedicated lipid and membrane packing sensors. Biol Chem 396(9–10):1043–1058
137. Ernst R, Ejsing CS, Antonny B (2016) Homeoviscous adaptation and the regulation of membrane lipids. J Mol Biol 428(24):4776–4791
138. Gawrisch K et al (1992) Membrane dipole potentials, hydration forces, and the ordering of water at membrane surfaces. Biophys J 61(5):1213–1223
139. Schamberger J, Clarke RJ (2002) Hydrophobic ion hydration and the magnitude of the dipole potential. Biophys J 82(6):3081–3088
140. Zheng C, Vanderkooi G (1992) Molecular origin of the internal dipole potential in lipid bilayers: calculation of the electrostatic potential. Biophys J 63(4):935–941
141. Clarke RJ (1997) Effect of lipid structure on the dipole potential of phosphatidylcholine bilayers. Biochim Biophys Acta Biomembr 1327(2):269–278
142. Starke-Peterkovic T, Clarke RJ (2009) Effect of headgroup on the dipole potential of phospholipid vesicles. Eur Biophys J 39(1):103–110
143. Szabo G (1974) Dual mechanism for the action of cholesterol on membrane permeability. Nature 252(5478):47–49

144. McIntosh TJ, Magid AD, Simon SA (1989) Repulsive interactions between uncharged bilayers. Hydration and fluctuation pressures for monoglycerides. Biophys J 55(5):897–904
145. Starke-Peterkovic T et al (2006) Cholesterol effect on the dipole potential of lipid membranes. Biophys J 90(11):4060–4070
146. Haldar S et al (2012) Differential effect of cholesterol and its biosynthetic precursors on membrane dipole potential. Biophys J 102(7):1561–1569
147. Cullis PR, De Kruijff B (1979) Lipid polymorphism and the functional roles of lipids in biological membranes. Biochim Biophys Acta Rev Biomembr 559(4):399–420
148. Sheetz MP, Singer SJ (1974) Biological membranes as bilayer couples. A molecular mechanism of drug-erythrocyte interactions. Proc Natl Acad Sci USA 71(11):4457–4461
149. Chernomordik LV, Kozlov MM (2008) Mechanics of membrane fusion. Nat Struct Mol Biol 15(7):675–683
150. Ailte I et al (2016) Addition of lysophospholipids with large head groups to cells inhibits Shiga toxin binding. Sci Rep 6:30336
151. McMahon HT, Boucrot E (2015) Membrane curvature at a glance. J Cell Sci 128(6):1065–1070
152. Holdbrook DA et al (2016) Dynamics of crowded vesicles: local and global responses to membrane composition. PLoS One 11(6):e0156963
153. Frisz JF et al (2013) Sphingolipid domains in the plasma membranes of fibroblasts are not enriched with cholesterol. J Biol Chem 288(23):16855–16861
154. Frisz JF et al (2013) Direct chemical evidence for sphingolipid domains in the plasma membranes of fibroblasts. Proc Natl Acad Sci USA 110(8):E613–E622
155. Sevcsik E et al (2015) GPI-anchored proteins do not reside in ordered domains in the live cell plasma membrane. Nat Commun 6:6969
156. Carquin M et al (2016) Recent progress on lipid lateral heterogeneity in plasma membranes: from rafts to submicrometric domains. Prog Lipid Res 62:1–24
157. Takatori S, Mesman R, Fujimoto T (2014) Microscopic methods to observe the distribution of lipids in the cellular membrane. Biochemistry 53(4):639–653
158. Maekawa M, Fairn GD (2014) Molecular probes to visualize the location, organization and dynamics of lipids. J Cell Sci 127(22):4801–4812
159. Skocaj M et al (2013) The sensing of membrane microdomains based on pore-forming toxins. Curr Med Chem 20(4):491–501
160. Maekawa M, Yang Y, Fairn GD, Perfringolysin O (2016) Theta toxin as a tool to monitor the distribution and inhomogeneity of cholesterol in cellular membranes. Toxins (Basel) 8(3):67
161. Montes LR et al (2008) Ceramide-enriched membrane domains in red blood cells and the mechanism of sphingomyelinase-induced hot-cold hemolysis. Biochemistry 47(43):11222–11230
162. Cai M et al (2012) Direct evidence of lipid rafts by in situ atomic force microscopy. Small 8(8):1243–1250
163. Hao M, Mukherjee S, Maxfield FR (2001) Cholesterol depletion induces large scale domain segregation in living cell membranes. Proc Natl Acad Sci USA 98(23):13072–13077
164. Grassme H et al (2003) Host defense against Pseudomonas aeruginosa requires ceramide-rich membrane rafts. Nat Med 9(3):322–330
165. Stancevic B, Kolesnick R (2010) Ceramide-rich platforms in transmembrane signaling. FEBS Lett 584(9):1728–1740
166. Kaiser HJ et al (2009) Order of lipid phases in model and plasma membranes. Proc Natl Acad Sci USA 106(39):16645–16650
167. Pataraia S et al (2014) Effect of cytochrome c on the phase behavior of charged multicomponent lipid membranes. Biochim Biophys Acta 1838(8):2036–2045
168. de la Serna JB et al (2009) Segregated phases in pulmonary surfactant membranes do not show coexistence of lipid populations with differentiated dynamic properties. Biophys J 97(5):1381–1389

169. Hammond AT et al (2005) Crosslinking a lipid raft component triggers liquid ordered-liquid disordered phase separation in model plasma membranes. Proc Natl Acad Sci 102(18):6320–6325
170. Fidorra M et al (2006) Absence of fluid-ordered/fluid-disordered phase coexistence in ceramide/POPC mixtures containing cholesterol. Biophys J 90(12):4437–4451
171. Bagatolli LA (2006) To see or not to see: lateral organization of biological membranes and fluorescence microscopy. Biochim Biophys Acta Biomembr 1758(10):1541–1556
172. Levental I, Grzybek M, Simons K (2011) Raft domains of variable properties and compositions in plasma membrane vesicles. Proc Natl Acad Sci 108(28):11411–11416
173. Girard P et al (2004) A new method for the reconstitution of membrane proteins into Giant Unilamellar vesicles. Biophys J 87(1):419–429
174. Veatch SL et al (2008) Critical fluctuations in plasma membrane vesicles. ACS Chem Biol 3(5):287–293
175. Bouwstra JA et al (2003) Structure of the skin barrier and its modulation by vesicular formulations. Prog Lipid Res 42(1):1–36
176. Downing DT (1992) Lipid and protein structures in the permeability barrier of mammalian epidermis. J Lipid Res 33(3):301–313
177. Mileykovskaya E, Dowhan W (2000) Visualization of phospholipid domains in Escherichia coli by using the Cardiolipin-specific fluorescent dye 10-N-Nonyl Acridine Orange. J Bacteriol 182(4):1172–1175
178. Kawai F et al (2004) Cardiolipin domains in Bacillus subtilis marburg membranes. J Bacteriol 186(5):1475–1483
179. Lopez D (2015) Molecular composition of functional microdomains in bacterial membranes. Chem Phys Lipids 192:3–11
180. Bramkamp M, Lopez D (2015) Exploring the existence of lipid rafts in bacteria. Microbiol Mol Biol Rev 79(1):81–100
181. Huijbregts RP, de Kroon AI, de Kruijff B (2000) Topology and transport of membrane lipids in bacteria. Biochim Biophys Acta 1469(1):43–61
182. Lopez-Lara IM, Geiger O (2016) Bacterial lipid diversity. Biochim Biophys Acta
183. Parsons JB, Rock CO (2013) Bacterial lipids: metabolism and membrane homeostasis. Prog Lipid Res 52(3):249–276
184. Huang Z, London E (2016) Cholesterol lipids and cholesterol-containing lipid rafts in bacteria. Chem Phys Lipids 199:11–16
185. Lin M, Rikihisa Y (2003) Ehrlichia chaffeensis and Anaplasma phagocytophilum lack genes for lipid a biosynthesis and incorporate cholesterol for their survival. Infect Immun 71(9):5324–5331
186. Saenz JP et al (2012) Functional convergence of hopanoids and sterols in membrane ordering. Proc Natl Acad Sci USA 109(35):14236–14240
187. Doughty DM et al (2014) Probing the subcellular localization of Hopanoid lipids in bacteria using NanoSIMS. PLoS One 9(1):e84455
188. Donovan C, Bramkamp M (2009) Characterization and subcellular localization of a bacterial flotillin homologue. Microbiology 155(6):1786–1799
189. López D, Kolter R (2010) Functional microdomains in bacterial membranes. Genes Dev 24(17):1893–1902
190. Malinska K et al (2003) Visualization of protein compartmentation within the plasma membrane of living yeast cells. Mol Biol Cell 14(11):4427–4436
191. Klose C et al (2010) Yeast lipids can phase-separate into micrometer-scale membrane domains. J Biol Chem 285(39):30224–30232
192. Spira F et al (2012) Patchwork organization of the yeast plasma membrane into numerous coexisting domains. Nat Cell Biol 14(8):890–890
193. Kabeche R et al (2015) Eisosomes regulate phosphatidylinositol 4,5-bisphosphate (PI(4,5)P2) cortical clusters and mitogen-activated protein (MAP) kinase signaling upon osmotic stress. J Biol Chem 290(43):25960–25973

194. Guiney EL et al (2015) Calcineurin regulates the yeast synaptojanin Inp53/Sjl3 during membrane stress. Mol Biol Cell 26(4):769–785
195. Toulmay A, Prinz WA (2013) Direct imaging reveals stable, micrometer-scale lipid domains that segregate proteins in live cells. J Cell Biol 202(1):35–44
196. Zachowski A (1993) Phospholipids in animal eukaryotic membranes: transverse asymmetry and movement. Biochem J 294(1):1–14
197. Goodman SR et al (2013) The proteomics and interactomics of human erythrocytes. Exp Biol Med (Maywood) 238(5):509–518
198. D'Auria L et al (2011) Segregation of fluorescent membrane lipids into distinct micrometric domains: evidence for phase compartmentation of natural lipids? PLoS One 6(2):e17021
199. Ekyalongo RC et al (2015) Organization and functions of glycolipid-enriched microdomains in phagocytes. Biochim Biophys Acta 1851(1):90–97
200. Makino A et al (2015) Visualization of the heterogeneous membrane distribution of sphingomyelin associated with cytokinesis, cell polarity, and sphingolipidosis. FASEB J 29(2):477–493
201. Abe M et al (2012) A role for sphingomyelin-rich lipid domains in the accumulation of phosphatidylinositol-4,5-bisphosphate to the cleavage furrow during cytokinesis. Mol Cell Biol 32(8):1396–1407
202. Mizuno H et al (2011) Fluorescent probes for superresolution imaging of lipid domains on the plasma membrane. Chem Sci 2:1548–1553
203. Kiyokawa E et al (2005) Spatial and functional heterogeneity of sphingolipid-rich membrane domains. J Biol Chem 280(25):24072–24084
204. Leonard et al (2017) Science reports 2017. Sci Rep 7(1):4264. doi:10.1038/s41598-017-04388-z
205. Agbani EO et al (2015) Coordinated membrane ballooning and Procoagulant spreading in human platelets. Circulation 132(15):1414–1424
206. Heemskerk JW et al (1997) Collagen but not fibrinogen surfaces induce bleb formation, exposure of phosphatidylserine, and procoagulant activity of adherent platelets: evidence for regulation by protein tyrosine kinase-dependent Ca2+ responses. Blood 90(7):2615–2625
207. Gousset K et al (2002) Evidence for a physiological role for membrane rafts in human platelets. J Cell Physiol 190(1):117–128
208. Bali R et al (2009) Macroscopic domain formation during cooling in the platelet plasma membrane: an issue of low cholesterol content. Biochim Biophys Acta 1788(6):1229–1237
209. Pierini LM et al (2003) Membrane lipid organization is critical for human neutrophil polarization. J Biol Chem 278(12):10831–10841
210. Sonnino S et al (2009) Role of very long fatty acid-containing glycosphingolipids in membrane organization and cell signaling: the model of lactosylceramide in neutrophils. Glycoconj J 26(6):615–621
211. Jackman N, Ishii A, Bansal R (2009) Oligodendrocyte development and myelin biogenesis: parsing out the roles of glycosphingolipids. Physiology (Bethesda) 24:290–297
212. Aggarwal S, Yurlova L, Simons M (2011) Central nervous system myelin: structure, synthesis and assembly. Trends Cell Biol 21(10):585–593
213. Boggs JM, Wang H (2004) Co-clustering of galactosylceramide and membrane proteins in oligodendrocyte membranes on interaction with polyvalent carbohydrate and prevention by an intact cytoskeleton. J Neurosci Res 76(3):342–355
214. Boggs JM et al (2010) Participation of galactosylceramide and sulfatide in glycosynapses between oligodendrocyte or myelin membranes. FEBS Lett 584(9):1771–1778
215. Ozgen H et al (2014) The lateral membrane organization and dynamics of myelin proteins PLP and MBP are dictated by distinct galactolipids and the extracellular matrix. PLoS One 9(7):e101834
216. Decker L, French-Constant C (2004) Lipid rafts and integrin activation regulate oligodendrocyte survival. J Neurosci 24(15):3816–3825
217. Meder D et al (2006) Phase coexistence and connectivity in the apical membrane of polarized epithelial cells. Proc Natl Acad Sci USA 103(2):329–334

218. Ikenouchi J et al (2012) Lipid polarity is maintained in absence of tight junctions. J Biol Chem 287(12):9525–9533
219. Umeda K et al (2006) ZO-1 and ZO-2 independently determine where claudins are polymerized in tight-junction strand formation. Cell 126(4):741–754
220. Fanning AS, Van Itallie CM, Anderson JM (2012) Zonula occludens-1 and -2 regulate apical cell structure and the zonula adherens cytoskeleton in polarized epithelia. Mol Biol Cell 23(4):577–590
221. Orsini F et al (2012) Atomic force microscopy imaging of lipid rafts of human breast cancer cells. Biochim Biophys Acta 1818(12):2943–2949
222. Dinic J et al (2015) The T cell receptor resides in ordered plasma membrane nanodomains that aggregate upon patching of the receptor. Sci Rep 5:10082
223. Gomez-Mouton C et al (2001) Segregation of leading-edge and uropod components into specific lipid rafts during T cell polarization. Proc Natl Acad Sci USA 98(17):9642–9647
224. Paparelli L et al (2016) Inhomogeneity based characterization of distribution patterns on the plasma membrane. PLoS Comput Biol 12(9):e1005095
225. Golebiewska U et al (2008) Diffusion coefficient of fluorescent phosphatidylinositol 4,5-bisphosphate in the plasma membrane of cells. Mol Biol Cell 19(4):1663–1669
226. Perkins RG, Scott RE (1978) Plasma membrane phospholipid, cholesterol and fatty acyl composition of differentiated and undifferentiated L6 myoblasts. Lipids 13(5):334–337
227. Aresta-Branco F et al (2011) Gel domains in the plasma membrane of Saccharomyces cerevisiae: highly ordered, ergosterol-free, and sphingolipid-enriched lipid rafts. J Biol Chem 286(7):5043–5054
228. Fujita A, Cheng J, Fujimoto T (2009) Segregation of GM1 and GM3 clusters in the cell membrane depends on the intact actin cytoskeleton. Biochim Biophys Acta 1791(5):388–396
229. Das A et al (2013) Use of mutant 125I-perfringolysin O to probe transport and organization of cholesterol in membranes of animal cells. Proc Natl Acad Sci USA 110(26):10580–10585
230. Jin H, McCaffery JM, Grote E (2008) Ergosterol promotes pheromone signaling and plasma membrane fusion in mating yeast. J Cell Biol 180(4):813–826
231. Chierico L et al (2014) Live cell imaging of membrane/cytoskeleton interactions and membrane topology. Sci Rep 4:6056
232. Veatch SL, Keller SL (2003) Separation of liquid phases in giant vesicles of ternary mixtures of phospholipids and cholesterol. Biophys J 85(5):3074–3083
233. Machta BB, Veatch SL, Sethna JP (2012) Critical Casimir forces in cellular membranes. Phys Rev Lett 109(13):138101–138101
234. Lee I-H et al (2015) Live cell plasma membranes do not exhibit a miscibility phase transition over a wide range of temperatures. J Phys Chem B 119(12):4450–4459
235. Adami C (1995) Self-organized criticality in living systems. Phys Lett A 2013:29–32
236. Jensen H (1998) Self-organized criticality: emergent complex behavior in physical and biological systems (Cambridge lecture notes in Physics), Cambridge University Press,Cambridge
237. Heberle FA et al (2013) Bilayer thickness mismatch controls domain size in model membranes. J Am Chem Soc 135(18):6853–6859
238. Samsonov AV, Mihalyov I, Cohen FS (2001) Characterization of cholesterol-sphingomyelin domains and their dynamics in bilayer membranes. Biophys J 81(3):1486–1500
239. D'Auria L et al (2013) Surfactins modulate the lateral organization of fluorescent membrane polar lipids: a new tool to study drug:membrane interaction and assessment of the role of cholesterol and drug acyl chain length. Biochim Biophys Acta 1828(9):2064–2073
240. Kuzmin PI et al (2005) Line tension and interaction energies of membrane rafts calculated from lipid splay and tilt. Biophys J 88(2):1120–1133
241. Frolov VAJ et al (2006) "Entropic Traps" in the Kinetics of Phase Separation in Multicomponent Membranes Stabilize Nanodomains. Biophys J 91(1):189–205
242. García-Sáez AJ, Schwille P (2010) Stability of lipid domains. FEBS Lett 584(9):1653–1658
243. Simons K, Vaz WLC (2004) Model systems, lipid rafts, and cell membranes. Annu Rev Biophys Biomol Struct 33(1):269–295

244. Castro BM, Prieto M, Silva LC (2014) Ceramide: a simple sphingolipid with unique biophysical properties. Prog Lipid Res 54:53–67
245. Sot J et al (2006) Detergent-resistant, ceramide-enriched domains in sphingomyelin/ceramide bilayers. Biophys J 90(3):903–914
246. Koldso H et al (2014) Lipid clustering correlates with membrane curvature as revealed by molecular simulations of complex lipid bilayers. PLoS Comput Biol 10(10):e1003911
247. Patel DS et al (2016) Influence of ganglioside GM1 concentration on lipid clustering and membrane properties and curvature. Biophys J 111(9):1987–1999
248. Mukherjee S, Soe TT, Maxfield FR (1999) Endocytic sorting of lipid analogues differing solely in the chemistry of their hydrophobic tails. J Cell Biol 144(6):1271–1284
249. Mukherjee S, Maxfield FR (2000) Role of membrane organization and membrane domains in endocytic lipid trafficking. Traffic (Copenhagen, Denmark) 1(3):203–211
250. Roux A et al (2005) Role of curvature and phase transition in lipid sorting and fission of membrane tubules. EMBO J 24(8):1537–1545
251. Parthasarathy R, Yu C-h, Groves JT (2006) Curvature-modulated phase separation in lipid bilayer membranes. LANGMUIR 22(11):5095–5099
252. Baumgart T, Hess ST, Webb WW (2003) Imaging coexisting fluid domains in biomembrane models coupling curvature and line tension. Nature 425(6960):821–824
253. Ursell TS, Klug WS, Phillips R (2009) Morphology and interaction between lipid domains. Proc Natl Acad Sci USA 106(32):13301–13306
254. Veatch SL, Keller SL (2005) Seeing spots: complex phase behavior in simple membranes. Biochimica et Biophysica Acta (BBA) Mol Cell Res 1746(3):172–185
255. Kiessling V, Crane JM, Tamm LK (2006) Transbilayer effects of raft-like lipid domains in asymmetric planar bilayers measured by single molecule tracking. Biophys J 91(9):3313–3326
256. Fujiwara T et al (2002) Phospholipids undergo hop diffusion in compartmentalized cell membrane. J Cell Biol 157(6):1071–1082
257. Nawaz S et al (2009) Phosphatidylinositol 4,5-bisphosphate-dependent interaction of myelin basic protein with the plasma membrane in Oligodendroglial cells and its rapid perturbation by elevated calcium. J Neurosci 29(15):4794
258. May S et al (2009) Trans-monolayer coupling of fluid domains in lipid bilayers. Soft Matter 5(17):3148–3148
259. Leibler S, Andelman D (1987) Ordered and curved meso-structures in membranes and amphiphilic films. J Phys 48(11):2013–2018
260. Merkel R, Sackmann E, Evans E (1989) Molecular friction and epitactic coupling between monolayers in supported bilayers. J Phys 50(12):1535–1555
261. Collins MD, Keller SL (2008) Tuning lipid mixtures to induce or suppress domain formation across leaflets of unsupported asymmetric bilayers. Proc Natl Acad Sci 105(1): 124–128
262. Brewer J et al (2017) Enzymatic studies on planar supported membranes using a widefield fluorescence LAURDAN generalized polarization imaging approach. Biochim Biophys Acta 1859(5):888–895
263. Kusumi A, Koyama-Honda I, Suzuki K (2004) Molecular dynamics and interactions for creation of stimulation-induced stabilized rafts from small unstable steady-state rafts. Traffic 5(4):213–230
264. Gri G et al (2004) The inner side of T cell lipid rafts. Immunol Lett 94(3):247–252
265. Pyenta PS, Holowka D, Baird B (2001) Cross-correlation analysis of inner-leaflet-anchored green fluorescent protein co-redistributed with IgE receptors and outer leaflet lipid raft components. Biophys J 80(5):2120–2132
266. Lee KYC, McConnell HM (1993) Quantized symmetry of liquid monolayer domains. J Phys Chem 97(37):9532–9539
267. Travesset A (2006) Effect of dipolar moments in domain sizes of lipid bilayers and monolayers. J Chem Phys 125(8):084905–084905

268. Inoue I, Kobatake Y, Tasaki I (1973) Excitability, instability and phase transitions in squid axon membrane under internal perfusion with dilute salt solutions. Biochim Biophys Acta Biomembr 307(3):471–477
269. Melamed-Harel H, Reinhold L (1979) Hysteresis in the responses of membrane potential, membrane resistance, and growth rate to cyclic temperature change. Plant Physiol 63(6):1089–1094
270. Herman P et al (2015) Depolarization affects the lateral microdomain structure of yeast plasma membrane. FEBS J 282(3):419–434
271. Malinsky J, Tanner W, Opekarova M (2016) Transmembrane voltage: Potential to induce lateral microdomains. Biochimica et Biophysica Acta (BBA) – Mol Cell Biol Lipids 1861(8):806–811
272. Rossy J, Ma Y, Gaus K (2014) The organisation of the cell membrane: do proteins rule lipids? Curr Opin Chem Biol 20:54–59
273. Mayor S, Rao M (2004) Rafts: scale-dependent, active lipid Organization at the Cell Surface. Traffic 5(4):231–240
274. Sharma P et al (2004) Nanoscale organization of multiple GPI-anchored proteins in living cell membranes. Cell 116(4):577–589
275. Nicolini C et al (2006) Visualizing association of N-Ras in lipid microdomains: influence of domain structure and interfacial adsorption. J Am Chem Soc 128(1):192–201
276. García-Sáez AJ et al (2007) Pore formation by a Bax-derived peptide: effect on the line tension of the membrane probed by AFM. Biophys J 93(1):103–112
277. Jensen MØ, Mouritsen OG (2004) Lipids do influence protein function—the hydrophobic matching hypothesis revisited. Biochim Biophys Acta Biomembr 1666(1–2):205–226
278. Mitra K et al (2004) Modulation of the bilayer thickness of exocytic pathway membranes by membrane proteins rather than cholesterol. Proc Natl Acad Sci USA 101(12):4083–4088
279. Larson DR et al (2005) Temporally resolved interactions between antigen-stimulated IgE receptors and Lyn kinase on living cells. J Cell Biol 171(3):527–536
280. Gaus K et al (2005) Condensation of the plasma membrane at the site of T lymphocyte activation. J Cell Biol 171(1):121–131
281. Kusumi A et al (2005) Paradigm shift of the plasma membrane concept from the two-dimensional continuum fluid to the partitioned fluid: high-speed single-molecule tracking of membrane molecules. Annu Rev Biophys Biomol Struct 34:351–378
282. Kay JG et al (2012) Phosphatidylserine dynamics in cellular membranes. Mol Biol Cell 23(11):2198–2212
283. Liu AP, Fletcher DA (2006) Actin polymerization serves as a membrane domain switch in model lipid bilayers. Biophys J 91(11):4064–4070
284. Honigmann A et al (2014) A lipid bound actin meshwork organizes liquid phase separation in model membranes. Elife 3:e01671
285. Arumugam S, Petrov EP, Schwille P (2015) Cytoskeletal pinning controls phase separation in multicomponent lipid membranes. Biophys J 108(5):1104–1113
286. Honigmann A, Pralle A (2016) Compartmentalization of the cell membrane. J Mol Biol 428(24):4739–4748
287. Kusumi A et al (2012) Dynamic organizing principles of the plasma membrane that regulate signal transduction: commemorating the fortieth anniversary of singer and Nicolson's fluid-mosaic model. Annu Rev Cell Dev Biol 28:215–250
288. Kusumi A et al (2012) Membrane mechanisms for signal transduction: the coupling of the meso-scale raft domains to membrane-skeleton-induced compartments and dynamic protein complexes. Semin Cell Dev Biol 23(2):126–144
289. Kusumi A et al (2011) Hierarchical mesoscale domain organization of the plasma membrane. Trends Biochem Sci 36(11):604–615
290. Janmey PA, Lindberg U (2004) Cytoskeletal regulation: rich in lipids. Nat Rev Mol Cell Biol 5(8):658–666
291. Rao M, Mayor S (2014) Active organization of membrane constituents in living cells. Curr Opin Cell Biol 29:126–132

292. Sheetz MP, Sable JE, Dobereiner HG (2006) Continuous membrane-cytoskeleton adhesion requires continuous accommodation to lipid and cytoskeleton dynamics. Annu Rev Biophys Biomol Struct 35:417–434
293. de la Serna JB et al (2016) There is no simple model of the plasma membrane organization. Front Cell Dev Biol 4:106
294. Johnson SA et al (2010) Temperature-dependent phase behavior and protein partitioning in giant plasma membrane vesicles. Biochim Biophys Acta Biomembr 1798(7):1427–1435
295. Lingwood D et al (2008) Plasma membranes are poised for activation of raft phase coalescence at physiological temperature. Proc Natl Acad Sci USA 105(29):10005–10010
296. Windschiegl B et al (2009) Lipid reorganization induced by Shiga toxin clustering on planar membranes. PLoS One 4(7):e6238–e6238
297. Zhao H et al (2013) Membrane-sculpting BAR domains generate stable lipid microdomains. Cell Rep 4(6):1213–1223
298. Picas L et al (2014) BIN1/M-Amphiphysin2 induces clustering of phosphoinositides to recruit its downstream partner dynamin. Nat Commun 5:5647
299. Levental I et al (2009) Calcium-dependent lateral organization in phosphatidylinositol 4,5-bisphosphate (PIP2)- and cholesterol-containing monolayers. Biochemistry 48(34):8241–8248
300. Sarmento MJ et al (2014) Ca(2+) induces PI(4,5)P2 clusters on lipid bilayers at physiological PI(4,5)P2 and Ca(2+) concentrations. Biochim Biophys Acta 1838(3):822–830
301. Carvalho K et al (2008) Giant unilamellar vesicles containing phosphatidylinositol(4,5)bisphosphate: characterization and functionality. Biophys J 95(9):4348–4360
302. Laux T et al (2000) GAP43, MARCKS, and CAP23 modulate PI(4,5)P(2) at plasmalemmal rafts, and regulate cell cortex actin dynamics through a common mechanism. J Cell Biol 149(7):1455–1472
303. McLaughlin S, Murray D (2005) Plasma membrane phosphoinositide organization by protein electrostatics. Nature 438(7068):605–611
304. Milovanovic D et al (2016) Calcium promotes the formation of Syntaxin 1 mesoscale domains through phosphatidylinositol 4,5-bisphosphate. J Biol Chem 291(15):7868–7876
305. Shi X et al (2013) Ca2+ regulates T-cell receptor activation by modulating the charge property of lipids. Nature 493(7430):111–115
306. Li L et al (2014) Ionic protein-lipid interaction at the plasma membrane: what can the charge do? Trends Biochem Sci 39(3):130–140
307. Visser D et al (2013) TRPM7 triggers Ca2+ sparks and invadosome formation in neuroblastoma cells. Cell Calcium 54(6):404–415
308. Wei C et al (2009) Calcium flickers steer cell migration. Nature 457(7231):901–905
309. Wu M, Wu X, De Camilli P (2013) Calcium oscillations-coupled conversion of actin travelling waves to standing oscillations. Proc Natl Acad Sci USA 110(4):1339–1344
310. Foret L et al (2005) A simple mechanism of raft formation in two-component fluid membranes. Europhy Lett (EPL) 71(3):508–514
311. Turner MS, Sens P, Socci ND (2005) Nonequilibrium raftlike membrane domains under continuous recycling. Phys Rev Lett 95(16):168301
312. Schmid F (2016) Physical mechanisms of micro- and nanodomain formation in multicomponent lipid membranes. Biochim Biophys Acta 1859:509
313. Fan J, Sammalkorpi M, Haataja M (2010) Formation and regulation of lipid microdomains in cell membranes: theory, modeling, and speculation. FEBS Lett 584(9):1678–1684
314. Bernal P et al (2007) A pseudomonas putida cardiolipin synthesis mutant exhibits increased sensitivity to drugs related to transport functionality. Environ Microbiol 9(5): 1135–1145
315. Tocheva EI, Li Z, Jensen GJ (2010) Electron cryotomography. Cold Spring Harb Perspect Biol 2(6):a003442–a003442
316. Huang KC, Ramamurthi KS (2010) Macromolecules that prefer their membranes curvy. Mol Microbiol 76(4):822–832

317. Hirschberg CB, Kennedy EP (1972) Mechanism of the enzymatic synthesis of cardiolipin in Escherichia coli. Proc Natl Acad Sci USA 69(3):648–651
318. Huang KC et al (2006) A curvature-mediated mechanism for localization of lipids to bacterial poles. PLoS Comput Biol 2(11):e151–e151
319. Mukhopadhyay R, Huang KC, Wingreen NS (2008) Lipid localization in bacterial cells through curvature-mediated microphase separation. Biophys J 95(3):1034–1049
320. Lin T-Y et al (2015) A Cardiolipin-deficient mutant of Rhodobacter sphaeroides has an altered cell shape and is impaired in biofilm formation. J Bacteriol 197(21):3446–3455
321. Ursell TS et al (2014) Rod-like bacterial shape is maintained by feedback between cell curvature and cytoskeletal localization. Proc Natl Acad Sci USA 111(11):E1025–E1034
322. Derganc J (2007) Curvature-driven lateral segregation of membrane constituents in Golgi cisternae. Phys Biol 4(4):317–324
323. Akimov SA et al (2007) Lateral tension increases the line tension between two domains in a lipid bilayer membrane. Phys Rev E Stat Nonlinear Soft Matter Phys 75(1):011919
324. Mohandas N, Gallagher PG (2008) Red cell membrane: past, present, and future. Blood 112(10):3939–3948
325. Lutz HU, Bogdanova A (2013) Mechanisms tagging senescent red blood cells for clearance in healthy humans. Front Physiol 4:387
326. Antonelou MH, Kriebardis AG, Papassideri IS (2010) Aging and death signalling in mature red cells: from basic science to transfusion practice. Blood Transfus 8(Suppl 3):s39–s47
327. Fricke K, Sackmann E (1984) Variation of frequency spectrum of the erythrocyte flickering caused by aging, osmolarity, temperature and pathological changes. Biochim Biophys Acta 803(3):145–152
328. Gov NS, Safran SA (2005) Red blood cell membrane fluctuations and shape controlled by ATP-induced cytoskeletal defects. Biophys J 88(3):1859–1874
329. Edwards CL et al (2005) A brief review of the pathophysiology, associated pain, and psychosocial issues in sickle cell disease. Int J Behav Med 12(3):171–179
330. Willekens FL et al (2008) Erythrocyte vesiculation: a self-protective mechanism? Br J Haematol 141(4):549–556
331. Stewart A et al (2005) The application of a new quantitative assay for the monitoring of integrin-associated protein CD47 on red blood cells during storage and comparison with the expression of CD47 and phosphatidylserine with flow cytometry. Transfusion 45(9):1496–1503
332. Bogdanova A et al (2013) Calcium in red blood cells-a perilous balance. Int J Mol Sci 14(5):9848–9872
333. Salzer U et al (2002) Ca(++)-dependent vesicle release from erythrocytes involves stomatin-specific lipid rafts, synexin (annexin VII), and sorcin. Blood 99(7):2569–2577
334. Li H, Lykotrafitis G (2015) Vesiculation of healthy and defective red blood cells. Phys Rev E Stat Nonlinear Soft Matter Phys 92(1):012715
335. Vind-Kezunovic D et al (2008) Line tension at lipid phase boundaries regulates formation of membrane vesicles in living cells. Biochim Biophys Acta 1778(11):2480–2486
336. Del Conde I et al (2005) Tissue-factor-bearing microvesicles arise from lipid rafts and fuse with activated platelets to initiate coagulation. Blood 106(5):1604–1611
337. Scheiffele P et al (1999) Influenza viruses select ordered lipid domains during budding from the plasma membrane. J Biol Chem 274(4):2038–2044
338. Maddock JR, Shapiro L (1993) Polar location of the chemoreceptor complex in the Escherichia coli cell. Science (New York, NY) 259(5102):1717–1723
339. Lutkenhaus J (2002) Dynamic proteins in bacteria. Curr Opin Microbiol 5(6):548–552
340. Mileykovskaya E et al (1998) Localization and function of early cell division proteins in filamentous Escherichia coli cells lacking phosphatidylethanolamine. J Bacteriol 180(16):4252–4257
341. Renner LD, Weibel DB (2011) Cardiolipin microdomains localize to negatively curved regions of Escherichia coli membranes. Proc Natl Acad Sci USA 108(15):6264–6269

342. Martin SW, Konopka JB (2004) Lipid raft polarization contributes to hyphal growth in Candida albicans. Eukaryot Cell 3(3):675–684
343. Takeshita N et al (2008) Apical sterol-rich membranes are essential for localizing cell end markers that determine growth directionality in the filamentous fungus Aspergillus nidulans. Mol Biol Cell 19(1):339–351
344. Wachtler V, Rajagopalan S, Balasubramanian MK (2003) Sterol-rich plasma membrane domains in the fission yeast Schizosaccharomyces pombe. J Cell Sci 116(5):867–874
345. Bagnat M, Simons K (2002) Cell surface polarization during yeast mating. Proc Natl Acad Sci USA 99(22):14183–14188
346. Sun Y et al (2000) Sli2 (Ypk1), a homologue of mammalian protein kinase SGK, is a downstream kinase in the sphingolipid-mediated signaling pathway of yeast. Mol Cell Biol 20(12):4411–4419
347. Proszynski TJ et al (2006) Plasma membrane polarization during mating in yeast cells. J Cell Biol 173(6):861–866
348. Garrenton LS et al (2010) Pheromone-induced anisotropy in yeast plasma membrane phosphatidylinositol-4,5-bisphosphate distribution is required for MAPK signaling. Proc Natl Acad Sci USA 107(26):11805–11810
349. Vernay A et al (2012) A steep phosphoinositide bis-phosphate gradient forms during fungal filamentous growth. J Cell Biol 198(4):711–730
350. Iwamoto K et al (2004) Local exposure of phosphatidylethanolamine on the yeast plasma membrane is implicated in cell polarity. Genes Cells 9(10):891–903
351. Ng MM, Chang F, Burgess DR (2005) Movement of membrane domains and requirement of membrane signaling molecules for cytokinesis. Dev Cell 9(6):781–790
352. Emoto K et al (2005) Local change in phospholipid composition at the cleavage furrow is essential for completion of cytokinesis. J Biol Chem 280(45):37901–37907
353. Field SJ et al (2005) PtdIns(4,5)P2 functions at the cleavage furrow during cytokinesis. Curr Biol 15(15):1407–1412
354. Emoto K, Umeda M (2000) An essential role for a membrane lipid in cytokinesis. Regulation of contractile ring disassembly by redistribution of phosphatidylethanolamine. J Cell Biol 149(6):1215–1224
355. Kraft ML (2013) Plasma membrane organization and function: moving past lipid rafts. Mol Biol Cell 24(18):2765–2768
356. Sevcsik E, Schutz GJ (2016) With or without rafts? Alternative views on cell membranes. BioEssays 38(2):129–139

Chapter 6
Minimal Cellular Models for Origins-of-Life Studies and Biotechnology

Pasquale Stano

Abstract Minimal cellular models can be defined as those vesicle-based cell-like constructs that are assembled with the aim of (1) clarifying/understanding unknown aspects in origins-of-life research and hypotheses testing, (2) studying reconstituted biochemical pathways in a simplified system, (3) being exploited for potential biotechnological applications, and (4) developing novel concepts/technologies. These ‘synthetic cells’ are created by the bottom-up approach and within the synthetic/constructive paradigm. Here we shortly review the main ideas behind such novel usage of vesicles, and comment the experimental data collected in the past decades. An intriguing picture emerges, where technical progresses owing to the convergence of liposome, cell-free (and microfluidic) technologies lead to a fecund research area of great potential, which blends fundamental scientific question with the most modern and challenging facets of synthetic biology.

Keywords Protocells • Minimal cells • Synthetic biology • PURE system • Fatty acid vesicles • Synthetic cells • Bottom-up approach • Power law • Micro-compartmentalized reactions • Autopoiesis

6.1 Introduction

Not many scientific questions are so fascinating as the origin of life on Earth. This still unsolved conundrum permeated the history of science of all ages, but only in the twentieth century it became a central question of modern chemical investigation [1, 2], with the emergence of a chemical branch called prebiotic chemistry. Prebiotic chemistry typically focuses on the question “how complex chemical molecules originated from simple and primitively available building blocks?”. Several studies –

P. Stano (✉)
Department of Biological and Environmental Sciences and Technology (DiSTeBA), Università del Salento, Campus Ecotekne (S.P. 6 Lecce-Monteroni), 73100, Lecce, Italy
e-mail: pasquale.stano@unisalento.it

R.M. Epand, J.-M. Ruysschaert (eds.), *The Biophysics of Cell Membranes*, Springer Series in Biophysics 19, DOI 10.1007/978-981-10-6244-5_6

among which the famous Urey-Miller experiment [3] – have shown plausible paths for originating molecules as amino acids, sugars, nucleobases, and so on (for a review, see [4]).

However, the origin of complex chemical molecules is *not* the origin of life. Life does not reside in a particular molecule. Life is a *system* property, deriving from a coherent, cooperative, out-of-equilibrium, and orchestrated dynamics of several molecules which – as far as we know – are spatially and functionally organized as *cells*.

Therefore, in addition to understanding the chemical origin of those molecules, which later we will recognize as the biomolecules responsible for the emergence and propagation of all biological organisms (from unicellular ones to the largest ones), a key and unsolved question refers to the origin, the structure, and the functionality of primitive cells, or similar cell-like systems, and their contributions to the onset of life in our planet (Fig. 6.1).

Several cellular models have been proposed in order to mimic primitive cells. Martin Hanczyc has reviewed these models [5], describing how, in the past decades, scientists have focused on microcompartments of different nature for modeling primitive cells, such as sulphobes, coacervates, autocells, jeewanu, and

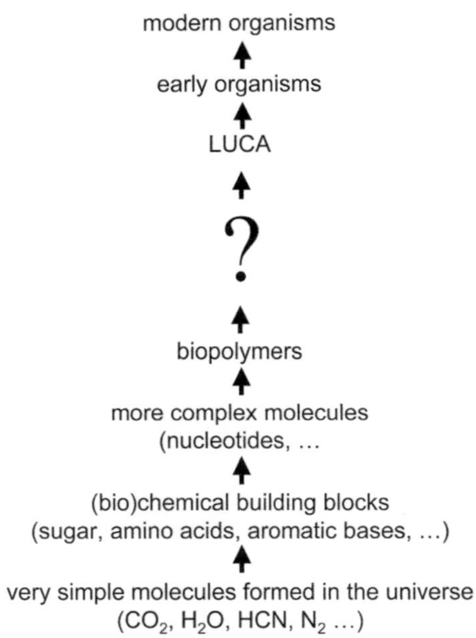

Fig. 6.1 Conceptual map describing the chemical (pre-biotic) and biological evolution, for the smallest molecules to modern biological cells and organisms. The research on primitive cells tries to fill the gap between understandings generated by classical prebiotic chemistry and backward evolutionary considerations ending to LUCA, the last universal common ancestor of all living organisms. The origins of life, according to the vision expressed by the author, should be investigated in the context of primitive cells origin, and not strictly related to the emergence of a particular molecule (e.g., a self-reproducing RNA)

microspheres. Although the interest in coacervates is still relevant, as witnessed by several recent publications [6–8], most of researchers now focus on *vesicles* as main cellular models.

Thus, from when in the 1960s Alec Bangham firstly reported on lipid vesicles (liposomes) [9, 10], the research on these tiny compartments, formed by self-assembly of lipids or other amphiphilic compounds in acqueous solutions, includes, among its several branches, the use of vesicles as cellular models, and in particular as primitive cell models (or protocell models). In particular, such models have been built from simple and primitive amphiphiles, such as fatty acids and some derivatives, but also from phospholipids and other synthetic compounds. The interest toward these models has increased a lot in the past years, till the point that the application of the same principles and methodologies developed for primitive cell models led to the novel perspective of assembling synthetic (or artificial) cells from scratch. In other words, the synthetic – or 'put-together' – approach which is so important in origins-of-life studies [11–13], has been extended to modern biotechnology aiming at synthesizing cells with minimal complexity. More specifically, the *bottom-up* construction of synthetic cells is one of the goal of synthetic biology [14], and it will serve advancements for biosensoring, in nanomedicine, for understanding cellular mechanisms in a simplified environment, and to design and build novel bionanomaterials [15–20].

Liposome-based cellular models represent, therefore, a promising wide field of inquiry, embracing origins-of-life and biotechnology. In many cases, the same technologies are used, the same analytical methods, the same operational (and sometimes conceptual) approaches. The goal of this chapter is to introduce the readers into this topic, presenting the simplest cases and the most recent reports. Although important steps have been recorded in the past two decades, the field is most at its beginning, and future efforts will certainly lead to exciting discoveries, useful knowledge, and new technological tools.

The chapter is organized in four parts. Firstly, there is a short survey on vesicles types and basic properties with respect to their use as cellular models. Secondly, a short epistemic remark to the "synthetic approach" is made, here intended not only as a methodology but also as a self-standing concept. Next, the use of vesicles as primitive cell model is presented. Finally, we will comment on the synthetic biology operations based on vesicles as artificial cells.

6.2 Types of Vesicles and Their Preparation

Vesicles are microscopic compartments, generally spherical, composed by a closed membrane. When the membrane consists of a lipid bilayer, vesicles are better called 'lipid vesicles' or *liposomes*. However, liposomes are not the only type of vesicles. Vesicles have been generated by several types of amphiphilic molecules, like fatty acids [21], terpenoids [22–25], block copolymers [26–29]. These compartments share similar features, like their formation by molecular self-assembly, a typically

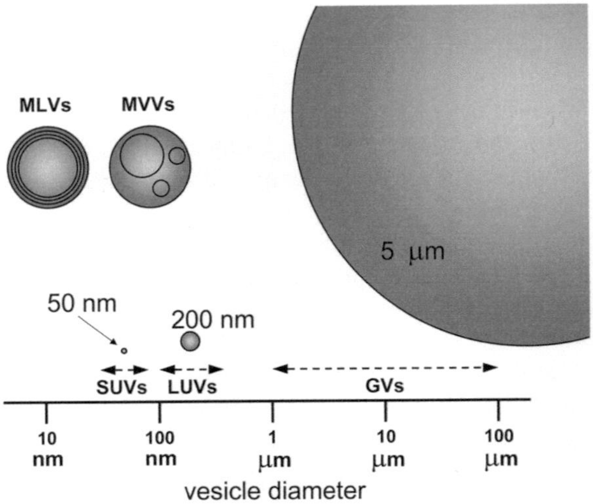

Fig. 6.2 Typical vesicle morphologies. Small Unilamellar Vesicles (SUVs), Large Unilamellar Vesicles (LUVs) and Giant Vesicles (GVs) (sometimes called Giant Unilamellar Vesicles, GUVs, if unilamellar) of diameters 50 nm, 200 nm, and 5 μm are represented approximately to scale. On the top, MultiLamellar Vesicles (MLVs) and Multi-Vesicular Vesicles (MVVs) (also called vesosomes) are schematically represented. MLVs and MVVs do not refers specifically to a determined size, rather to the vesicle morphology

semi-permeable membrane, or their reactivity dominated by surface forces (vesicles, after all, are colloidal particles). However, the specific chemical nature of the building blocks constituting the vesicle membrane is the first aspect that strongly impacts on other properties like stability, interaction with other vesicles or other molecules, methods of preparations, and compatibility with encapsulated material. Vesicles made of the above-mentioned chemicals (phospholipids, fatty acids, terpenoids, and sometimes block-copolymers) have been used in several instances as protocells.

The second aspect to keep into account is vesicle morphology (Fig. 6.2 and Table 6.1). In particular two vesicle types have been widely used, namely, large unilamellar vesicles (LUVs) and giant vesicles (GVs). There are good reasons to focus on these two vesicles types, which are equally important. LUVs (typical diameters: 100–400 nm) are easily produced by well-known and standardized methods. GVs (diameter 1–100 μm) are produced by a limited number of methods [30], but have the great advantage of being visible by light microscopy so that their behaviour can be directly assessed by visual inspection. A word of mention should be spent for another vesicle type, which consists in a more complex architecture, namely small vesicles inside a large vesicle. Technically, these vesicles are called multi-vesicular vesicles (MVVs) or vesosomes. Their construction is interesting because they somehow mimic eukaryotic cells with their intracellular organelles. The issue of internal organelle-like vesicles is intriguing, as it is plausible that sub-compartmentalization is a successful strategy for exploiting chemical gradients,

Table 6.1 Types of vesicles (cf. Fig. 6.2)

Abbreviation	Name	Specifications[a]
SUV	Small unilamellar vesicles	Whose diameter is typically < than 100 nm
LUV	Large unilamellar vesicles	Diameter in the ~100–400 nm range
MLV	Multilamellar vesicles	Several concentric vesicles, with variable size
MVV	Multivesicular vesicles[b]	Vesicles containing smaller non-concentric vesicles
GV	Giant vesicles	Diameter in the 1–100 μm range

[a] Note that commonly reported diameters are indicated. For example, 100 μm GVs are rare. MLVs and MVVs are terms not strictly related to size, rather to morphology
[b] Also called 'vesosomes'. Mimics of a cell with intracellular organelles

separating potentially interfering chemical paths, and providing at the same time a large surface for solute/membrane interactions. Although vesosomes have not been extensively used, it is foreseeable that future studies on artificial cells will be based on these structures, especially if methods for their systematic production will be optimized.

The third important factor is the method of preparation [30, 31]. Several methods have been employed, depending on the chemical nature of the vesicle building block, on the desired vesicle morphology, and on the compounds that need to be encapsulated or reconstituted in the vesicle. Actually it is not possible to give a general recipe and the choice must be deduced according to experimental restrictions and goal of the research. Such matter should also tuned on the basis of compatibility between the preparation method and the requirements of lipid/solute system - which in turn should be combined after considering chemical compatibility.

6.2.1 Chemical Nature of Lipids

To be more specific, some general considerations about the chemical nature of lipids are reported below (a detailed discussion can be also find in [32]):

1. Phospholipid and fatty acids (Fig. 6.3a, b) are the two main compounds which have been used to construct protocellular models. Phospholipids, however, owing to their chemical complexity, cannot be considered as very primitive chemicals, and therefore fatty acid vesicles are better suited when the focus is on the membrane structure, behavior, chemical and inter-vesicle reactivity. On the other hand, phospholipid vesicles (liposomes) can still be used as primitive cell models if the focus is on intra-vesicle reactions, or on other aspects, providing that the nature of the vesicle membrane is not a conceptual issue.
2. More in detail, also fatty acids can be differentiated on the basis of primitive plausibility. In this context, short chain saturated fatty acids, such as the decanoic acid/decanoate system (C10:0) [33], probably represents better the nature of

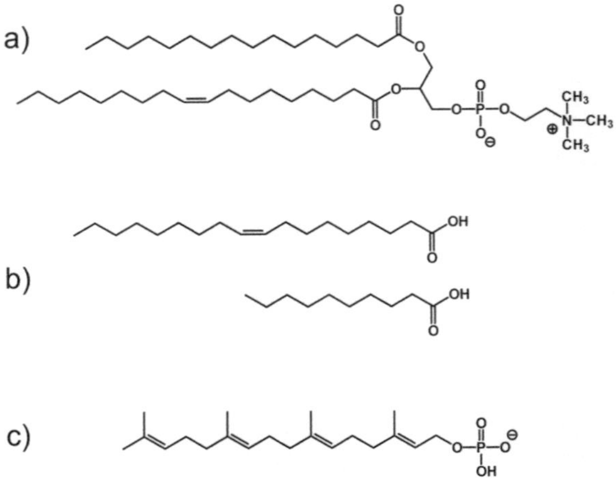

Fig. 6.3 Chemical structures of amphiphiles used for assembling vesicles as cellular models. (**a**) Phospholipids; here one of the most used lipid is shown, namely a phosphatidylcholine (lecithin) named 1-palmitoyl-2-oleoyl-*sn*-glycero-3-phosophatidylcholine (POPC); (**b**) fatty acids; in particular the protonated forms of oleic acid (C18:1) and decanoic acid (C10:0) are shown – note that only the second has a realistic primitive relevance; (**c**) isoprenoids, in particular a partially ionized geranylgeranylphosphate is shown

primitive cells membrane. However, most of the published studies have been carried out with oleic acid/oleate vesicles (C18:1) [34–42] and myristoleic acid/myristoleate (C14:1) [43–46].

3. Membranes composed by only one chemical species, on the other hand, are not realistic, and mixtures of diverse amphiphilic molecules better represent primitive membranes [44, 45, 47]. Studies on pure compounds are nevertheless useful to preliminarily decipher the properties of individual compounds, before engaging with the study of mixtures.
4. In contrast to fatty acid vesicles, which have been investigated at a considerable extent, isoprenoid compounds (Fig. 6.3c), such as polyprenyl phosphates (alone or as mixtures with polyprenols), have been studied only in a very few cases [22–25, 48, 49]. The ionizable phosphate head group implies a pH-dependence in their self-assembly properties. Linear and branched polyprenyl compounds can form vesicles whose properties are only partially known. This contrasts with the importance of isoprenoids in modern cells. *Archaea* membranes are made of isoprenoid-derivatives monolayers; whereas cholesterol, ergosterol and lanosterol are typically found in *Eukarya* cells.
5. Pure fatty acid vesicles, being composed in their 'stable' form by about 50% carboxylate (typically as sodium salts) are sensitive to important multivalent cations such as Fe^{2+}, Fe^{3+}, Ca^{2+} and Mg^{2+} [50] (at relatively low concentration) and to monovalent cations (at high concentration) [51]. Note also that H^+ destabilizes fatty acid vesicles by binding to carboxylate ($RCOO^- + H^+ \rightleftarrows RCOOH$). Indeed, the limited pH-range of existence of pure fatty acid vesicles (generally

between 7–7.5 and 9–9.5) should be always considered. Sensitivity to Mg^{2+} (an important cation due to its interaction with nucleic acids and their precursors) has been improved by mixed systems composed of fatty acids and monoacylglycerols [44, 50–53] or by chelating Mg^{2+} by citrate [54].

6. Most of liposomes-based work has been carried out with the zwitterionic phosphatidylcholine (lecithin), which self-assemble in a very stable membrane and in a wide pH range. The other zwitterionic phospholipid, namely phosphatidylethanolamine does not form generally stable membranes, being characterized by an unfavorable packing parameter [55] ($v/al < 1$; v being the molecule volume, a the effective head group area, l the molecule length). Moreover the positive charge on the head amino group ($-NH_3^+$) is pH-dependent, whereas the phosphatidylcholine trimethylammonium group is not ($-NMe_3^+$). Other phospholipids have been generally used as a lipid mixtures (especially the anionic phosphatidylglycerol), with phosphatidylcholine being the main component, possibly also including cholesterol. The simplest phospholipid, namely, phosphatidic acid – which also form vesicles at intermediate pH – has not been deeply investigated in the context of origin of life.
7. Both in the case of fatty acids and phospholipids, care should be taken in order to be aware of the physical state of the hydrocarbon chains, i.e., solid-like or liquid-like. The transition temperature, T_m is an important parameter to consider for designing protocellular systems. The physical state of the membrane will impact on vesicle stability and small-solute permeability. Fatty acids are single-chain charged molecules, and their solubility can be high. Therefore in such systems, the critical aggregation concentration (c.a.c.) is an issue to consider.
8. Polymersomes, which have been occasionally employed to build synthetic cells, have no direct relevance for the origins of life. However, their development might be functional for specific biotechnological applications, in virtue of their great stability.

6.2.2 *Vesicle Type (Morphology)*

With respect to vesicle type (or morphology, see Fig. 6.2), vesicles are generally classified according to Table 6.1 entries. The most utilized vesicle types in origins of life studies are LUVs and GVs. LUVs have been probably the most common type of vesicles due to several reasons. Some are technical reasons and of-opportunity reasons, and it will be commented below. Another considerations – which might count contrasting opinion among investigators – focus on the idea (on the hypothesis) of how large where primitive cells. To be more detailed, we will see below that the spontaneous hydration of lipids generally brings about quite large vesicles, for example GVs. It is plausible then that in absence of strong shearing forces, large vesicles (in the micrometer range) are better candidates for representing protocells. Contemporary living cells also have similar sizes, from the smallest bacteria (ca. 1 μm) to large unicellular eukaryotes (10–50 μm). However, one should also consider the stringent conditions for realizing a living dynamics, in terms of number of molecules, local concentration, surface-to-volume requirements. These features are probably better embodied in LUVs (0.1–0.2 μm). An interesting discussion has been developed around the minimal requirements of life, also in terms of dimensions, as reported in the proceedings of a dedicated workshop [56], and experimental studies [57].

As mentioned, LUVs can be produced in highly reproducible way from a wide variety of lipids, thanks to standard procedures (film hydration, freeze-thawing, extrusion, purification by size exclusion chromatography/dialysis). The large part of a vesicle population prepared in this way is spherical and unilamellar. The procedure allows the entrapment of both water-soluble (dissolved in the aqueous buffer used to hydrate the lipid film) and lipid-soluble substances (dried together the lipids). Another advantage is that the resulting dispersion can be manipulated almost as a normal solution, and bulk measurements (absorption spectroscopy, fluorescence, etc.) can be applied. LUVs model small primitive cells, much smaller that bacteria. Their size is instead typical of viruses, and the size distribution is typically narrow (after extrusion). The fact that LUVs can be produced in such reproducible and homogeneous form (uniform with respect to size, shape, lamellarity) make LUVs a quite attractive model, especially if one is interested in average properties (averaged over the whole vesicle population). Fatty acid LUVs and phospholipid LUVs have been extensively used.

GVs, on the other hand, are also widely used. Their main feature is the very large size – in the 1–100 μm range (typically 5–20 μm), which allows their direct visualization by light microscopy in the form of aqueous suspension (whereas LUVs cannot). Importantly, GVs have high trapped volume and therefore contain a large number of solutes. GVs requires special preparation methods. The two main methods derive from the classical film hydration method that is used to produce LUVs and MLVs. These 'classical' methods are the so-called natural swelling method and the electroswelling method. The natural swelling method consists in hydrating thin lipid films without mechanical perturbation. The film, sometimes pre-hydrated by aqueous vapours, is left for a long time (hours) in contact with the aqueous solution without stirring, shaking, etc. Lipid films swell gently, creating GVs of various size and morphology, also multi-vesicular GVs. Electroswelling is essentially a way to accelerate this process, by application of an alternating electrical field. Lipids are stratified over wires or planar electrodes and alternating current is applied. Swelling occurs in shorter times (less than 1 h). Both methods work well for phosphatidylcholine GVs and low ionic strength buffers. This can be a limitation because in many cases physiological-like buffers are necessary. It has been shown that negatively charged GVs (e.g., including phosphatidylglycerol in their membrane) can be produced by the natural swelling method in the presence of high ionic strength buffers [58]. Note that fatty acid GVs have been produced only by modifications the film hydration/natural swelling method, not by electroswelling, whereas mixed phospholipid/fatty acid vesicles have been produced by the next-discussed droplet transfer method [59].

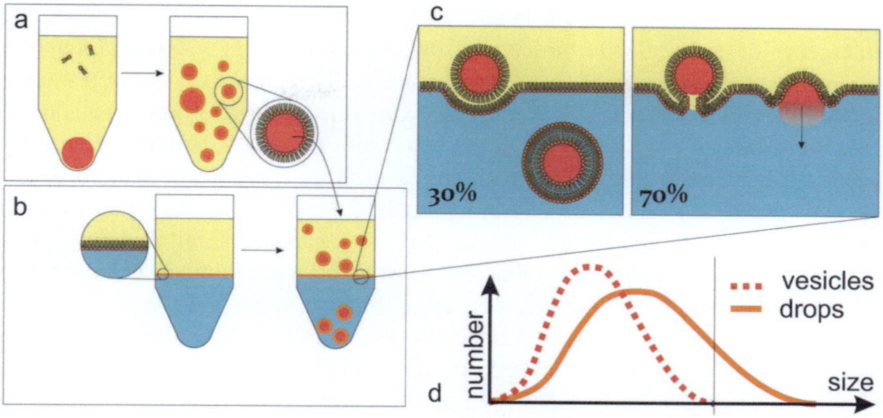

Fig. 6.4 GVs prepared by the droplet transfer method [60]. (**a**) Preparation of a water-in-oil macroemulsion with mineral oil, surfactants (POPC) and the inner-solution. (**b**) Preparation of an oil over outer-solution system with surfactants at the interface. In a second step the macroemulsion droplets are inserted in this system and sink down due to the density difference between the inner- and outer-solution. (**c**) As they wander through the interface they get a second layer of surfactants such that they now have a bilayer of phospholipids, i.e. they are vesicles now, if the conditions are good (according to our measurements, generally in about 30% of all cases) – otherwise they merge with the interface and the inner-solution is released into the outer-solution (which generally happens 70% of all cases). (**d**) This leads to a size distribution of the vesicles which does not allow vesicles greater then a critical size, even though the macroemulsion droplets generated in the first step were greater (Reproduced from [63] with the permission of Springer)

In addition to these two classical methods, which are very useful for studying the properties of lipid membranes, a novel GVs preparation protocol has been introduced recently [60]. This is based on the transformation of water-in-oil (w/o) lipid-stabilized droplets. W/o droplets are centrifuged across a lipid-containing interface and get covered by a second lipid monolayer, so to form GVs (Fig. 6.4). This method forms GVs with traces of the apolar solvent (the 'oil') used to prepare the w/o droplets and it is therefore questionable whether or not the resulting GVs can be used for accurate biophysical measurements of membrane properties. On the other hand, the strength of the method relies in its application to encapsulate molecules in the GVs lumen, especially macromolecules [61]. Moreover, asymmetric lipid membranes can be created with this method [62], and mixtures of lipids can be used, provided that the w/o droplets are sufficiently stabilized in the first step of the preparation. Mixed phosphatidylcholine/fatty acids GVs have been successfully prepared by the droplet transfer method [59]. Despite these advantages, the droplet transfer method (as well as the electroswelling method) cannot be considered of prebiotic relevance. Nevertheless it has been used in several cases to produce solute-filled vesicles which model primitive cells. The focus was therefore not on the mechanism of formation of such vesicles, but on their properties/dynamics/interaction with other vesicles, and so on.

The natural swelling method is therefore the preferred method to simulate the emergence of early cell-like structures from amphiphiles and water. It produces a heterogeneous population of vesicles that realistically represents a sort of primitive ecosystem, where synergies, cooperations, competitions and selections among these coexisting 'units' took place. In addition to these important features, especially when the realistic primitive cell modeling is desired, it should be reminded that the GVs prepared by the natural swelling method are characterized by an intrinsic diversity in size, lamellarity, and morphology. This makes difficult to define a sort of 'average' behavior. It follows that studies done on more homogeneous samples (LUVs, GVs prepared by the electroswelling method or droplet transfer method) and those done on spontaneously formed GVs by natural swelling complement each other. Moreover, it should be recalled that microfluidics offers a novel technological route for the construction of highly homogeneous GVs [64–71].

6.2.3 Preparation Methods and Solute Entrapment

It is worth to recall the interplay between lipid types, methods of preparation and an essential feature of lipid micro-compartments, namely, their capacity of entrapping water-soluble or lipid-soluble substances. Clearly, this is of vital importance when models of primitive cells are prepared. Water-soluble substances (inorganic salts, sugars, small polar molecules, proteins, nucleic acids, ribosomes, ...) are encapsulated inside vesicles in the moment of their formation (Fig. 6.5). Generally, such molecules have low permeability and their addition after vesicle formation does not bring about their internalization. An exception are small molecules which

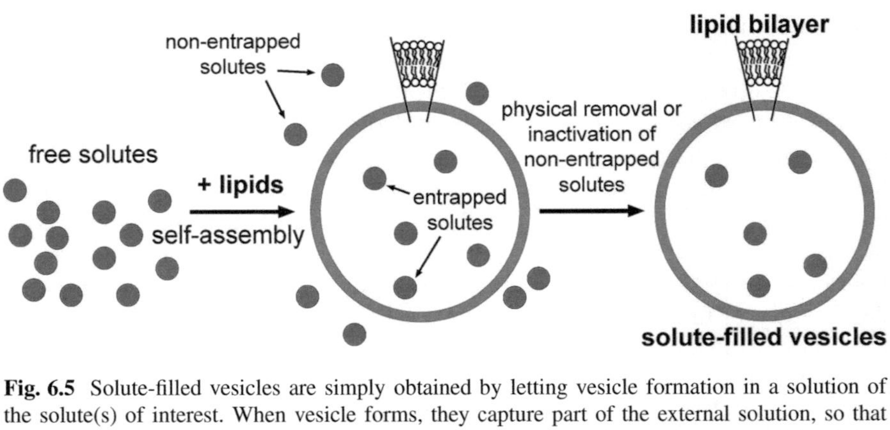

Fig. 6.5 Solute-filled vesicles are simply obtained by letting vesicle formation in a solution of the solute(s) of interest. When vesicle forms, they capture part of the external solution, so that solutes become encapsulated inside. Their permeability is low (as they are water-soluble) and thus they are not released. A purification step generally follows the entrapment, but in some cases, external solutes and their potential reactions are blocked/inhibited by the addition of another (non-permeable) agent

are present in solution in two forms (charged and non-charged), for example in the case of amines (RNH_3^+ and RNH_2), or other compounds with acid/base forms with a similar feature. In some cases, the neutral form can permeate the lipid membrane and be transformed in the charged non-permeable species inside the vesicle, due to a different pH value [72]. This procedure has been applied in the construction of doxorubicine-containing liposomes [73, 74] (for drug delivery) but a similar strategy can work in the case of other molecules of prebiotic interest [75–78].

Lipid vesicles originate, ultimately, from a closure mechanism whereby a curved lipid layer closed on itself capturing a portion of the aqueous solution. In ideal conditions, and in absence of strong solute-lipid interactions (as it could be, for example, when cationic lipids are allowed to form vesicles in the presence of nucleic acids), the entrapment process is equivalent to a random sampling. One can imagine that open lipid bilayers (homogeneously distributed in the solution) capture portions of the solution in random way. Probably the method that more closely matches with this description is the ethanol injection method [79]. In the other methods practical constraints prevent the occurrence of these ideal conditions, but nevertheless it is useful to describe what would be the ideal case and then compare it with the observations.

The fraction of whole volume V captured by a vesicle with volume v is v/V, and this ratio represents the entrapment probability p. The *average* number λ of entrapped solutes will be pN, where N is the total number of solutes in the whole solution. As $N = N_A C_0 V$, it results that $\lambda = pN = v/V \cdot N_A C_0 V = N_A C_0 v$ (where N_A is the Avogadro's number and C_0 the bulk solute concentration). As in all microscopic phenomena, stochastic events related to the randomness of microscopic conditions affect the local solute concentration; consequently the closure of open lipid bilayers produces a population of vesicles where the number n of solute entrapped in a certain vesicle differs from λ, the most probable one. Given a certain vesicle size, a solute occupancy distribution $\wp(n)$ is obtained, that can be typically described by a Poisson distribution (Fig. 6.6):

$$\wp(n) = \frac{\lambda^n e^{-\lambda}}{n!} \tag{6.1}$$

Note that the Poisson distribution becomes similar to the Gaussian distribution when λ is large. This simple analysis reveals that together with vesicles that encapsulate the expected number of solute molecules, there will be always vesicles with n smaller or larger than λ. This means that when vesicle forms, even in the most ideal conditions (negligible solute-solute and solute-lipid interactions, homogeneous and isotropic solutions, homogeneous distribution of lipids in the whole solution), the resulting vesicle population is – by definition – heterogeneous in terms of solute content.

This has two consequences. First, from the technical viewpoint, one has to distinguish among (i) the (very much used) *average* entrapment efficiency (or entrapment yield, or similar values) which considers the overall amount of entrapped

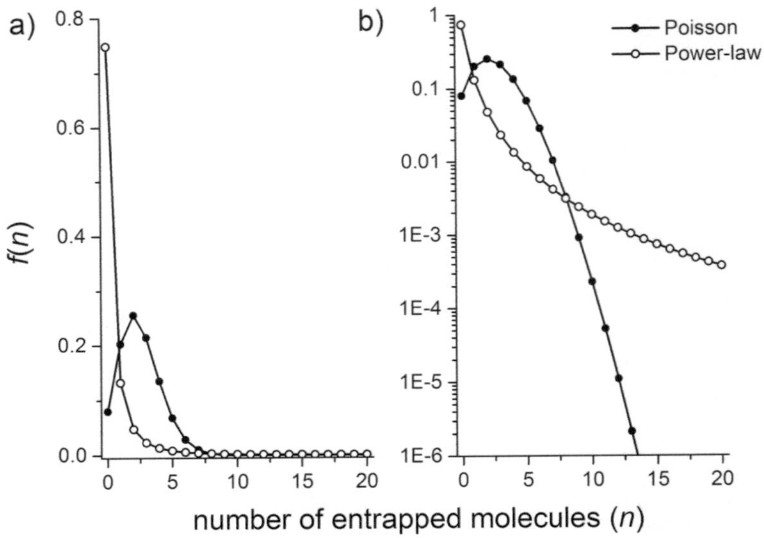

Fig. 6.6 Poisson and power-law distribution. Simulated Poisson and power-law distribution $f(n)$ as function of the number n of encapsulated molecules. The Poisson curve has been generated by using $\lambda = 2.52$, which corresponds to the expected value of encapsulated molecules when vesicles with diameter 200 nm are formed in a 1 μM solution. The power-law curve has been generated with $A = 0.75$ and $k = 2.5$ [82]. The plots refer to the same datasets but different y-axis: (**a**) linear, (**b**) logarithmic. The latter representation helps understanding that the encapsulation of high number of molecules (say, $n = 20$) when 2.52 are expected is highly improbable according to the Poisson distribution ($\sim 10^{-12}$), but 10^9 times more probable when a power-law is considered (10^{-3})

solute, divided by the number of vesicles (or by the lipid concentration), and (ii) the (less used) *individual* entrapment which instead measures the content of each vesicle, individually. The first measure is obtained by bulk measurements, the second, by techniques that allow the individual analysis of vesicles (e.g., microscopy [80–82], flow cytometry [83, 84]). Second, from the viewpoint of utilizing vesicles as primitive cell model, this fact evidences that populations of primitive cells, being formed by spontaneous lipid self-assembly into vesicles, and spontaneous entrapment of molecules, are ‘diverse’ also in terms of solute filling, in addition to size, lamellarity, morphology [85]. As the internalized solutes play the role of metabolic components, this implies that some protocells would function better than others, and being subjected to a proto-Darwinian selection, in virtue of better function, better reproduction rate, better stability (however, cooperation/synergy should not be neglected [59]). This has a profound implication for depicting realistic origins-of-life scenarios, as it has been indeed done in recent work, although referring to different behaviour [40, 86–88].

According to what has been discussed above, a vesicle population is characterized by a diversity in the inner solute content, which is spread around a certain average value λ. Such average solute number linearly depends on the product of vesicle volume v and bulk solute concentration C_0, as expected. Small vesicles display strong stochastic fluctuations in the number of encapsulated molecules. For instance, when LUVs (diameter 100 nm) are formed in the presence of 10 μM solute, $\lambda = 3.15$ meaning that ca. 60% of vesicles have entrapped 2, 3, or 4 solute molecules, whereas the others a quite different number. For example, the probability of finding a vesicle with 10 solute molecules is 0.1%. The 'long tail' on the right-side of the solute occupancy distribution mean (λ) is much more important than the left part. Here there are vesicles which are filled with solutes, much more than the expectation. Normally their number is very low, but – as detailed in the *'super-filled' vesicles* Box, there are cases that challenge the Poisson distribution.

'Super-filled' vesicles. Although the Poisson distribution is the expected solute-occupancy distribution for an ideal random entrapment process (demonstrated in [84, 89]), recent reports have shown an intriguing divergence from the expectations, as reviewed and discussed in [90]. In particular, a series of paper have revealed that when macromolecules are encapsulated in conventional submicrometer vesicles prepared by different methods, the experimentally observed solute occupancy distribution does not follow the Poisson distribution [82, 91–94]. Rather, it is shaped as a 'power law', namely $f(n) = A/(n+1)^k$, where A and k are positive parameters. This means that most of the vesicles are empty, and few of them are instead solute-filled (Fig. 6.6). However, because the power law function goes to zero slower than the Poisson function, it results that the amount of vesicles with high n is higher for a power law than for a Poisson process. In other words, actually there is a non-zero probability of finding solute-filled vesicles, with n well above the expected average λ. For example, the 0.1% of vesicles prepared in the presence of ferritin (or ribosomes) has an intra-vesicle solute concentration up to 10 times higher than the expected value. Clearly, when vesicles are formed in a mixture of essential macromolecules (essential for the sustainment of a network) some of the vesicles would be capable – against the expectations – of co-entrapping several copies of these solutes. In turn, such 'super-filled' vesicles could be very efficient in the realization of an internal reaction network, and so being favoured with respect to the empty or regularly filled one. Experimental evidences about this mechanism have been provided [82, 91, 92, 94]. A mechanism based on the perturbation of open bilayer closure rate has been suggested for accounting the observations (but for a counterexample, see [95, 96]).

The discussion on random encapsulation holds in the case of vesicles prepared by conventional methods, and when the solute encapsulation is the study focus. On the other hand, the above-mentioned droplet transfer method (Fig. 6.4) radically differ from the others and it is best suitable for a 'directed' encapsulation of water-soluble molecules. Due to recent developments, this method has become probably the best method for preparing solute-filled GVs intended as synthetic cells, when there is no intention to study the self-assembly phenomena underlying the spontaneous emergence of solute-filled vesicles (i.e., modeling the origin of primitive cells). The directed entrapment occurs because solutes are firstly emulsified in form of w/o droplet (achieving 100% compartmentalization), and droplets are transformed in vesicles. The remaining source of inter-vesicle variability is now only the solute partition among droplets while they form by coagulation/fragmentation steps [85].

A completely different discussion refers to lipid-soluble molecules. In contrast to water-soluble ones, they have the spontaneous tendency of binding to lipid membranes, and their inclusion in the vesicle structure is easy. The exceptions are membrane proteins. If a protocell model is designed in a way that it includes membrane proteins (enzymes, transport protein, signaling protein) care should be taken to design the lipid type and the vesicle preparation method (and these two things are somehow interlinked). It may seems that the reconstitution of complex membrane proteins in protocellular model is not a fully pertinent exercise. However, it is expected that primordial polypeptides with hydrophobic character could decorate early membranes complexifying their repertoire of functions (e.g., permeability changes, inter-vesicle interactions, vesicle-surface interactions, binding of free floating molecules). In this respect, the inclusion of lipid soluble or lipid-anchoring molecules is certainly relevant.

Moreover, when the protocell model is built to demonstrate the reliability of minimal transformation pathways, or their use for investigating protocell transformations, the reconstitution of modern membrane enzymes can be an essential step. From this perspective, also phospholipid membranes can be employed as primitive membrane models. An example is given by one of the early papers on vesicle self-reproduction. The four enzymes that convert lipid precursors into lipid molecules were reconstituted in lipid vesicles [97]. The goal of the work was to demonstrate that a lipid-producing liposome could grow like a cell, producing from within the building blocks for enlarging the membrane (according to the autopoiesis theory – see below).

In conclusion, when vesicles are intended as cell models, there are several aspects to consider, and these are somehow linked to each other. The type of lipids, the vesicle type, the preparation method and the solute entrapment are all connected, and unfortunately not all combinations are easily accessible or even already explored. Clearly, the choice will depend on the aim of a study. Table 6.2 offers a framework for such discussion. For example, allegedly primitive compounds,

Table 6.2 Different compounds and approaches in vesicles as cell model research

Component type	Scope	Examples
Allegedly primitive	Realistic protocell	Fatty acids, small peptides, ...
Modern biomolecules	Minimal functions	Enzymes, ribosomes, phospholipids, ...
Synthetic	Fully synthetic cells	Polymers, hybrid molecules, ...

as well as components extracted from modern cells (and synthetic molecules as well) have been successfully employed to build cellular models, and vesicles of very different types have been constructed. However, it should be useful, when discussing the choices underlying the constructed models, keep clear what is the aim of the study, and what the model aims at revealing. For example, it is evident that a realistic primitive cell model should be built from allegedly primitive compounds, such as simple membrane-forming lipids (fatty acids, isoprenoids, fatty alcohols, etc.), and molecules which mimic the early catalysts (such as small peptides or ribozymes). When the focus is on minimal cell function, irrespective from the actual molecular species that carry out a certain function, modern molecules have been used (enzymes, DNA, ribosomes, modern lipids, . . .). We have to distinguish two possible aims for the second-line approach of Table 6.2. One aim could be the construction of primitive cell models, but using modern molecules to test some hypotheses about minimal function. Another aim could be the construction of a biotechnological tool, a sort of synthetic cell, for some specific applications. Finally, one can imagine also fully-synthetic cell model that show how living-like functions can be achieved in 'orthogonal' way – namely – by using compounds not selected in the natural evolutionary pathway (synthetic polymers, ad hoc designed lipids, hybrid materials, and so on). Of course, the distinction between the approaches can become blurred in certain circumstances, and the possibility of hybridization should be considered also positively. Any advancement in this new field carry a scientific and technological potential that should not be necessarily restricted by classifications.

Despite these differences, the approaches indicated in Table 6.2 have a common ground, and this should be emphasized: *assembling a cell-like entity from separated parts*. This operational procedure, whose roots are in the chemical science (building a complex molecule from small parts), has been dubbed as 'synthetic' approach. It corresponds to an *operational bottom-up* approach, even when the protocell design might originate from a *conceptual top-down* approach (think, for instance, of designing a protocell by removing unnecessary components from a fully fledged cell, and then construct the protocell by assembling these essential parts). Describing and commenting this approach is the topic of the next section.

6.3 The Synthetic or Bottom-Up Approach

Let us remark better the concepts which lay at the basis of the vesicle usage as cell models. In the introduction it has been quickly said that the 'synthetic approach' (also called constructive approach) is typical of origins-of-life research. The philosophy of the synthetic approach can be summarized by the words of Liu and Fletcher [98], saying that "we are much better at taking cells apart than putting them together", evidencing how in modern science the analytic rather than the synthetic method has been largely applied for gaining knowledge about how biological entities work. Actually the analytical approach has been very successful.

In the field of primitive cell research, however, such an approach cannot be applied, and primitive cells – or better, their models – can be only constructed in the laboratory, by assembling molecular pieces and verifying hypotheses, checking which kind of constraints apply, building minimal metabolic pathways, and so on.

Understanding-by-building can be the motto of this practice. Clearly, difficulties exist due to the ignorance of the exact conditions that allowed the emergence of first cells from non-living molecules, and on the exact order of the steps that accompanied the success of cellular homeostasis and self-reproduction. Nevertheless the synthetic approach is the only one that might give scientific answers to this age-long question, and especially demonstrating that life emerges, as a result of out-of-equilibrium organization, yet according to the laws of physics and chemistry (we will see later that this special type of organization must necessarily be autopoietic).

Understanding-by-building, however, is also a motto of a modern biology branch, synthetic biology (SB). SB is a young discipline born by applying the engineering vision to biology [99]. However it differs from genetic engineering because it focuses on the engineering ('rewiring') of whole cells. SB generally aims at achieving concrete goal, typical of bioengineering, such as creating biosensors, letting biological cells (most often: bacteria) producing fine chemicals, or biofuels, or pharmaceuticals, constructing bacterial strains for bioremediations, and so on. On the other hand, from its very beginning, a special focus was reserved to the construction of *minimal cells*, i.e., living cells with minimal complexity. This goal is important for basic understanding of living cells (what is the minimal non-reducible complexity associated with life) and for biotechnological purposes (eliminating unnecessary circuitry is thought to be equivalent to optimize energy usage in cells).

Mainstream SB research looks at minimal synthetic cells as a product of genomic manipulation (with the already achieved goal of a cell that has been deprived of its native genome and transplanted with a minimal synthetic genome, according to the Venter's team approach [100, 101]). Such a path is now recognized as the 'top-down' approach to synthetic cell (using SB techniques). It is top-down because it starts from a pre-existing organism, and implies a minimization design done by the scientist. An alternative approach to minimal synthetic cells is based on the 'bottom-up' construction, from a minimal set of molecules, not from cells – as it has been described in this review. This second path would lead to the improvement of biological understanding of cellular processes once they are reconstituted from scratch in simplified systems, and when the noise due to other concurrent processes has been eliminated in the novel, minimized, cell-like design.

It is evident that the philosophy, the methods, the synthetic (constructive) approaches for the *bottom-up* construction of mimimal cells largely overlap with those which are typical of research on primitive cells. This leads to an unexpected, yet very fertile, common arena which is interesting also from the viewpoint of science epistemology [102, 103]. It is quite intriguing that one of the most ambitious biology branches shares with the research on the primitive cells its most fundamental conceptual and practical tools.

Bottom-up synthetic minimal cells and models of primitive cells, thus, are two different implementations of the synthetic bottom-up approach. Both aims at assembling cell-like systems from non-living molecular parts, and at understanding the function of cellular systems (understanding by building). Perhaps, the most obvious difference refers to the type of molecules used for building such minimalized cells, as already specified in Table 6.2. It should be noted that as the field is very young, the terminology is not yet crystallized. Different authors uses different terms like primitive cell, protocell, minimal cell, synthetic cell, artificial cell, semi-synthetic cell, semi-synthetic minimal cell. In contrary to their apparent diversity, all vesicle-based cell models built according to the bottom-up synthetic approach show a certain degree of similarity. Very different is the case of synthetic cells built by genetic manipulations of existing cells or by genome implanting, referred, as mentioned, as top-down approach. *Nota bene*: With respect to the bottom-up/top-down dichotomy, it is worth noting that the authentic bottom-up approach, based on non-designed self-organization and emergence is actually not really applied (*stricto sensu)*. Rather, the methodological bottom-up approach (assembling a cell-like structure from its parts, based on self-assembly of molecular systems) is often preceded by a design step (deciding what molecules include, foresee patterns, combining parts which function together), which is a typical top-down practice (designing a system with a final goal in mind).

6.4 Modelling Primitive Cells

One of the primary goal of origins-of-life research is creating model of primitive cells, those entities that originated from self-assembly of molecules which are supposed to be available in early times. These very basic structures were very different from the cells as we know them. They were simpler, performed worst than evolved cells, probably they were partially unstable and perhaps 'limping' [11]. As it has been recently discussed by us [104], the formation of self-reproducing protocells that are able to display essential features of biological autonomy marks the transition between non-life and life. One generally focuses on the notion of cells with minimal complexity, but what does 'minimal' life mean? The definition of life is still an open question that divides scholars [105, 106]. Most would agree on the fact that a living system displays homeostatic self-maintenance, self-reproduction and the capability to evolve. Autopoiesis (self-production) is a theory proposed by two biologists, Humberto Maturana and Francisco Varela [107, 108], that provides an operational definition of life as that process of self-bounded structures that produce their own components own to chemical transformations occurring within the autopoietic organization itself (Fig. 6.7). Autopoiesis does not explain how life originated, but tells us how a living system works – and therefore, how to construct it.

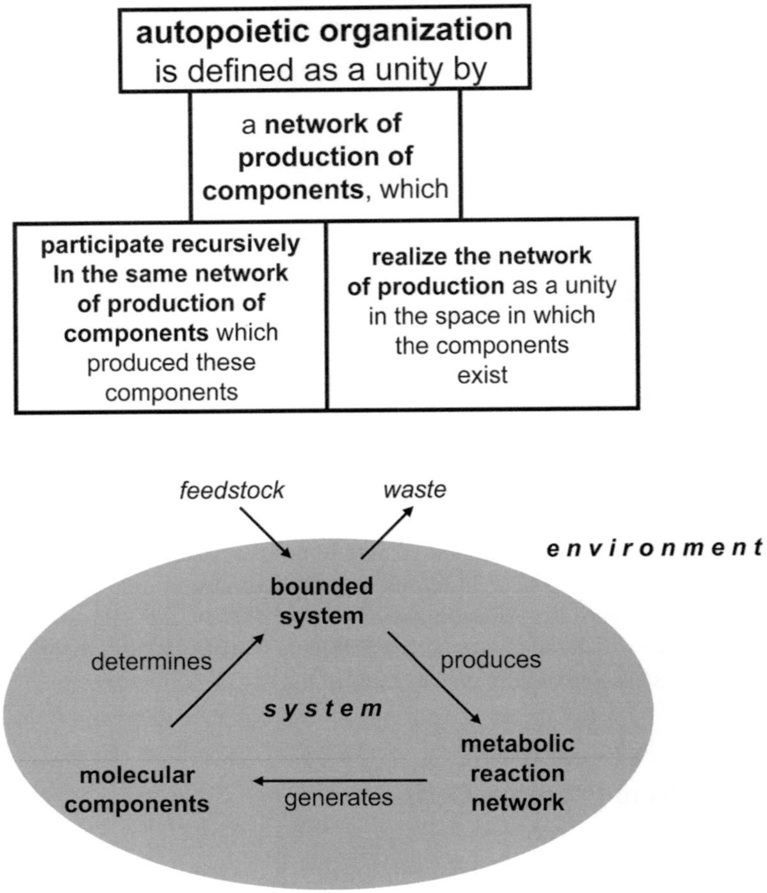

Fig. 6.7 Essentials of autopoiesis theory. *On the top*, the definition of autopoietic organization in terms of a network of production of components [107]; *on the bottom*, application of autopoietic theory to primitive cells, synthetic cells, etc. [109]

Here we present a brief overview of autopoietic theory [109], as recently discussed in our publication [104]. The central feature that characterises all living entities (and in particular unicellular organisms) is their self-maintenance. By self-maintenance, here we mean the self-generation of all components by chemical reactions, occurring within a boundary (e.g., the cell membrane). The boundary is also produced by the internal metabolic system. This self-generation is due to the peculiar form of chemical organization, that is, a dynamical organization typical of autopoietic systems. It is defined as a

(continued)

network of processes of productions of the molecular components, which (1) participate recursively in the network of processes for their own production and (2) occur in a defined region (physical space) delimited by a physical boundary (Fig. 6.7). The physical boundary also belongs to the autopoietic organization. The so-obtained physical entity is an autopoietic unit. Note that the autopoietic organization is a collective, distributed property of the whole system, and does not reside in any particular molecule. In autopoietic theory, the operational closure is often included in the discussion, to mean that the autopoietic unit reacts to environmental changes in order to maintain its own inner autopoietic organization, and that the unit tolerates only changes that can be accommodated within the autopoietic organization.

Based on autopoietic theory, some important experimental achievements have been obtained in the 1990s with fatty acid vesicles, namely, autopoietic self-reproduction of reverse micelles [110], micelles [111] and LUVs [34]. In all cases, fatty acids were used to construct these structures. The self-reproduction of fatty acid GVs has been also reported [42, 112]. Despite the historical and conceptual relevance of reverse micelle and micelle self-reproduction (a programmatic paper co-authored by Luisi and Varela appeared in 1989 [113] where chemical autopoiesis was firstly sketched in the reverse micelle system), in Sect. 6.4.1 only the self-reproduction of vesicles will be commented. Interested readers can find useful a recently published comprehensive review [114]. Next, selected examples showing the occurrence of reactions inside fatty acid vesicles will be presented (Sect. 6.4.3).

6.4.1 Vesicle Self-Reproduction

Autopoietic vesicle self-reproduction has been designed after inspiration to the autopoietic theory. Fatty acid vesicles have been used as model of self-reproducing vesicles. This was a somehow fortunate case, as there are easily accessible experimental routes for realizing a self-reproducing mechanism, *and* fatty acid vesicles are also the most plausible models of primitive cells. Pre-existing fatty acid vesicles (generally oleic acid/oleate vesicles, which are stable at slighly alkaline pH, i.e., 8.5) were provided with feeding material in form of externally added fatty acid precursors (water-insoluble fatty acid anhydride) or directly with free fatty acids in the form of micelles, which are easily deliverable in pseudo-homogeneous phase.

The rationale is the following. According to autopoiesis, an autopoietic cell takes up molecular precursors from its environment and transforms them in the components of its dynamical network (including assembling the physical boundary: the membrane, which limits and define the autopoietic cell against the background). Such a process is central to autopoietic organization, and involves by definition all components of the autopoietic network. However, the focus on the membrane allows a further advantageous implementation. Autopoiesis also means self-maintenance. Despite the concurrent and ever-present anabolic and catabolic processes, the autopoietic entity maintains its identity. Quantitatively this means that the rate of precursors uptake and components production (V_p) is balanced by the rate of components destruction/release in the environment (V_d), or $V_p = V_d$. No net component production or accumulation takes place. The autopoietic unit self-maintains in a *homeostatic* state. However, if $V_p > V_d$ the autopoietic system can grow, producing new components that can constitute a novel (daughter) system following a growth/division process (Fig. 6.8). Note that the reaction takes place within the boundary of structure/unit. This process, when applied to cells or their vesicle models, has been called autopoietic self-reproduction [35, 115].

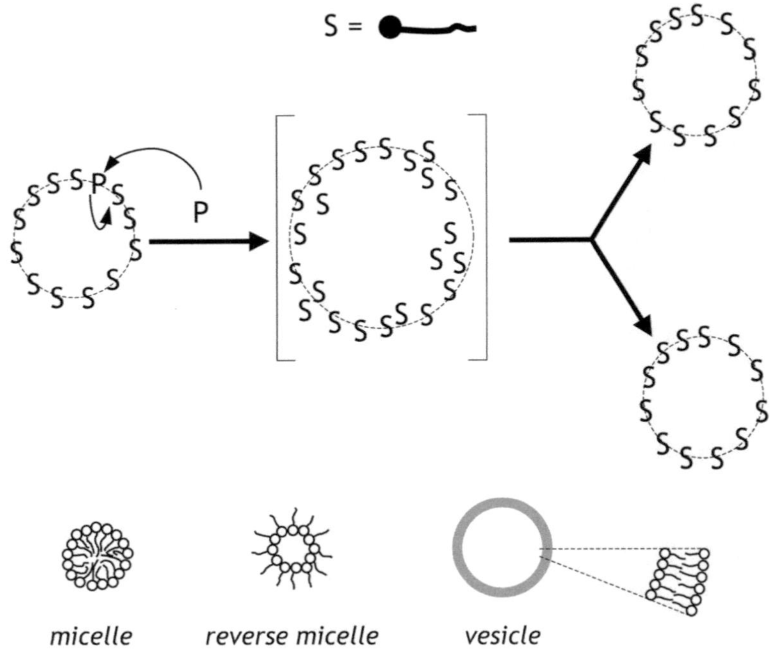

Fig. 6.8 General schemes for the self-reproduction of supramolecular structures. The uptake of a suitable precursor **P** by the preformed self-assembled structure, and its transformation to **S**, the membrane-forming compound, brings about the growth of the structure boundary. The growth induces a destabilization of the structure, which divides in two (or more) similar structures (not necessarily of the same size). The production of **S** proceeds with rate V_p, the destruction/release of **S** (not shown in the drawing) proceeds with rate V_d. If $V_p > V_d$ the structure grows; if $V_p = V_d$ the structure is in a (dynamic) homeostatic state; if $V_p < V_d$ the structure collapses. The autopoietic self-reproduction mechanism has been studied for micelles (*bottom, left*), reverse micelles (*bottom, centre*), and vesicles (*bottom, right*; not drawn to scale) (Reproduced from [116] with the permission of Springer)

6.4.1.1 Premise on Fatty Acids Dispersed in Water

Depending on pH, fatty acids can be found essentially in three different forms when dispersed in an aqueous solution. At low pH (for example, below ca. 7.5) fatty acids are protonated ($R - COOH$) and separate from the solution. Intramolecular interactions are dominantly hydrophobic. At high pH (for example above ca. 9.5) fatty acids are deprotonated, in the form of soaps ($R - COO^{-}Na^{+}$) and due to their large hydrated head group they self-assembly into micelles, as expected from the Israelachvili-Mitchell-Ninham generalization [55]. At intermediate pH values, both forms of fatty acids are significantly present, so that hydrogen bonding can occurs between the carboxylic form and the carboxylate ($R - COO^{-} \ldots HOOC{-}R$). The p$K_a$ is within this pH range. The average head group area decreases when compared to fully deprotonated molecules, and fatty acids self-assemble as bilayers (and therefore as vesicles). It has been proposed that a dynamic network of hydrogen bonds characterizes fatty acid membrane surface at these intermediate pH values [117]. Note that when compared to short-chain carboxylic acids, the pK_a of fatty acids in their associated form is ca. 3 units higher. This is because the carboxylate charges that originate after deprotonation are near each other in the membrane, making difficult the extraction of further protons.

6.4.1.2 Biphasic System

Oleic anhydride is not soluble in water. When oleic anhydride is stratified over a pH 8.5–9.0 buffer, basic hydrolysis occurs at the macroscopic interface between the oleic anhydride phase (or droplets if the system is dispersed) and the aqueous phase. The number of anhydride molecules hydrolized in this way is very small and the hydrolysis occurs very slowly. If oleic acid/oleate vesicles are instead present in the aqueous phase, the oleic anhydride molecules are taken up by the vesicles, which solubilize the anhydride in the membrane (Fig. 6.9a). Anhydride molecules are then easily and rapidly hydrolized to form fatty acids, the vesicles' building blocks. The net result is the accretion of the 'parent' vesicles due to in situ synthesis of their membrane components, reaching an unstable state, then split to give rise to 'daughter' vesicles.

Such an approach has been studied in a number of cases [35, 118], which also include intra-vesicle reactions (see Sect. 6.4.3).

6.4.1.3 Homogeneous Sytems

At high pH, fatty acids in aqueous solution form micelles. Micelles, in contrary to vesicles, as very small spherical assemblies (the diameter of a small oleate micelles can be estimated to be 3.5–4 nm, the sum of the length of two extended oleate molecules) (for a discussion on the co-existence of vesicles and micelles, see [119]). Due to the small micelle size, a micelle solution is transparent. Micelles are dynamic systems, whose fatty acid (FA) components are in rapid equilibrium with the monomer form in solution, $FA_n \rightleftarrows nFA$.

Most experiments have been carried out with the oleic acid/oleate system. When a small aliquot of oleate micelles (high pH) are added to a pH 8.5 buffer (e.g., bicine or borate buffer), the protonation of oleate molecules brings about a structural rearrangement of the micelles which transform into bilayer and then into vesicles. The whole process takes several minutes to occur, as evidenced by turbidity measurements. If, however, pre-existing oleic acid/oleate vesicles are present in the buffered solution, an alternative (and faster) path is available. Pre-existing vesicles uptake oleate molecules from the micelles (or even include the micelles) in their membrane. The membrane area increases due to the insertion of new oleic acid/oleate molecules (Fig. 6.9b). Consequently, the vesicle grow (plausibly in non-spherical way [120]), reach an unstable state, and divide into two daughter vesicles. Many studies have been carried out on this system, especially with oleic acid/oleate

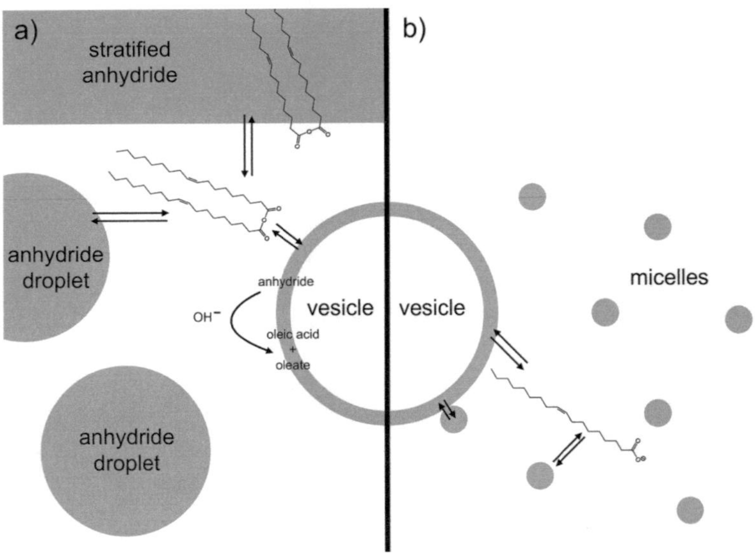

Fig. 6.9 Two different feeding methods for achieving fatty acid vesicles growth and division (cf. Fig. 6.8). (**a**) Oleic acid/oleate vesicles are added to a solution in the presence of oleic anhydride. The anhydride is stratified over the aqueous phase and dispersed in form of oil-in-water droplets under mechanical agitation. Anhydride molecules are taken up by the vesicles, which solubilize the anhydride molecules in their bilayer, where it is hydrolized by hydroxy ions to give oleic acid and oleate – the membrane-forming compound. In this way, the parent vesicle increases its membrane area, reach an unstable state (not shown) and divide in two or more 'daughter' vesicles (not shown). (**b**) Oleic anhydride can be substituted by oleate micelles, which are added, in a small volume, to an oleic acid/oleate vesicle suspension. Oleate micelles deliver oleate molecules to the vesicle bilayer either via the monomeric form, either after vesicle/micelle collision. As in the case (**a**), the vesicle growth and division are not shown

LUVs, which has the advantage of being homogeneous (in contrary to the anhydride method, which occurs in a two-phases system), and therefore it can be studied by spectroscopic methods [37, 39–41, 121]. In addition to the increased rate of micelle-to-vesicle transformation due to the uptake-growth-division mechanism, it results that the size distribution of parent and daughter vesicles are approximately similar. This has been called 'matrix effect' [36–39]. A possible vesicle intermediate has been visualized by cyro-transmission electronmicroscopy [120]. Moreover, it has been recently extended to oleic acid/oleate GVs. In the latter case, direct observation by light microscopy has shown that GVs uptake oleate micelles, elongate to form tubular vesicles, then fragment into many new vesicles by a pearling/breakage mechanism [42, 122, 123].

6.4.2 Relevance of Vesicle Autopoietic Self-Reproduction

The above-mentioned observations have great relevance in origins-of-life scenario, because they show that the proliferation of cell-like structures is possible also in absence of the complex macromolecular machineries that characterize modern cell growth-division. The only requirement is that adequate precursors are available in the environment of the protocells. Moreover it demonstrates that a quite complex chemical system, hold together solely by non-covalent interactions, can behave in coordinate manner displaying some of the features of biological systems. As we will see in the next section, if a chemical reaction is occurring inside the self-reproducing vesicles, this corresponds to a minimal model of cell. One of the still missing goal is the substitution of the *externally-added* lipids with the *internally-produced* ones. A first attempt was done in 1991, with the incorporation of four lipid-producing enzymes in phosphatidylcholine liposomes [97]. More recent studies have followed this idea [124, 125] but the desired pattern has not been observed yet (mainly because the limited amount of newly-produced lipids). Indeed, the realization of compartmentalized reactions inside self-reproducing vesicles is the closest way for modeling primitive cells.

It is important to emphasize that the vesicle self-reproduction has been discovered by using fatty acids. The 'substrate' vesicles have been fatty acid LUVs or GVs, or phosphatidylcholine LUVs [39] (for a study on phosphatidylcholine GVs and fatty acid micelles, see [126]). Studies on other chemical systems have been also reported. In one case, an *ad hoc* designed artificial surfactant was shown to display GVs growth and division [127, 128]. In another case, lipids have been generated by chemical ligation [129] (an interesting case even if self-reproduction was not observed). Whereas fatty acids are prominent candidates for the first membrane-forming compounds [130], terpenoids, which can also have primitive origin [22], have been investigated much less with respect to the dynamical properties of their assemblies. The Murchison meteorite, in addition, also contains a suite of alkyl dicarboxylic acids up to 18 carbons [131, 132], whose self-assembly properties are not well known (C. Thomas and P. L. Luisi, unpublished results).

6.4.3 Reactions and Other Patterns

The second issue for the construction of primitive cellular models from the bottom-up is the realization of reactions inside vesicles. Most of work is done with enzyme-based pathways, which model primitive pathways. The latter are difficult to identify and carry out. Examples of these allegedly primitive pathways are the formose reaction (producing sugars from formaldehyde in basic conditions [133]), or the various enzyme-free paths for oligonucleotide synthesis from activated mononucleotides, or primitive condensation reactions to form oligopeptides, and so on. Enzyme-based pathways, on the other hand, are easy to implement as purified enzymes are available from several sources, or house-made, or synthesized *in situ* by cell-free protein synthesis.

The first two studies (1994/1995) of compartmentalized reactions deserve special mention, as the reactions occurred inside self-reproducing fatty acid vesicles. Both systems focused on the production of nucleic acids inside oleic acid/oleate LUVs: (1) the so-called Oparin reaction (oligomerization of ADP to give poly(A), under catalysis of polynucleotide phosphorylase (PNPase)); and (2) the RNA replication catalized by Qβ-replicase. These two systems are sketched in Fig. 6.10. In the first case, ADP was added to PNPase-containing vesicles. ADP permeates into the vesicles and is polymerized inside the aqueous lumen, with the production of inorganic phosphate [35, 134]. In the second case, a (+)-strand RNA template, nucleotides triphosphate and Qβ-replicase have being co-entrapped inside vesicles, with the result of producing the complementary RNA (–)-strand [118]. These reactions were simultaneous to vesicle growth and division, so that a simplified model of RNA-producing primitive cell was realized.

Next (1995–1999), the polymerase chain reaction was carried out in phosphatidylcholine LUVs (non self-reproducing) [135] (but see also [136, 137]) and

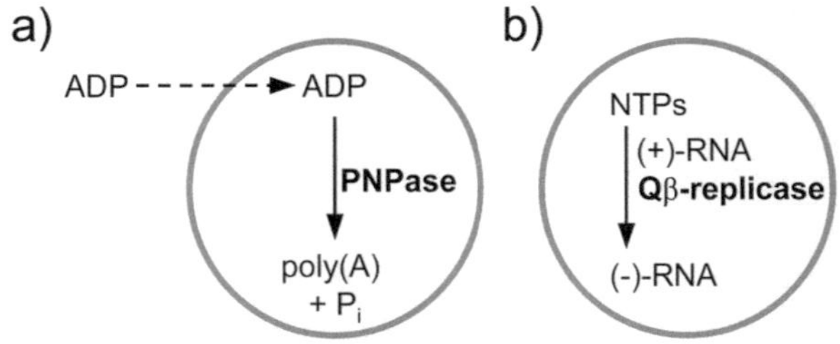

Fig. 6.10 Reactions inside vesicles. (**a**) The Oparin reaction consists in the oliogomerization/polymerization of ADP operated by PNPase. The reaction is interesting from the viewpoint of origins-of-life because it produce a RNA molecule without a template. The Oparin reaction was carried out in DMPC vesicles [134] or in oleic acid/oleate vesicles [35]. In both cased ADP was added externally and ADP firstly diffuses from the environment to the vesicle core. (**b**) Qβ-replicase is another interesting enzyme as it is a RNA-dependent RNA polymerase, capable of replicating RNA without need of DNA. All components required for the reaction were co-entrapped inside oleic acid/oleate vesicles [118]. Note that in works [35, 118] autopoietic vesicle self-reproduction occurred simultaneously to internalized reactions

few years later the first production of a long peptide – poly(Phe) – by translation was similarly observed [138]. In particular, the latter example signifies a transition towards an approach that later was recognized as typical of bottom-up synthetic biology (i.e., the synthesis of proteins in the vesicle lumen). This promising and flourishing topic will be specifically discussed in Sect. 6.5. Note, however, that protein synthesis inside fatty acid vesicles has not been reported yet.

A number of studies have been carried out on reactivity inside fatty acid vesicles, in particular for showing the oligomerization of activated nucleotides in the vesicle lumen [44], to the function of encapsulated ribozymes [46, 139], and to face the Mg^{2+}-induced destabilization of fatty acids [54, 88, 140]. Of particular relevance are those studies where ribozymes are encapsulated inside fatty acid vesicles because these systems model quite closely the primitive cell-like structures that played a role in the RNA-world hypothesis [141]. On the other hand, the intra-vesicle synthesis of short peptides, catalyzed by Ser-His [142], has been also reported, and because the reaction product migrates to the vesicle membrane, a mechanism of competition among vesicles emerges [87].

In conclusion, a number of reactions have been reported as occurring inside fatty acid vesicles. However, due to the relatively high solubility of fatty acids (it depends on the chain length), when compared with double-chain phospholipids, they can potentially interfere with the reactions. The field of micro-compartmentalized reaction is certainly richer in examples when phospholipid vesicles (liposomes) are used. We should comment, however, on the necessity of developing experimental cases based on primitive membranes, like fatty acids, terpenoids, or mixtures of different surfactants in order to test hypotheses on primordial compartmentalization and robustness of early chemical pathways.

6.5 Semi-synthetic Minimal Cells

The point of junction between the synthetic approach typical of origins-of-life and the advancing synthetic biology (SB) derives from a common program which focuses on the synthesis of proteins inside vesicles. In origins-of-life perspective, this is a key step that would allow the construction, from a bottom-up approach, of primitive cell models displaying functions for the realization of an autopoietic cell. In synthetic biology, cell-free protein synthesis (CFPS) represents a tool for engineering man-made devices, such as functional synthetic cells for a variety of applications (understanding biochemical paths, for biosensoring, as proof-of-concept, etc., up to nanomedicine vehicles).

Such goals can be reached by constructing the so-called '*semi-synthetic minimal cells*' (Fig. 6.11), that can be defined as those cell-like structures built by co-encapsulation of the minimal number of biochemicals (DNA, RNA, enzymes, ...) inside lipid vesicles, in order to achieve a certain function (ultimately, being alive). Here, for brevity, semi-synthetic minimal cells will be referred simply as synthetic cells.

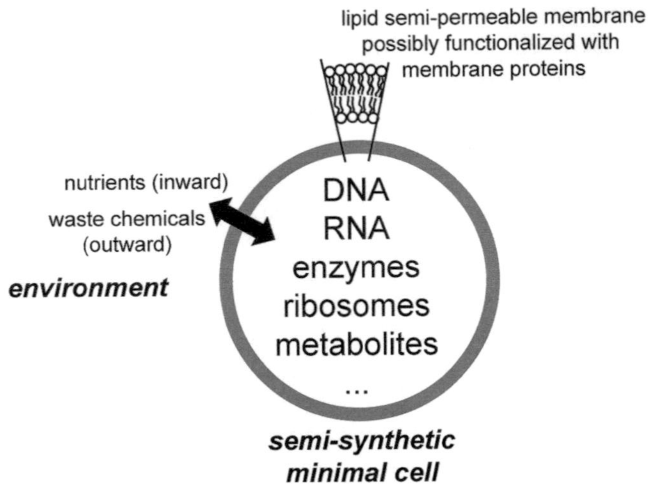

Fig. 6.11 Semi-synthetic minimal cells. The minimal number of genes, enzymes, RNAs, and low molecular-weight compounds are encapsulated into synthetic lipid vesicles. The membrane acts as a boundary to confine the interacting internalized molecules, and thanks to its semi-permeability allows the material exchange between the semi-synthetic minimal cell and its environment (nutrients uptake, waste release). More elaborated model include membrane proteins for transportation, sensing, catalysis

The bottom-up constructive approach perfectly fits with SB philosophy based on the concept of standard biological parts (http://parts.igem.org), of 'biobricks' and on the idea of letting a structure/function emerge as the result of molecular parts interaction in the form of a molecular system. Moreover, it is well represented by the concepts of orthogonality, modularity, programmability that are typical of SB. However, subtle differences in the epistemology of the research on assembling synthetic cells under the bottom-up SB view and origins-of-life perspectives have been revealed [143].

The bottom-up construction of synthetic 'cells' – not necessarily alive – for practical application is an essentially unexplored field where most of concepts, techniques, usages, advantages and limitations have to be discovered. Up to now, such kind of research has been essentially done by scholars more interested in basic science, and especially in using bottom-up synthetic cells as cellular models in search of biochemical/biophysical understanding. Most of researchers working in traditional fields of biochemistry, molecular biology, physiology, biotechnology, and so on can express, especially if working in multidisciplinary teams, a great potential to fully exploit these structures.

The intersection between SB and the practice of assembling synthetic cells has been discussed elsewhere, in most of its conceptual and practical aspects [11, 17, 144, 145]. Here we would like to recall only two of these facets. The first is the so-called *minimal genome*. The minimal genome can be defined as the minimal set of genes that encode the proteins (enzymes) capable of supporting autopoietic cellular self-maintenance under permissive conditions. By permissive conditions we mean that most of the low molecular weight molecules required for cell metabolism are available in the cell environment and do not require the corresponding intracellular synthetic step. Comparative genomics have revealed that, based on the smallest prokaryotes a minimal gene set could include about 200–300 genes (reviewed in [11]). A study focused on the endosymbiont *Buchnera* set the figure to 206 genes [146], most of which ($\sim$50%) are devoted to transcription-translation (TX-TL). As it will become clear in the next section, the goal of creating a bottom-up synthetic cell by including 206 genes is currently beyond the actual possibility. Moreover, this is not the central point in contemporary research, which is instead focused at understanding the interplay between microcompartmentalized reactions and the compartment feature, their interplay in terms of physics and chemistry [17, 85]. The second facet is a technical one, and refers to the development of *microfluidics* for the assembly of synthetic cells – a process that is currently just a potential one, but it can quickly become a realistic procedure. As we have stressed in an early report [144], microfluidic fabrication of solute-filled vesicles is a goal that pave the way to fully programmable systems (Fig. 6.12). Microfluidic devices would reduce both the size heterogeneity and the wide solute occupancy distribution that typically characterize spontaneous vesiculation. By reducing these two stochastic phenomena one achieves vesicle populations (cell models) with uniform and programmable features, that can be very advantageous for practical applications (note that, in contrary, deciphering and exploiting stochastic phenomena is a key value for primitive cell modeling). As mentioned in Sect. 6.2, vesicle-producing microfluidic devices have been increasingly developed in the past years [64–71].

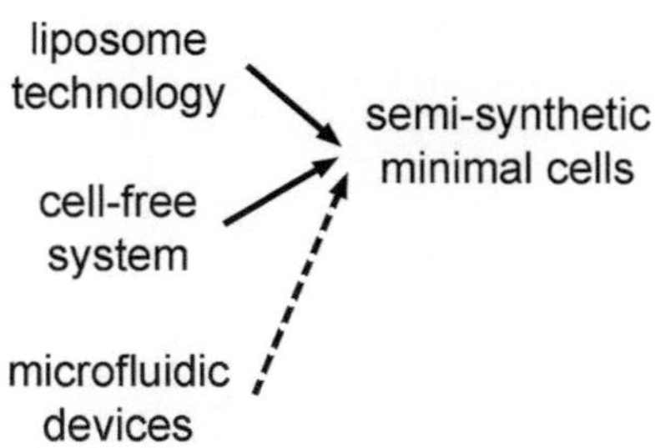

Fig. 6.12 Operative aspects of semi-synthetic minimal cell construction. The techniques currently involved for the production of minimal cells, mainly based on cell-free systems and liposome technology can be possibly improved by the future use of microfluidic devices. Some reports pointing to this direction have been already published

6.5.1 Protein Synthesis Inside Vesicles

As it has been mentioned, the core of current synthetic cell research is the CFPS inside lipid vesicles. After the pioneer report on ribosomal poly(U) translation inside phosphatidylcholine LUVs [138], in 2001 the Yomo group published the first example of a functional protein synthesis, the green fluorescent protein (GFP) in a heterogeneous vesicle population prepared by the natural swelling method [147]. This paper was followed by a 2002 short note on the enhanced-GFP (eGFP) synthesis inside liposomes prepared by the ethanol injection method [148]. From that moment the number of reports on CFPS inside liposomes increased constantly, and a certain number of proteins have been successfully produced in their correct fold, as witnessed by their functionality (reviewed in [144]).

CFPS micro-compartmentalized reactions occurs after co-encapsulating all components ($\sim$80) of CFPS 'kits', and an encoding DNA (or RNA) sequence, inside vesicle (Fig. 6.13). Typically a DNA plasmid is mixed with a CFPS kit, kept at low temperature to prevent the reaction start, vesicles are formed, and membrane-impermeable killers of the reaction are added externally in order to block the reaction outside vesicles (protease or RNase are often used). The reaction is often started just by increasing the temperature. Therefore, protein synthesis inside vesicles is a combination of liposome technology and CFPS methods. Lipids and CFPS kits must be chemically compatible, capable of forming good vesicles, and matching with the needs of the synthesized protein (think to membrane enzymes). A practical example of how these three constraints have been overcome can be found in [124].

The approach proposed by Noieraux and Libchaber (2004) focused on the expression of α-hemolysin (αHL) in GVs [61]. Thanks to the pore generated by the self-assembly of αHL in the vesicle membrane, it was possible to 'feed' the vesicle bioreactor for 4 days, as the energy-rich compounds were added to vesicles and permeate in the vesicles core *via* the αHL pore. Notably, the pore also allowed the release of by-products from the vesicles. After this report, αHL has been often used for this aim.

A quite interesting CFPS kit for the bottom-up SB approach is the PURE system (see the grey-box and Table 6.3).

The PURE system (Protein synthesis Using Recombinant Elements) is a partially recombinant, cell-free, protein-synthesis system reconstituted solely from those essential elements of the *Escherichia coli* translation system [149, 150]. As shown in Table 6.3, it is composed of 36 individually purified His-tagged proteins, purified ribosomes, and tRNAs mix (overall, 83 macromolecular components). It can be considered as a standard chassis for synthetic biology. The PURE system performs CFPS by combining T7-RNA-

(continued)

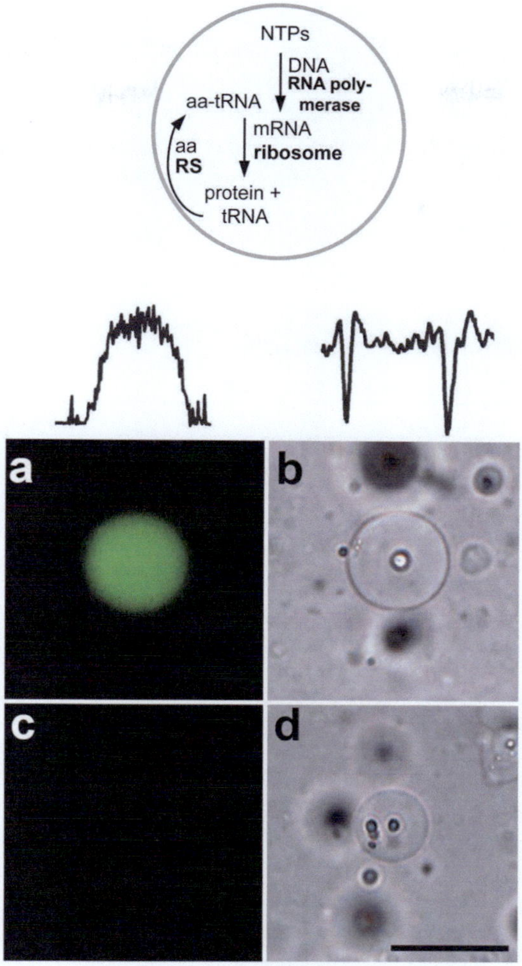

Fig. 6.13 Protein synthesis inside lipid vesicles. *Top*: Reaction scheme. NTPs are polymerized to give mRNA using a DNA template and RNA polymerase. The resulting mRNA is a template for the protein synthesis, fuelled by aa-tRNAs. tRNA are re-charged by aminoacyl-tRNA synthetases (RS) and amino acids. These three modules consume energy (ATP, GTP), which is re-formed inside the vesicle by a fourth module (not shown) at expenses of creatine phosphate. *Bottom*: eGFP synthesis in GVs produced by the droplet transfer method [144]. Panels (**a,b**): cell-free extracts plus DNA; panels (**c,d**): cell-free extracts without DNA (negative control); panels (**a,c**): fluorescence imaging; panels (**b,d**): bright-field imaging. *On the top*, equatorial profile of micrographs' pixel luminosity (panels **a,b**). Size bar represents 50 μm (The *bottom* part (panels **a**), (**b**), (**c**), (**d**) have been reproduced from Ref. [144] with permission from The Royal Society of Chemistry)

polymerase transcription and *E. coli* translation. With respect to traditional CFPS kits, the PURE system has a lower yield (ca. one third [151]), but its usage adds to the synthetic cell research from the viewpoint of the potential of a full design and modularity. Indeed, one can imagine of modifying the PURE system composition at will, or substitute some of its components with others, and so on.

Table 6.3 Composition of the PURE system [150]

Component	Concentration	Component	Concentration
IF1	2.70 μM	IF2	0.40 μM
IF3	3.50 μM	EF-G	0.26 μM
EF-Tu	0.92 μM	EF-Ts	0.96 μM
RF1	0.25 μM	RF2	0.24 μM
RF3	0.17 μM	RRF	0.50 μM
AlaRS	1900 U/mL	ArgRS	2500 U/mL
AsnRS	20 mg/mL	AspRS	2500 U/mL
CysRS	630 U/mL	GlnRS	1300 U/mL
GluRS	1900 U/mL	GlyRS	500 U/mL
HisRS	630 U/mL	IleRS	2500 U/mL
LeuRS	3800 U/mL	LysRS	3800 U/mL
MetRS	6300 U/mL	PheRS	1300 U/mL
ProRS	1300 U/mL	SerRS	1900 U/mL
ThrRS	1300 U/mL	TrpRS	630 U/mL
TysRS	630 U/mL	ValRS	3100 U/mL
MTF	4500 U/mL	Ribosomes	1.2 μM
Creatine kinase	4 μg/mL	Myokinase	3 μg/mL
NDP kinase	1.1 μg/mL	Pyrophosphatase	2.0 U/mL
T7 RNA polymerase	10 μg/mL	Creatine phosphate	20 mM
ATP, GTP	2.0 mM	CTP, UTP	1 mM
HEPES-KOH (pH 7.6)	50 mM	Potassium glutamate	100 mM
Magnesium acetate	13 mM	Spermidine	2.0 mM
DTT	1.0 mM	10 amino acids	0.3 mM
tRNA mix	56 A260/mL	10-formyl-5,6,7,8-tetrahydrofolic acid	10 mg/mL

The research on protein synthesis inside liposomes greatly advances. A number of proteins have been produced inside vesicles in addition to GFP (or other fluorescent proteins or β-galactosidase and β-glucuronidase as reporter proteins [152, 153]). For example, T7 RNA polymerase to realize a two-steps genetic cascade [154], αHL to create a pore in the membrane [61], lipid-synthesizing enzymes [124, 125], Qβ-replicase in order to replicate RNA [152]. More recent applications

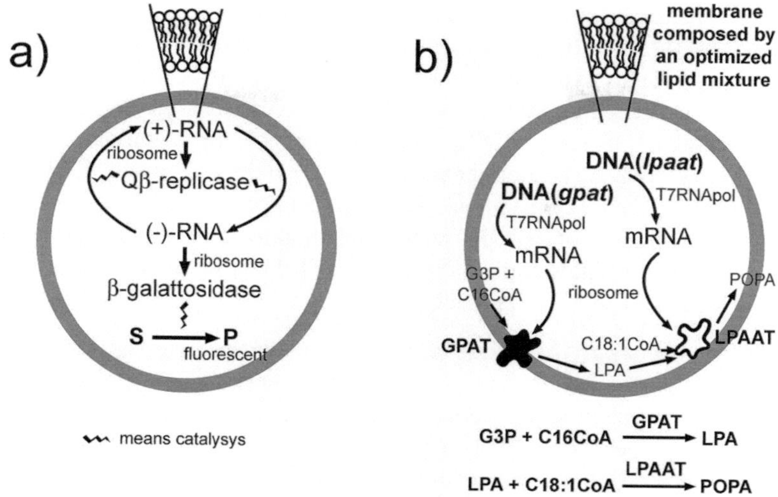

Fig. 6.14 Two remarkable examples of semi-synthetic minimal cell construction. (**a**) Starting from Qβ-replicase-encoding (+)-RNA strand and cell-free protein expression system, the production of Qβ-replicase was carried out inside liposomes, so that the complementary (−)-RNA strand is produced from nucleotides and (+)-RNA template. In turn, (−)-RNA acts as a template for the Qβ-replicase catalyzed (+)-RNA strand synthesis. The correct production of (−)-RNA is confirmed by the fact that it encodes for β-galactosidase (that successfully catalyzes the formation of a fluorogenic substrate [152]. (**b**) The two genes encoding for glycerol-3-phosphate acyltransferase (GPAT) and lysophosphatidic acid acyltransferase (LPAAT) needs to be co-encapsulated inside lipid vesicles, together with a CFPS kit (e.g., the PURE system). Actually, the two proteins require different redox conditions, so that they were actually synthesized in two different vesicle populations. In two steps, these membrane enzymes produce phosphatidic acid starting from glycerol-3-phosphate and an acyl donor. Note that the vesicle membrane composition is a key factor for realizing a functional system POPC 50.8%, POPE 35.6%, POPG 11.5%, CL 2.1% (POPC 1-palmitoyl-2-oleoyl-*sn*-glycero-3-phosphatidylcholine, POPE 1-palmitoyl-2-oleoyl-*sn*-glycero-3- phosphatidylethanolamine, POPG 1-palmitoyl-2-oleoyl-*sn*-glycero-3-phosphatidylglycerol, CL cardiolipin) [124]

refer to: σ^{70} transcription factor (in order to realize a two-stage genetic cascade) [155]; MreB (bacterial cytoskeleton filaments) [156]; EmrE (a transporter protein) [157, 158]; BmOR1 and BmOrco (olfactory receptors and co-receptors) [159]; Sec translocon (mediator of membrane translocation of single- and multi-span membrane proteins) [160].

For example, Yomo and collaborators [152] assembled a self-encoding replicase system as it follows (Fig. 6.14a). Messenger RNA encoding the RNA-dependent RNA replicase (Qβ-replicase), was encapsulated inside liposomes together with essential transcription-translation components. The Qβ-replicase coding sequence was successfully translated and the resultant Qβ-replicase enzyme was functional. It therefore replicated the RNA gene, producing its complementary strand. In turn, the complementary RNA strand also encoded for β-galactosidase, whose successful translation was detected by measuring the conversion of fluorogenic substrate.

Another instance comes from the attempts of synthesizing phospholipids inside lipid vesicles (Fig. 6.14b). In particular, starting from glycerol-3-phosphate it in principle possible to synthesize phosphatidic acid by two successive acylation steps (using acyl-CoA as acyl donor). These two reactions are catalyzed by the integral membrane enzyme G3PAT [2.3.1.15] and by the membrane associated enzyme LPAAT [2.3.1.51]. Both enzymes can be synthesized inside liposomes by CFPS (PURE system). Due to some issues about redox conditions the enzymes were produced in two different liposome population to carry out phosphatidic acid synthesis. A more recent version of this approach led to a significant understanding, in several details, of such an approach [125].

Why protein synthesis? As it can be understood by the flourishing research on CFPS inside liposomes, dominating micro-compartmentalized protein synthesis is an essential conceptual and practical goal for the bottom-up assembly of cell-like particles, either intended as primitive cell models or synthetic cells for SB/biotechnology. It is a conceptual advancement because, as we have specified, about 50% of the minimal genome refers to protein synthesis. Currently, TX-TL components, purified from biological organisms, are inserted in liposomes, whereas only a handful of genes is expressed. In future, one can imagine of realizing a PURE system-producing synthetic cells, where all (∼80) PURE system components are synthesized *in situ* starting from the corresponding DNA sequences and PURE system. This recursive logic is typical of biological systems. On the other hand, CFPS is important from practical reasons as it paves the way to synthetic cell functionalization with pores, receptor, enzymes, transducers, cytoskeletal elements, and so on. All this will allow a stepwise increase of complexity and a better understanding of biochemical systems by their full reconstitution.

6.5.2 Toward More Complex Functionalization

Some of the most interesting scenarios can be summarized by looking at the contemporary trends, reports, and to the open questions [161].

1. MVV (vesosomes, see Fig. 6.2) are interesting structures consisting of small vesicles contained inside a larger one. This structure is interesting because it models a cell with subcellular organelles. Some reports have shown how to produce such structure, but their use in synthetic cell research, to the best of our knowledge, has not been reported [162–164].
2. The production of biochemical energy (e.g., ATP) is another open issue. A possible way is the conversion of light energy in the form of a chemical gradient (pH gradient), then exploiting such a gradient for fostering ATP synthesis via ATP synthase. Photosynthetic proteins (or bacteriorhodopsin) are membrane proteins, as ATP synthase. Clearly, the synthesis (or the reconstitution) of correctly folded and correctly oriented membrane proteins inside liposomes is

thus very important. We have been recently involved in the reconstitution of the photosynthetic reaction centre in phospholipid GVs [165].

3. Movement is a property that is not strictly necessary for the definition of living beings, but it is intriguing the construction of a synthetic cell capable of autonomous motility. Similarly, cytoskeletal elements, division elements (such as the bacterial *Z*-ring), and other similar proteins may be accessible targets for the bottom-up approach.
4. From one to many. Most of current research focuses on individual cells, but it is well known that even the simplest organisms often live in community. An attempt to investigate the cooperation aspects of primitive cells has been done by associating GVs in the form of colonies, on a flat support [59]. Intriguingly, the colonies displayed several emerging properties that are not found in individual vesicles. Other approaches based on molecular recognition of DNA-decorated vesicles have been presented [166]. In some cases, GFP was produced in the associated vesicles [167]. Note that such assemblies can be used as tissue models.
5. Mimicking inter-cellular communication is probably one of the most fascinating goals (Fig. 6.15). Inspired by a hypothesis paper on the application of Turing test to artificial cells [168], a first paper appeared in 2009 on liposomes sending a signal molecule to bacteria [133], but the generation of the signal molecule was not under genetic control. The idea was further elaborated within the realm of bottom-up SB approaches (the combination of CFPS and liposome technology), and an experimental plan for the implementation of synthetic-to-synthetic and synthetic-to-natural communciation (and vice versa) was devised by Stano and Rampioni [18] (for other approaches see [169–171]). It was realized also that such an approach would bring about minimal cells as tools for the emerging field of bio-chem information and communication technologies (bio-chem ICTs [172, 173]). From then, important papers appeared on experimental realization of this approach [174–178], that eventually goes in the direction described in the next final point.

6.5.2.1 An Intriguing Perspective: Minimal Cognition in the Chemical Domain

The reference here is to the so-called "embodied approach" to the study of cognition [179], which aims at overcoming the mind-body dichotomy by bringing into focus the fundamental role(s) the biological body plays in cognition. Despite this consideration mainly refers to a high-level cognition, its extrapolation to the minimal terms of unicellular systems can be interesting and perhaps fruitful. Synthetic cell research contributes to the embodied approach to (minimal) cognition thanks to the approaches originally introduced as proto-cybernetics, which are coherent with those explained above (understanding-by-building, constructive/synthetic approach, bottom-up). In particular the construction of artificial systems as models of cognitive and biological processes serves to test and develop scientific theories about the mechanism underlying the natural processes. This is common to all sciences that study natural processes through artificial models [180], and include software, hardware, and – as a result of origins-of-life/SB bottom-up approaches – also *wetware* models (synthetic cells) [102].

We have seen above that primitive cell models have been built under the paradigm of autopoietic theory [108, 109]. However, according to the proponents of autopoiesis, minimal autopoietic systems are also minimal cognitive systems, by actively maintaining a dynamical coupling with their environment [181], i.e., being capable of perceiving external events and actively reacting to them in a conservative way (generation of an internal meaning to external perturbations).

This theoretical perspective offers to synthetic cell research the possibility of playing a role in the *avant-gard* of artificial intelligence, in form of embodied artificial intelligence. The synthetic cells, in fact, do not process external stimuli under a representational scheme, but being guided by the physico-chemical constraints referring to their intra-cellular dynamical/functional inter- and supra-molecular relationships (embodiment). This new frontier is dense of possible developments [182].

6.6 Concluding Remarks

No questions are more fascinating and difficult to face than unveiling the mechanisms that led to the origin of life on Earth more than 3.5 billions years ago. The scientific search for the physico-chemical basis of such an important event – which *is* within the realm of experimental study, started less than 100 years ago, and it

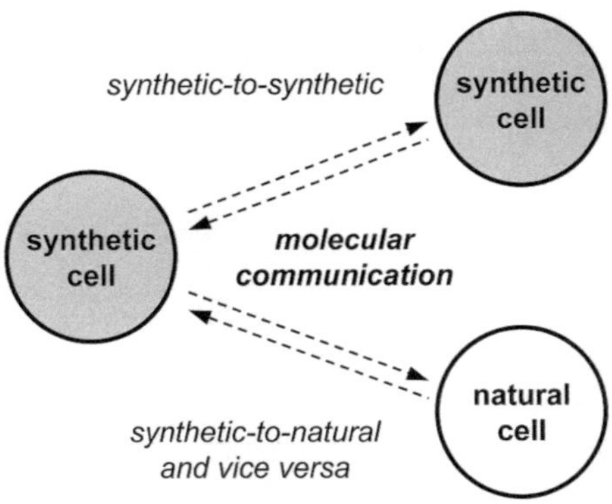

Fig. 6.15 Molecular communication between synthetic cells and between synthetic and natural cells [18]. By a proper design it is possible to engineer synthetic cells for being capable of communicating with other synthetic cells or with biological cells via chemical signaling. This approaches paves the way to nanomedicine – imagine for example an 'intelligent' drug delivery vehicle consisting of a minimal cells capable of interacting with biological cells in an organism and synthesize a drug in situ only when necessary

is continuously progressing thanks to technical and conceptual advancements. An important branch of origins-of-life investigations is dedicated to understanding the formation, the structure, the function of primitive cells, filling the gap between pure prebiotic chemistry (the chemistry of molecules) and the last universal common ancestor (LUCA). Is this the realm of the chemistry of molecular systems (*systems chemistry*), where ensembles of molecules behaves and should be considered as a whole.

Cell models based on vesicles are central to this arena, and it is significant that the two faces of this approach look at primitive times, but simultaneously to modern synthetic biology. In this chapter we have presented both directions evidencing the common grounding despite the apparent differences. We are convinced that future developments of these fields will contribute to a *qualitative* jump destined to modify scientific and technological knowledge.

Hopefully, together with enforcing a new and fruitful wave in synthetic biology, cell models will finally demonstrate that living systems emerge from non-living matter without additional requirements. Or, with the words of Eschenmoser and Kisakürek [183]:

> The aim of an experimental aetiological chemistry is not primarily to delineate the pathway along which our ('natural') life on earth *could* have originated, but to provide decisive experimental evidence, through the realization of model systems ('artificial chemical life') that life *can* arise as a result of the organization of the organic matter.

Acknowledgements The author is grateful to Pier Luigi Luisi (ETH Zürich, Switzerland and Uniroma3, Italy) for inspiring discussions over the past 15 years. Luisa Damiano (Univ. Messina and CERCO, Univ. of Bergamo) is acknowledged for the discussion presented in Sect. 6.5.2.1. Synthetic cell research has been developed with the framework of two European COST Actions, namely CM-1304 *Emergence and Evolution of Complex Chemical Systems* and TD-1308 *Origins and evolution of life on Earth and in the Universe (ORIGINS)*.

References

1. Haldane JBS (1929) The origin of life. Ration Annu 148:3–10
2. Oparin AI (1953) The origin of life, 2nd edn. (Tras: Morgulis S). Dover Publications, New York
3. Miller SL (1953) A production of amino acids under possible primitive earth conditions. Science 117(3046):528–529
4. Ruiz-Mirazo K, Briones C, de la Escosura A (2014) Prebiotic systems chemistry: new perspectives for the origins of life. Chem Rev 114(1):285–366
5. Hanczyc MM (2009) The early history of protocells – the search for the recipe of life. In: Rasmussen S, Bedau MA, Chen L, Deamer D, Krakauer DC, Packard NH, Stadler PF (eds) Protocells: bridging nonliving and living matter. MIT Press, Cambridge, pp 3–18
6. Jia TZ, Hentrich C, Szostak JW (2014) Rapid RNA exchange in aqueous two-phase system and coacervate droplets. Orig Life Evol Biosph 44:1–12
7. van Swaay D, Dora Tang T-Y, Mann S, de Mello A (2015) Microfluidic formation of membrane-free aqueous coacervate droplets in water. Angew Chem Int Ed Engl 54(29):8398–8401

8. Dora Tang T-Y, van Swaay D, deMello A, Ross Anderson JL, Mann S (2015) In vitro gene expression within membrane-free coacervate protocells. Chem Commun (Camb) 51(57):11429–11432
9. Bangham AD, Horne RW (1964) Negative staining of phospholipids and their structural modification by surface-active agents as observed in the electron microscope. J Mol Biol 8:660–668
10. Bangham AD, Standish MM, Watkins JC (1965) Diffusion of univalent ions across the lamellae of swollen phospholipids. J Mol Biol 13(1):238–252
11. Luisi PL, Ferri F, Stano P (2006) Approaches to semi-synthetic minimal cells: a review. Naturwissenschaften 93(1):1–13
12. Mansy SS, Szostak JW (2009) Reconstructing the emergence of cellular life through the synthesis of model protocells. Cold Spring Harb Symp Quant Biol 74:47–54
13. Ichihashi N, Yomo T (2016) Constructive approaches for understanding the origin of self-replication and evolution. Life (Basel) 6(3)
14. de Lorenzo V, Danchin A (2008) Synthetic biology: discovering new worlds and new words. EMBO Rep 9(9):822–827
15. Pohorille A, Deamer D (2002) Artificial cells: prospects for biotechnology. Trends Biotechnol 20(3):123–128
16. Noireaux V, Maeda YT, Libchaber A (2011) Development of an artificial cell, from self-organization to computation and self-reproduction. PNAS 108(9):3473–3480
17. Stano P (2011) Minimal cells: relevance and interplay of physical and biochemical factors. Biotechnol J 6(7):850–859
18. Stano P, Rampioni G, Carrara P, Damiano L, Leoni L, Luisi PL (2012) Semi-synthetic minimal cells as a tool for biochemical ICT. BioSystems 109(1):24–34
19. Ichihashi N, Aita T, Motooka D, Nakamura S, Yomo T (2015) Periodic pattern of genetic and fitness diversity during evolution of an artificial cell-like system. Mol Biol Evol 32(12):3205–3214
20. Xu C, Hu S, Chen X (2016) Artificial cells: from basic science to applications. Mater Today 19(9):516–532
21. Gebicki Jm, Hicks M (1973) Ufasomes are stable particles surrounded by unsaturated fatty-acid membranes. Nature 243(5404):232–234
22. Ourisson G, Nakatani Y (1994) The terpenoid theory of the origin of cellular life: the evolution of terpenoids to cholesterol. Chem Biol 1(1):11–23
23. Pozzi G, Birault V, Werner B, Dannenmuller O, Nakatani Y, Ourisson G, Terakawa S (1996) Single-chain polyprenyl phosphates form primitive membranes. Angew Chem Int Ed Engl 35(2):177–180
24. Streiff S, Ribeiro N, Wu Z, Gumienna-Kontecka E, Elhabiri M, Albrecht-Gary AM, Ourisson G, Nakatani Y (2007) "Primitive" membrane from polyprenyl phosphates and polyprenyl alcohols. Chem Biol 14(3):313–319
25. Nakatani Y, Ribeiro N, Streiff S, Gotoh M, Pozzi G, Dsaubry L, Milon A (2014) Search for the most primitive membranes and their reinforcers: a review of the polyprenyl phosphates theory. Orig Life Evol Biosph 44(3):197–208
26. Discher BM, Won YY, Ege DS, Lee JCM, Bates FS, Discher DE, Hammer DA (1999) Polymersomes: Tough vesicles made from diblock copolymers. Science 284(5417):1143–1146
27. Choi H-J, Montemagno CD (2005) Artificial organelle: ATP synthesis from cellular mimetic polymersomes. Nano Lett 5(12):2538–2542
28. Chandrawati R, Caruso F (2012) Biomimetic liposome- and polymersome-based multicompartmentalized assemblies. Langmuir 28(39):13798–13807
29. Martino C, Shin-Hyun Kim, Horsfall L, Abbaspourrad A, Rosser SJ, Cooper J, Weitz DA (2012) Protein expression, aggregation, and triggered release from polymersomes as artificial cell-like structures. Angew Chem-Int Edit 51(26):6416–6420
30. Walde P, Cosentino K, Engel H, Stano P (2010) Giant vesicles: preparations and applications. Chembiochem 11(7):848–865

31. Walde P (2004) Preparation of vesicles (liposomes). In: Encyclopedia of nanoscience and nanotechnology, vol 9. H. S. Nalwa, American Scientific Publishers, pp 43–79
32. Walde P (2006) Surfactant assemblies and their various possible roles for the origin(s) of life. Orig Life Evol Biosph 36(2):109–150
33. Namani T, Walde P (2005) From decanoate micelles to decanoic acid/dodecylbenzenesulfonate vesicles. Langmuir 21(14):6210–6219
34. Walde P, Wick R, Fresta M, Mangone A, Luisi PL (1994) Autopoietic self-reproduction of fatty-acid vesicles. J Am Chem Soc 116(26):11649–11654
35. Walde P, Goto A, Monnard Pa, Wessicken M, Luisi Pl (1994) Oparins reactions revisited – enzymatic-synthesis of poly (adenylic Acid). J Am Chem Soc 116(17):7541–7547
36. Blöchliger E, Blocher M, Walde P, Luisi PL (1998) Matrix effect in the size distribution of fatty acid vesicles. J Phys Chem B 102(50):10383–10390
37. Lonchin S, Luisi PL, Walde P, Robinson BH (1999) A matrix effect in mixed phospholipid/fatty acid vesicle formation. J Phys Chem B 103(49):10910–10916
38. Berclaz N, Blöchliger E, Müller M, Luisi PL (2001) Matrix effect of vesicle formation as investigated by cryotransmission electron microscopy. J Phys Chem B 105(5):1065–1071
39. Rasi S, Mavelli F, Luisi PL (2003) Cooperative micelle binding and matrix effect in oleate vesicle formation. J Phys Chem B 107(50):14068–14076
40. Chen IA, Roberts RW, Szostak JW (2004) The emergence of competition between model protocells. Science 305(5689):1474–1476
41. Chen IA, Szostak JW (2004) A kinetic study of the growth of fatty acid vesicles. Biophys J 87(2):988–998
42. Zhu TF, Szostak JW (2009) Coupled growth and division of model protocell membranes. J Am Chem Soc 131(15):5705–5713
43. Fujikawa SM, Chen IA, Szostak JW (2005) Shrink-wrap vesicles. Langmuir 21(26):12124–12129
44. Mansy SS, Schrum JP, Krishnamurthy M, Tobé S, Treco DA, Szostak JW (2008) Template-directed synthesis of a genetic polymer in a model protocell. Nature 454(7200):122–U10
45. Mansy SS, Szostak JW (2008) Thermostability of model protocell membranes. Proc Natl Acad Sci USA 105(36):13351–13355
46. Engelhart AE, Adamala KP, Szostak JW (2016) A simple physical mechanism enables homeostasis in primitive cells. Nat Chem 8(5):448–453
47. Rendón A, Carton DG, Sot J, García-Pacios M, Montes L-R, Valle M, Arrondo J-LR, Goñi FM, Ruiz-Mirazo K (2012) Model systems of precursor cellular membranes: long-chain alcohols stabilize spontaneously formed oleic acid vesicles. Biophys J 102(2): 278–286
48. Birault V, Pozzi G, Plobeck N, Eifler S, Schmutz M, Palanché T, Raya J, Brisson A, Nakatani Y, Ourisson G (1996) Di(polyprenyl) phosphates as models for primitive membrane constituents: synthesis and phase properties. Chem Eur J 2(7):789–799
49. Gotoh M, Miki A, Nagano H, Ribeiro N, Elhabiri M, Gumienna-Kontecka E, Albrecht-Gary A-M, Schmutz M, Ourisson G, Nakatani Y (2006) Membrane properties of branched polyprenyl phosphates, postulated as primitive membrane constituents. Chem Biodivers 3(4):434–455
50. Monnard PA, Deamer DW (2002) Membrane self-assembly processes: steps toward the first cellular life. Anat Rec 268(3):196–207
51. Namani T, Deamer DW (2008) Stability of model membranes in extreme environments. Orig Life Evol Biosph, 38(4):329–341
52. Monnard P-A, Apel CL, Kanavarioti A, Deamer DW (2002) Influence of ionic inorganic solutes on self-assembly and polymerization processes related to early forms of life: implications for a prebiotic aqueous medium. Astrobiology 2(2):139–152
53. Maurer SE, Deamer DW, Boncella JM, Monnard P-A (2009) Chemical evolution of amphiphiles: glycerol monoacyl derivatives stabilize plausible prebiotic membranes. Astrobiology 9(10):979–987

54. Adamala K, Szostak JW (2013) Nonenzymatic template-directed RNA synthesis inside model protocells. Science 342(6162):1098–1100
55. Israelachvili JN, John Mitchell D, Ninham BW (1976) Theory of self-assembly of hydrocarbon amphiphiles into micelles and bilayers. J Chem Soc Faraday Trans 2 72(0): 1525–1568
56. (1999) Steering Group for the workshop on size limits of very small microorganisms National Research Council. In: Knoll A, Osborn MJ, Baross J, Berg HC, Pace NR, Sogin M (eds) Size limits of very small microorganisms: proceedings of a workshop. National Academic Press, Washington, DC
57. Pereira de Souza T, Stano P, Luisi PL (2009) The minimal size of liposome-based model cells brings about a remarkably enhanced entrapment and protein synthesis. Chembiochem 10(6):1056–1063
58. Akashi K, Miyata H, Itoh H, Kinosita K (1996) Preparation of giant liposomes in physiological conditions and their characterization under an optical microscope. Biophys J 71(6):3242–3250
59. Carrara P, Stano P, Luisi PL (2012) Giant vesicles "colonies": a model for primitive cell communities. Chembiochem 13(10):1497–1502
60. Pautot S, Frisken BJ, Weitz DA (2003) Production of unilamellar vesicles using an inverted emulsion. Langmuir 19(7):2870–2879
61. Noireaux V, Libchaber A (2004) A vesicle bioreactor as a step toward an artificial cell assembly. Proc Natl Acad Sci USA 101(51):17669–17674
62. Pautot S, Frisken BJ, Weitz DA (2003) Engineering asymmetric vesicles. Proc Natl Acad Sci USA 100(19):10718–10721
63. Stano P, Wodlei F, Carrara P, Ristori S, Marchettini N, Rossi F (2014) Approaches to molecular communication between synthetic compartments based on encapsulated chemical oscillators. In: Pizzuti C, Spezzano G (eds) Advances in artificial life and evolutionary computation. WIVACE 2014. Volume 445 of communications in computer and information science. Springer, Cham
64. Sugiura S, Kuroiwa T, Kagota T, Nakajima M, Sato S, Mukataka S, Walde P, Ichikawa S (2008) Novel method for obtaining homogeneous giant vesicles from a monodisperse water-in-oil emulsion prepared with a microfluidic device. Langmuir 24(9):4581–4588
65. Matosevic S, Paegel BM (2011) Stepwise synthesis of giant unilamellar vesicles on a microfluidic assembly line. J Am Chem Soc 133(9):2798–2800
66. Hu PC, Li S, Malmstadt N (2011) Microfluidic fabrication of asymmetric giant lipid vesicles. ACS Appl Mater Interfaces 3(5):1434–1440
67. Nishimura K, Suzuki H, Toyota T, Yomo T (2012) Size control of giant unilamellar vesicles prepared from inverted emulsion droplets. J Colloid Interface Sci 376(1):119–125
68. Matosevic S (2012) Synthesizing artificial cells from giant unilamellar vesicles: state-of-the art in the development of microfluidic technology. Bioessays 34(11):992–1001
69. Shiomi H, Tsuda S, Suzuki H, Yomo T (2014) Liposome-based liquid handling platform featuring addition, mixing, and aliquoting of femtoliter volumes. PLoS ONE 9(7):e101820
70. Karamdad K, Law RV, Seddon JM, Brooks NJ, Ces O (2015) Preparation and mechanical characterisation of giant unilamellar vesicles by a microfluidic method. Lab Chip 15(2): 557–562
71. Morita M, Onoe H, Yanagisawa M, Ito H, Ichikawa M, Fujiwara K, Saito H, Takinoue M (2015) Droplet-shooting and size-filtration (DSSF) method for synthesis of cell-sized liposomes with controlled lipid compositions. Chembiochem 16(14):2029–2035
72. Deamer DW, Prince RC, Crofts AR (1972) The response of fluorescent amines to pH gradients across liposome membranes. Biochim Biophys Acta 274(2):323–335
73. Haran G, Cohen R, Bar LK, Barenholz Y (1993) Transmembrane ammonium sulfate gradients in liposomes produce efficient and stable entrapment of amphipathic weak bases. Biochim Biophys Acta 1151(2):201–215
74. Cullis PR, Hope MJ, Bally MB, Madden TD, Mayer LD, Fenske DB (1997) Influence of pH gradients on the transbilayer transport of drugs, lipids, peptides and metal ions into large unilamellar vesicles. Biochim Biophys Acta 1331(2):187–211

75. Chakrabarti AC, Deamer DW (1992) Permeability of lipid bilayers to amino acids and phosphate. Biochim Biophys Acta 1111(2):171–177
76. Chakrabarti AC (1994) Permeability of membranes to amino acids and modified amino acids: mechanisms involved in translocation. Amino Acids 6:213–229
77. Chakrabarti AC, Deamer DW (1994) Permeation of membranes by the neutral form of amino acids and peptides: relevance to the origin of peptide translocation. J Mol Evol 39(1):1–5
78. Monnard PA, Deamer DW (2001) Nutrient uptake by protocells: a liposome model system. Orig Life Evol Biosph 31(1-2):147–155
79. Batzri S, Korn ED (1973) Single bilayer liposomes prepared without sonication. Biochimica et Biophysica Acta (BBA) – Biomembranes 298(4):1015–1019
80. Dominak LM, Keating CD (2007) Polymer encapsulation within giant lipid vesicles. Langmuir 23(13):7148–7154
81. Lohse B, Bolinger P-Y, Stamou D (2008) Encapsulation efficiency measured on single small unilamellar vesicles. J Am Chem Soc 130(44):14372–14373
82. Luisi PL, Allegretti M, de Souza TP, Steiniger F, Fahr A, Stano P (2010) Spontaneous protein crowding in liposomes: a new vista for the origin of cellular metabolism. ChemBioChem 11(14):1989–1992
83. Nishimura K, Hosoi T, Sunami T, Toyota T, Fujinami M, Oguma K, Matsuura T, Suzuki H, Yomo T (2009) Population analysis of structural properties of giant liposomes by flow cytometry. Langmuir 25(18):10439–10443
84. Sakakura T, Nishimura K, Suzuki H, Yomo T (2012) Statistical analysis of discrete encapsulation of nanomaterials in colloidal capsules. Anal Methods 4(6):1648–1655
85. Stano P, de Souza TP, Carrara P, Altamura E, D'Aguanno E, Caputo M, Luisi PL, Mavelli F (2015) Recent biophysical issues about the preparation of solute-filled lipid vesicles. Mech Adv Mater Struct 22(9):748–759
86. Cheng Z, Luisi PL (2003) Coexistence and mutual competition of vesicles with different size distributions. J Phys Chem B 107(39):10940–10945
87. Adamala K, Szostak JW (2013) Competition between model protocells driven by an encapsulated catalyst. Nat Chem 5(6):495–501
88. Adamala KP, Engelhart AE, Szostak JW (2016) Collaboration between primitive cell membranes and soluble catalysts. Nat Commun 7:11041
89. Boukobza E, Sonnenfeld A, Haran G (2001) Immobilization in surface-tethered lipid vesicles as a new tool for single biomolecule spectroscopy. J Phys Chem B 105(48):12165–12170
90. Pereira de Souza T, Fahr A, Luisi PL, Stano P (2014) Spontaneous encapsulation and concentration of biological macromolecules in liposomes: an intriguing phenomenon and its relevance in origins of life. J Mol Evol 1–14
91. de Souza TP, Steiniger F, Stano P, Fahr A, Luisi PL. Spontaneous crowding of ribosomes and proteins inside vesicles: a possible mechanism for the origin of cell metabolism. Chembiochem 12(15):2325–2330 (2011)
92. Stano P, Erica D'Aguanno, Jürgen Bolz, Fahr A, Luisi PL (2013) A remarkable self-organization process as the origin of primitive functional cells. Angew Chem Int Ed Engl 52(50):13397–13400
93. Mavelli F, Stano P (2015) Experiments on and numerical modeling of the capture and concentration of transcription-translation machinery inside vesicles. Artif Life 21(4): 445–463
94. D'Aguanno E, Altamura E, Mavelli F, Fahr A, Stano P, Luisi PL (2015) Physical routes to primitive cells: an experimental model based on the spontaneous entrapment of enzymes inside micrometer-sized liposomes. Life 5(1):969–996
95. van Hoof B, Markvoort AJ, van Santen RA, Peter Hilbers AJ (2012) On protein crowding and bilayer bulging in spontaneous vesicle formation. J Phys Chem B 116(42): 12677–12683
96. van Hoof B, Markvoort AJ, van Santen RA, Peter Hilbers AJ (2014) Molecular simulation of protein encapsulation in vesicle formation. J Phys Chem B 118(12):3346–3354

97. Schmidli PK, Schurtenberger P, Luisi PL (1991) Liposome-mediated enzymatic synthesis of phosphatidylcholine as an approach to self-replicating liposomes. J Am Chem Soc 113(21):8127–8130
98. Liu AP, Fletcher DA (2009) Biology under construction: in vitro reconstitution of cellular function. Nat Rev Mol Cell Biol 10(9):644–650
99. Endy D (2005) Foundations for engineering biology. Nature 438(7067):449–453
100. Gibson DG, Glass JI, Lartigue C, Noskov VN, Chuang R-Y, Algire MA, Benders GA, Montague MG, Ma L, Moodie MM, Merryman C, Vashee S, Krishnakumar R, Assad-Garcia N, Andrews-Pfannkoch C, Denisova EA, Young L, Qi Z-Q, Segall-Shapiro TH, Calvey CH, Parmar PP, Hutchison CA, 3rd, Smith HO, Venter JC (2010) Creation of a bacterial cell controlled by a chemically synthesized genome. Science 329(5987):52–56
101. Hutchison CA, Chuang R-Y, Noskov VN, Assad-Garcia N, Deerinck TJ, Ellisman MH, Gill J, Kannan K, Karas BJ, Ma L, Pelletier JF, Qi Z-Q, Alexander Richter R, Strychalski EA, Sun L, Suzuki Y, Tsvetanova B, Wise KS, Smith HO, Glass JI, Merryman C, Gibson DG, Venter JC (2016) Design and synthesis of a minimal bacterial genome. Science 351(6280):aad6253
102. Damiano L, Hiolle A, Canamero L (2011) Grounding synthetic knowledge. In: Lenaerts T, Giacobini M, Bersini H, Bourgine P, Dorigo M, Doursat R (eds) Advances in artificial life, ECAL 2011. MIT press, Cambridge, pp 200–207
103. Dumouchel P, Damiano L (2017) Living with robots. Harvard University Press, Boston
104. Stano P, Luisi PL (2016) Theory and construction of semi-synthetic minimal cells. In: Nesbeth DN (ed) Synthetic biology handbook. CRC Press, pp 209–257
105. Luisi PL (1998) About various definitions of life. Orig Life Evol Biosph 28(4–6):613–622
106. Szostak JW (2012) Attempts to define life do not help to understand the origin of life. J Biomol Struct Dyn 29(4):599–600
107. Varela FG, Maturana HR, Uribe R (1974) Autopoiesis: the organization of living systems, its characterization and a model. Biosystems 5(4):187–196
108. Maturana HR, Varela FJ (1980) Autopoiesis and cognition: the realization of the living, 1st edn. D. Reidel Publishing Company
109. Luisi PL (2003) Autopoiesis: a review and a reappraisal. Naturwissenschaften 90(2):49–59
110. Bachmann PA, Walde P, Luisi PL, Lang J (1990) Self-replicating reverse micelles and chemical autopoiesis. J Am Chem Soc 112(22):8200–8201
111. Bachmann PA, Luisi PL, Lang J (1992) Autocatalytic self-replicating micelles as models for prebiotic structures. Nature 357(6373):57–59
112. Wick R, Walde P, Luisi PL (1995) Light-microscopic investigations of the autocatalytic self-reproduction of giant vesicles. J Am Chem Soc 117(4):1435–1436
113. Luisi PL, Varela FJ (1989) Self-replicating micelles – a chemical version of a minimal autopoietic system. Orig Life Evol Biosph 19(6):633–643
114. Stano P, Luisi PL (2010) Achievements and open questions in the self-reproduction of vesicles and synthetic minimal cells. Chem Commun (Camb) 46(21):3639–3653
115. Luisi PL (1994) The chemical implementation of autopoiesis. In: Fleischaker GR, Colonna S, Luisi PL (eds) Self-production of supramolecular structures. From synthetic structures to models of minimal living systems. Number 446 in nato science series C, 1st edn. Kluver Academic Publisher, pp 197–179
116. Stano P (2010) Synthetic biology of minimal living cells: primitive cell models and semi-synthetic cells. Syst Synth Biol 4(3):149–156
117. Haines TH (1983) Anionic lipid headgroups as a proton-conducting pathway along the surface of membranes – a hypothesis. Proc Natl Acad Sci USA 80(1):160–164
118. Oberholzer T, Wick R, Luisi PL, Biebricher CK (1995) Enzymatic RNA replication in self-reproducing vesicles: an approach to a minimal cell. Biochem Biophys Res Commun 207(1):250–257
119. Dejanović B, Mirosavljević K, Noethig-Laslo V, Pecar S, Sentjurc M, Walde P (2008) An ESR characterization of micelles and vesicles formed in aqueous decanoic acid/sodium decanoate systems using different spin labels. Chem Phys Lipids 156(1–2):17–25
120. Stano P, Wehrli E, Luisi PL (2006) Insights into the self-reproduction of oleate vesicles. J Phys: Condens Matter 18(33):S2231

121. Chen IA, Szostak JW (2004) Membrane growth can generate a transmembrane pH gradient in fatty acid vesicles. Proc Natl Acad Sci USA 101(21):7965–7970
122. Budin I, Debnath A, Szostak JW (2012) Concentration-driven growth of model protocell membranes. J Am Chem Soc 134(51):20812–20819
123. Hentrich C, Szostak JW (2014) Controlled growth of filamentous fatty acid vesicles under flow. Langmuir 30(49):14916–14925
124. Kuruma Y, Stano P, Ueda T, Luisi PL (2009) A synthetic biology approach to the construction of membrane proteins in semi-synthetic minimal cells. Biochim Biophys Acta 1788(2): 567–574
125. Scott A, Noga MJ, de Graaf P, Westerlaken I, Yildirim E, Danelon C (2016) Cell-free phospholipid biosynthesis by gene-encoded enzymes reconstituted in liposomes. PLoS ONE 11(10):e0163058
126. Peterlin P, Arrigler V, Kogej K, Svetina S, Walde P (2009) Growth and shape transformations of giant phospholipid vesicles upon interaction with an aqueous oleic acid suspension. Chem Phys Lipids 159(2):67–76
127. Takakura K, Toyota T, Sugawara T (2003) A novel system of self-reproducing giant vesicles. J Am Chem Soc 125(27):8134–8140
128. Kurihara K, Tamura M, Shohda K-I, Toyota T, Suzuki K, Sugawara T (2011) Self-reproduction of supramolecular giant vesicles combined with the amplification of encapsulated DNA. Nat Chem 3(10):775–781
129. Brea RJ, Cole CM, Devaraj NK (2014) In situ vesicle formation by native chemical ligation. Angew Chem Int Ed Engl 53(51):14102–14105
130. Deamer DW, Dworkin JP (2005) Chemistry and physics of primitive membranes. In: Walde P (ed) Prebiotic chemistry. Number 259 in topics in current chemistry. Springer, Berlin/Heidelberg, pp 1–27
131. Lawless JG, Zeitman B, Pereira WE, Summons RE, Duffield AM (1974) Dicarboxylic acids in the Murchison meteorite. Nature 251(5470):40–42
132. Pizzarello S, Shock E (2010) The organic composition of carbonaceous meteorites: the evolutionary story ahead of biochemistry. Cold Spring Harb Perspect Biol 2(3):a002105
133. Gardner PM, Winzer K, Davis BG (2009) Sugar synthesis in a protocellular model leads to a cell signalling response in bacteria. Nat Chem 1(5):377–383
134. Chakrabarti AC, Breaker RR, Joyce GF, Deamer DW (1994) Production of RNA by a polymerase protein encapsulated within phospholipid vesicles. J Mol Evol 39(6):555–559
135. Oberholzer T, Albrizio M, Luisi PL (1995) Polymerase chain reaction in liposomes. Chem Biol 2(10):677–682
136. Mason JT, Xu L, Sheng Z-m, O'Leary TJ (2006) A liposome-PCR assay for the ultrasensitive detection of biological toxins. Nat Biotech 24(5):555–557
137. Lee S, Koo H, Na JH, Lee KE, Jeong SY, Choi K, Kim SH, Kwon IC, Kim K (2014) DNA amplification in neutral liposomes for safe and efficient gene delivery. ACS Nano 8(5): 4257–4267
138. Oberholzer T, Nierhaus KH, Luisi PL (1999) Protein expression in liposomes. Biochem Biophys Res Commun 261(2):238–241
139. Chen IA, Salehi-Ashtiani K, Szostak JW (2005) RNA catalysis in model protocell vesicles. J Am Chem Soc 127(38):13213–13219
140. Anella F, Danelon C (2014) Reconciling ligase ribozyme activity with fatty acid vesicle stability. Life (Basel) 4(4):929–943
141. Gilbert W (1986) Origin of life: the RNA world. Nature 319(6055):618–618
142. Gorlero M, Wieczorek R, Adamala K, Giorgi A, Schininà ME, Stano P, Luisi PL (2009) Ser-His catalyses the formation of peptides and PNAs. FEBS Lett 583(1):153–156
143. Luisi PL (2011) The synthetic approach in biology: epistemological notes for synthetic biology. In: Luisi PL, Chiarabelli C (eds) Chemical synthetic biology. Wiley, Chichester, pp 343–362

144. Stano P, Carrara P, Kuruma Y, de Souza TP, Luisi PL (2011) Compartmentalized reactions as a case of soft-matter biotechnology: synthesis of proteins and nucleic acids inside lipid vesicles. J Mater Chem 21(47):18887–18902
145. Stano P, Luisi PL (2013) Semi-synthetic minimal cells: origin and recent developments. Curr Opin Biotechnol 24(4):633–638
146. Gil R, Silva FJ, Peret J, Moya A (2004) Determination of the core of a minimal bacterial gene set. Microbiol Mol Biol Rev 68(3):518–537. Table of contents
147. Yu W, Sato K, Wakabayashi M, Nakaishi T, Ko-Mitamura EP, Shima Y, Urabe I, Yomo T (2001) Synthesis of functional protein in liposome. J Biosci Bioeng 92(6):590–593
148. Oberholzer T, Luisi PL (2002) The use of liposomes for constructing cell models. J Biol Phys 28(4):733–744
149. Shimizu Y, Inoue A, Tomari Y, Suzuki T, Yokogawa T, Nishikawa K, Ueda T (2001) Cell-free translation reconstituted with purified components. Nat Biotechnol 19(8):751–755
150. Shimizu Y, Kanamori T, Ueda T (2005) Protein synthesis by pure translation systems. Methods 36(3):299–304
151. Hillebrecht JR, Chong S (2008) A comparative study of protein synthesis in in vitro systems: from the prokaryotic reconstituted to the eukaryotic extract-based. BMC Biotechnol 8:article no 58
152. Kita H, Matsuura T, Sunami T, Hosoda K, Ichihashi N, Tsukada K, Urabe I, Yomo T (2008) Replication of genetic information with self-encoded replicase in liposomes. Chembiochem 9(15):2403–2410
153. Hosoda K, Sunami T, Kazuta Y, Matsuura T, Suzuki H, Yomo T (2008) Quantitative study of the structure of multilamellar giant liposomes as a container of protein synthesis reaction. Langmuir 24(23):13540–13548
154. Ishikawa K, Sato K, Shima Y, Urabe I, Yomo T (2004) Expression of a cascading genetic network within liposomes. FEBS Lett 576(3):387–390
155. Shin J, Noireaux V (2012) An E. coli cell-free expression toolbox: application to synthetic gene circuits and artificial cells. ACS Synth Biol 1(1):29–41
156. Maeda YT, Nakadai T, Shin J, Uryu K, Noireaux V, Libchaber A (2012) Assembly of MreB filaments on liposome membranes: a synthetic biology approach. ACS Synth Biol 1(2):53–59
157. Soga H, Fujii S, Yomo T, Kato Y, Watanabe H, Matsuura T (2014) In vitro membrane protein synthesis inside cell-sized vesicles reveals the dependence of membrane protein integration on vesicle volume. ACS Synth Biol 3(6):372–379
158. Uyeda A, Nakayama S, Kato Y, Watanabe H, Matsuura T (2016) Construction of an in vitro gene screening system of the E. coli EmrE transporter using liposome display. Anal Chem 88(24):12028–12035
159. Hamada S, Tabuchi M, Toyota T, Sakurai T, Hosoi T, Nomoto T, Nakatani K, Fujinami M, Kanzaki R (2014) Giant vesicles functionally expressing membrane receptors for an insect pheromone. Chem Commun (Camb) 50(22):2958–2961
160. Matsubayashi H, Kuruma Y, Ueda T (2014) In vitro synthesis of the E. coli sec translocon from DNA. Angew Chem Int Ed 53(29):7535–7538
161. Stano P, Rampioni G, Francesca D'Angelo, Altamura E, Mavelli F, Marangoni R, Rossi F, Damiano L (2018) Current directions in synthetic cell research. In: Piotto S, Rossi F, Concilio S, Reverchon E, Cattaneo G (eds) Advances in bionanomaterials. Lecture notes in bioengineering. Springer, pp 141–154
162. Walker SA, Kennedy MT, Zasadzinski JA (1997) Encapsulation of bilayer vesicles by self-assembly. Nature 387(6628):61–64
163. Kisak ET, Coldren B, Evans CA, Boyer C, Zasadzinski JA (2004) The vesosome– a multicompartment drug delivery vehicle. Curr Med Chem 11(2):199–219
164. Paleos CM, Pantos A (2014) Molecular recognition and organizational and polyvalent effects in vesicles induce the formation of artificial multicompartment cells as model systems of eukaryotes. Acc Chem Res 47(5):1475–1482
165. Altamura E, Milano F, Tangorra RR, Trotta M, Hassan Omar O, Stano P, Mavelli F (2017) Highly oriented photosynthetic reaction centres generate a proton gradient in synthetic protocells. Proc Natl Acad Sci USA 114:3837–3842

166. Hadorn M, Eggenberger Hotz P (2010) DNA-mediated self-assembly of artificial vesicles. PLoS ONE 5(3):e9886
167. Hadorn M, Boenzli E, Sørensen KT, De Lucrezia D, Hanczyc MM, Yomo T (2013) Defined DNA-mediated assemblies of gene-expressing giant unilamellar vesicles. Langmuir 29(49):15309–15319
168. Cronin L, Krasnogor N, Davis BG, Alexander C, Robertson N, Steinke JHG, Schroeder SLM, Khlobystov AN, Cooper G, Gardner PM, Siepmann P, Whitaker BJ, Marsh D (2006) The imitation game–a computational chemical approach to recognizing life. Nat Biotechnol 24(10):1203–1206
169. Tomasi R, Noël J-M, Zenati A, Ristori S, Rossi F, Cabuil V, Kanoufi F, Abou-Hassan A (2014) Chemical communication between liposomes encapsulating a chemical oscillatory reaction. Chem Sci 5(5):1854–1859
170. Liu Y, Wu H-C, Chhuan M, Terrell JL, Tsao C-Yu, Bentley WE, Payne GF (2015) Functionalizing soft matter for molecular communication. ACS Biomater Sci Eng 1(5):320–328
171. Liu Y, Tsao C-Yu, Kim E, Tschirhart T, Terrell JL, Bentley WE, Payne GF (2017) Using a redox modality to connect synthetic biology to electronics: hydrogel-based chemo-electro signal transduction for molecular communication. Adv Healthc Mater 6(1):article no 1600908
172. Nakano T, Moore M, Enomoto A, Suda T (2011) Molecular communication technology as a biological ICT. In: Sawai H (ed) Biological functions for information and communication technologies. Number 320 in studies in computational intelligence. Springer, Berlin/Heidelberg, pp 49–86
173. Nakano T, Eckford AW, Haraguchi T (2013) Molecular communications. Cambridge University Press, Cambridge
174. Rampioni G, Damiano L, Messina M, D'Angelo F, Leoni L, Stano P (2013) Chemical communication between synthetic and natural cells: a possible experimental design. Electron Proc Theor Comput Sci 130:14–26
175. Rampioni G, Mavelli F, Damiano L, DAngelo F, Messina M, Leoni L, Stano P (2014) A synthetic biology approach to bio-chem-ICT: first moves towards chemical communication between synthetic and natural cells. Nat Comput 13:1–17
176. Lentini R, Santero SP, Chizzolini F, Cecchi D, Fontana J, Marchioretto M, Del Bianco C, Terrell JL, Spencer AC, Martini L, Forlin M, Assfalg M, Dalla Serra M, Bentley WE, Mansy SS (2014) Integrating artificial with natural cells to translate chemical messages that direct E. coli behaviour. Nat Commun 5:4012
177. Adamala KP, Martin-Alarcon DA, Guthrie-Honea KR, Boyden ES (2017) Engineering genetic circuit interactions within and between synthetic minimal cells. Nat Chem 9:431–439. Advance online publication
178. Lentini R, Martín NY, Forlin M, Belmonte L, Fontana J, Cornella M, Martini L, Tamburini S, Bentley WE, Jousson O, Mansy SS (2017) Two-way chemical communication between artificial and natural cells. ACS Cent Sci 3:117–123
179. Shapiro L (2007) The embodied cognition research programme. Philos Compass 2:338–346
180. Simon HA (1996) The sciences of the artificial. MIT Press, Cambridge
181. Bich L, Damiano L (2012) Life, autonomy and cognition: an organizational approach to the definition of the universal properties of life. Orig Life Evol Biosph 42:389–397
182. Damiano L, Kuruma Y, Stano P (2016) What can synthetic biology offer to artificial intelligence (and vice versa)? BioSystems 148:1–3
183. Eschenmoser A, Volkan Kisakuerek M (1996) Chemistry and the origin of life. Helv Chim Acta 79(5):1249–1259

MIX
Papier aus verantwortungsvollen Quellen
Paper from responsible sources
FSC® C105338

If you have any concerns about our products,
you can contact us on
ProductSafety@springernature.com

In case Publisher is established outside the EU,
the EU authorized representative is:
Springer Nature Customer Service Center GmbH
Europaplatz 3, 69115 Heidelberg, Germany

Printed by Libri Plureos GmbH
in Hamburg, Germany